PDE Toolbox Primer for Engineering Applications with MATLAB® Basics

PDE Toolbox Primer for Engineering Applications with MATLAB® Basics

Leonid Burstein

CRC Press
Taylor & Francis Group
Boca Raton London New York

CRC Press is an imprint of the
Taylor & Francis Group, an **informa** business

First edition published 2022
by CRC Press
6000 Broken Sound Parkway NW, Suite 300, Boca Raton, FL 33487-2742

and by CRC Press
4 Park Square, Milton Park, Abingdon, Oxon, OX14 4RN

CRC Press is an imprint of Taylor & Francis Group, LLC

ISBN: 978-1-032-05997-6 (hbk)
ISBN: 978-1-032-06022-4 (pbk)
ISBN: 978-1-003-20035-2 (ebk)

DOI: 10.1201/9781003200352

Typeset in Times
by codeMantra

To my dear family: wife Inna and son Dmitri,

and to my friends,

for memory…

Contents

Preface

Scientists, engineers, and students of technical and scientific fields need computer software, first of all, to solve actual problems arising in the design, development, and manufacturing of products and technological systems. In many cases, it is necessary to solve the problems described by differential equations – ordinary and especially partial differential equations (ODEs and PDEs). This book introduces the basics of programming tools using examples from fields such as mechanics, heat and mass transfer, electricity, tribology (surfaces, friction, and lubrication), materials science, technical physics, and biotechnology. In these and many other engineering areas, ODEs and PDEs play a key role as they describe technological phenomena and processes and are used to analyze, design, and model optimal technical products. Among the wide range of programming tools available, the Partial Differential Equation Toolbox™ included in the MATLAB® software is an application for implementing spatial and transient PDE solutions. The programmatic and PDE Modeler tools of this toolbox, together with some of the core MATLAB® commands for solving PDEs, are recognized as convenient and effective tools for solving the topical problems of modern science and technology. Thus, mastering its latest versions and practical solutions with their help are increasingly essential for the creation of new products in mechanics, electronics, chemistry, life sciences, and modern industry as a whole. Specialists in modern technology widely use computers and some special programs, but they need a universal tool for solving, simulating, and modeling specific problems from their field. Numerous MATLAB® tutorials have been written, but none has been specially designed for teaching and mustering engineering problem with the PDE Toolbox. The proposed book covers this gap.

Problem solving with the differential equations involves calculations of mechanical parts, machine elements, production process, quality assurance, fluid mechanics parameters, thermodynamic and rheological properties of the materials, state equations, lubrication and tribo-characteristics of rubbing parts, bacteria population, and dissociation and reaction-diffusion kinetics. The book is a primer in which MATLAB® and PDE toolbox are introduced as an essential foundation for solving both traditional and emerging engineering PDE problems. The real benefit of the book is a variety of illustrative examples from a broad range of engineering fields, thus making understanding the presented material easier.

The first chapter presents the objectives of the book, the topics covered, and the structure of the chapters and outlines the problems that are solved in the book chapters by the MATLAB® software and its PDE Toolbox.

The next four chapters describe the computational and graphical tools with examples of various practical applications. For program writing, the regular and live editors are described. This chapter illustrates numerical methods and ode- and bvp-solvers applied to solve problems.

The sixth chapter presents the types of partial differential equations and boundary conditions to be solved using the PDE Toolbox programmatic tools.

The Modeler graphical interface is described in the seventh chapter. These two chapters consider two-dimensional PDEs. The eighth chapter deals with one-dimensional PDEs and its pdepe solver.

In the ninth chapter, solutions of the couplet PDEs with the PDE Modeler and programmatic solutions of the three-dimensional PDEs are presented and applied to scientific and engineering problems.

The final chapter depicts biotechnology problems that are described with differential equations and solved using the ode-, bvp-, pdepe-solvers, and PDE software tools.

The appendix contains a collection of the commands and functions presented in the book for solving actual problems of engineering. It also lists the problems resolved in each chapter.

The book accumulates experience of many years studying and teaching in courses for students, practitioners, and scientists specializing in the area in question (at the Technion – IIT, Haifa University, the ORT Braude College of Engineering, and the Kinneret Academic College).

I thank The MathWorks Inc.*, which kindly granted permission to use certain materials.

I am grateful to IGI Global for granting permission to use materials from my book *MATLAB® with Applications in Mechanics and Tribology* (2021) – used specifically in Chapters 2–4, 7, and 8.

I am also grateful to Elsevier Inc. for permission to use the certain text tables, figures, and screenshots from my book *A MATLAB® Primer for Technical Programming in Materials Science and Engineering* (2020) – used in Chapters 2, 3, and 8.

I would like to express my appreciation to Stephen Rifkind, an ITA Recognized Translator, who edited parts of the book.

I hope the book will prove useful to engineers, scientists, and students in all areas of engineering and enable them to work with MATLAB® and PDE Toolbox™.

The author will gratefully accept any reports of bugs or errata as well as comments and suggestions for improving the book.

Leonid Burstein
Independent researcher
Nesher, Israel
March – November 2021

*MATLAB® is a registered trademark of The MathWorks, Inc. For product information, please contact:

The MathWorks, Inc.
3 Apple Hill Drive
Natick, MA 01760-2098 USA
Tel: 508-647-7000
Fax: 508-647-7001
E-mail: info@mathworks.com
Web: www.mathworks.com

Author

Leonid Burstein, PhD, taught as an Associate Staff Member (ret) of the Software Engineering Department at the ORT Braude College of Engineering for 17 years. He also worked for many years as a lecturer at large and taught in the Quality Assurance Department of the Kinneret Academic College and at Technion – ITI and was regularly invited as a lecturer at a number of leading universities and high schools of the Western and Lower Galilee.

Following an MA studying in thermophysics at the Lomonosov Technological Institute in Odessa, Ukraine, he undertook doctoral studies at All-Union Scientific Research Institute of Physical-Technical and Radiotechnical Measurements at Mendeleevo, Moscow region, Russia. Burstein earned a PhD in thermophysical properties of materials from the Heat and Mass Transfer Institute of the Belarusian Academy of Sciences, Minsk, in 1974.

After a short period of work in Russia and Belarus, Burstein continued his carrier at the Piston Ring Institute in Odessa, Ukraine, where he worked from 1974 to 1990 as Head of Projects and Head of the CAD/CAM group. Since 1991, he has worked at the Technion – IIT, first at the ME Faculty, then at the Quality Assurance and Reliability Program at the Faculty of IEM and later at the Taub Computer Centre as a software science consultant. At the time, he also worked at the Technion Research and Development Foundation as the principal researcher in funded projects in various areas, including diesel tribology and environment control. During this period, he taught various courses at Haifa University, the Technion, the Kinneret Academic College, and other institutions.

Dr Burstein is also the editor of a handbook published by IGI Global and author of several patents. He is the author and contributor to published textbooks as well as an editorial board member and reviewer for a number of international scientific journals.

1 Introduction

1.1 PREAMBLE

This book provides the basics of MATLAB® and the Partial Differential Equation (PDE) Toolbox™ accompanied by examples taken from various engineering areas, including mechanics, electricity, heat and mass transfer, tribology, materials science, technical physics, and biotechnology. In these as in many other engineering fields, differential equations, in particular PDEs, play a key role as they describe technological phenomena and processes and are vital for the analysis, design, and modeling of optimal technical products. The solutions of the spatial and transient PDEs are implemented by the PDE Toolbox, which is included in the MATLAB® software. The PDE Modeler of this toolbox is known as a convenient and effective tool for solving actual problems of modern science and technology. Thus, mastering its latest versions is becoming more and more relevant for finding practical solutions and creating new products in mechanics, electronics, chemistry, life sciences, and modern industry in general. Modern technology specialists widely use computers and some special programs but require some universal tool for solving, simulating, and modeling specific problems in their area. Numerous primers on MATLAB® have been written, but none have been designed especially for teaching and solving engineering problem with the PDE Toolbox. This book covers the gap.

MATLAB® is introduced in the book as an essential foundation for PDE Modeler and programmatic tool of the PDE Toolbox with appropriate explanatory solutions applied to both traditional and emerging engineering problems. One benefit of the book is a variety of explanatory examples from a broad range of engineering fields, including modern and classical mechanics, material sciences, mass-heat transfer, and biotechnology, which will facilitate the understanding of the presented material.

1.2 A BIT OF HISTORY AND ADVANTAGES OF THE SOFTWARE PRESENTED IN THIS BOOK

The foundations and language of MATLAB® were established in the 1970s by mathematician Cleve Moler. Later, the language was rewritten in C and improved by specialists that had joined the founder. Initially, the language was intended for students to adapt LINPACK and EISPACK mathematical packages, but it was then applied to control engineering. In a fairly short period, researchers and engineers recognized MATLAB® as an effective and convenient tool for solving not only mathematical but also many technological problems. Early commercial

DOI: 10.1201/9781003200352-1

versions of MATLAB® appeared in the general software market in the mid-1980s. Later, graphics and special engineering-oriented means, such as user-oriented interfaces and toolboxes, begin to be incorporated into MATLAB®, giving it its modern form. Among them, the PDEs Toolbox appeared in 1995, adding some commands and, most importantly for practical use, an interface for solving spatially two-dimensional (2D) transient PDEs. The toolbox has constantly evolved and is intended today for the fourth type of PDEs, specifically elliptical, parabolic, hyperbolic, and eigenvalue. In the mid-90s, a software called FEMLAB was separated from the PDE Toolbox and is now developing independently. Possibilities of the PDE Toolbox are constantly expanding and, since the early 2010s, are able to solve some spatially three-dimensional (3D) problems using its programmatic tool. As a whole, today's MATLAB® and its PDE Toolbox are a unique assembly of implemented classical and modern numerical methods and specialized interfaces for engineering calculations designed for specialists of various fields. With other programming tools for various computations, this software has a valuable place among them as a technical computing software. Without going into details, the following abilities and their interactions explain the sustainable preference for PDE Toolbox and MATLAB® as a whole:

- PDE tool:
  - Provides a powerful and flexible environment for exploring and solving PDEs in two and three spatial dimensions and time;
  - Assures a user-friendly graphical interface allowing the user to draw the geometry of the real technical part, specify its boundaries, select the appropriate PDE type, solve the problem, and visualize the calculated results;
  - Supplies a wide range of commands to efficiently create programs solving one or more 2D or 3D PDEs.
- MATLAB® itself:
  - Ensures substantial universality and the ability to solve both simple and complex scientific and technical problems using a large set of various all-general and specialized commands;
  - The proposed practical applicability in various fields of technology and science is provided by a wide variety of problem-oriented tools called toolboxes;
  - Provides convenience means for visualizing the achieved solutions of the general or specific scientific and technological problems;
  - Has quick access to built-in help and well-organized extensive documentation.

To these characteristics should be added the innovation of recent years – the Live Editor – which provides the ability to display text, images, and codes together with the resulting tables and graphs in one window, with the results immediately visible after entering the codes.

1.3 THE GOALS OF THE BOOK AND ITS AUDIENCE

The book has two goals:

- First, to provide researchers, engineers, machine designers, teachers, and students with guidance to teach them how to use the PDE Toolbox™ as well as the MATLAB® solvers for 1D PDEs and initial value and boundary value problems (IVP and BVP, respectively) for ordinate differential equations (ODE).
- Second, following from the first aim, to provide a basic, simple, and comprehensive guide to MATLAB® without which it is impossible to use the PDE Toolbox for solving differential equations (DE).

It is assumed that the reader has no programming experience and will be using the software for the first time. Therefore, the book provides MATLAB® basics to make PDE Toolbox available to the widest possible audience. To make the basic steps of programming and make the commands understandable to the target audience, the book provides examples of problems from various fields of science and engineering. As the basic programming knowledge of the reader is increased, the problems become more complex and are solved with special solvers, ode and bvp. Then, the PDE Toolbox is sequentially introduced with its programmatic and interface tools for spatially 2D PDEs, followed by 3D options.

MATLAB® and its toolboxes are updated and improved in parallel with the development of modern technologies. Thus, the technical analysis and calculations that can be conducted in MATLAB® and in PDE Toolbox in particular have contributed to the fact that the scientific community has recognized it as a convenient and effective tool for use in modern science and engineering. Thus, mastering its latest versions and practical solutions with their help is increasingly essential for the creation of new products in mechanics, electronics, chemistry, life sciences, and modern industry as a whole. Specialists in these areas, among others, widely use computers and some special programs but also require some universal tool for solving, simulating, and modeling specific problems in their area. This also applies to the usage of the PDE Toolbox, which is frequently not used due to the ignorance of its capabilities and lack of familiarity. However, the problems that are described by differential equations and can be solved with the PDE Toolbox cover a wide range of phenomena. These include the strength and durability of mechanical parts, machine elements, production processes, quality assurance, fluid mechanics parameters, thermodynamic and rheological properties of the materials, state equations, lubrication and tribo-characteristics of rubbing parts and descriptive statistics as well as bacteria population, dilution, dissociation, and reaction-diffusion kinetics. Thus, knowledge of the available apparatus for solving such equation is critically important for effective solutions of many real engineering problems. This book is oriented to the reader with a modest mathematical background and introduces the programming or technical concepts using some simplifications of the traditional approach. A variety of examples from a broad range of modern and classical engineering help solidify the understanding of the presented material and show specialists the options for using the software in their specific fields. As a whole, the book can serve as a guide for two categories of users:

- Scientists, engineers, and students that would like to see how to use the PDE Toolbox in their specific areas;
- Researchers and technicians wanting to learn and apply MATLAB® in their industry.
- Summarizing the above, the principal audiences of the book include:
- Scientists, engineers, and specialists that seek to solve their problems and search for similar problems that were solved by computer;
- Non-programmer professionals and the academic community dealing with modeling and simulation machinery and processes in areas of technique and technology from mechanics or electricity to the biotechnology;
- Students, engineers, managers, and teachers from academic and university communities in the field of technology;
- Instructors and their audiences in study courses where PDE Toolbox and/or MATLAB® is used as a supplemental but required tool;
- Staff, students, and non-programmers as well as self-taught readers for quick mastering of the programs for their needs;
- Freshmen and participants in advanced scientific and engineering courses, seminars, or workshops where MATLAB® is taught;
- Researchers and professionals using a computer for modeling calculations to solve actual engineering and chemical/bioengineering problems applying the book as a reference.

1.4 ABOUT THE MATERIAL IN THE CHAPTERS

The material in the chapters is based on nearly 25 years of research and 18 years of multiple MATLAB® authoring courses in the fields of mechanics, mathematics, quality assurance, and biotechnology. The topics in the chapters are presented so that a beginner can gradually move from one topic to another from topics presented in MATLAB®-introductory sections to sections on PDE Toolbox programming and its interface, with previously acquired material to be used as a basis for each subsequent chapter.

This chapter, the first of ten chapters of this book, outlines the objectives of the book, the topics covered, and the structure of the chapters and outlines engineering problems that can be solved by the MATLAB® software and its PDE Toolbox.

The next three chapters describe the computational and graphics tools with examples of various practical applications. The most important, basic MATLAB® features are introduced in the second chapter, which describes the software desktop, toolbars, and main windows. It discusses elementary functions, input and output commands, numbers and strings, vectors, matrices/arrays, their manipulations, and flow control commands as well as relational and logical operators.

The commands of this chapter are intended to enable beginners to write, perform, and display simple calculations interactively and directly in the Command Window. Chapter 3 introduces the user-defined functions and presents the regular Editor window for writing program scripts and user-defined functions and then the Live Editor window for writing live scripts and functions. All commands, regular and live scripts/functions, are explained with examples from engineering fields. The visualization means for generating 2D and 3D plots are described in Chapter 4. It describes the formatting commands for inserting labels, titles, text, and symbols into a plot as well as the color, marker, and line qualifiers. How to develop graphs containing more than one curve and graphs with multiple plots on one page is explained. Accompanying applications demonstrate how to generate 2D and 3D graphs for water surface tension, bandpass filter, and other practical applications. Understanding the material of the second, third, and fourth chapters allows the reader to generate rather complex programs applying technical calculations and their graphical presentation.

Chapter 5 presents more advanced topics but still refers to basic MATLAB®, namely ordinary differential equations, ODE and solvers for initial and boundary value problems, IVP and BVP. The finite difference method is also explained here. The chapter presents applied solutions to IVP and BVP problems such as RLC series current and heater wire temperature distribution.

In the next two chapters, the sixth and seventh, PDE Toolbox programming and modeling tools are described and applied to scientific and engineering problems that are modeled by PDEs. Chapter 6 illustrates the finite element collocation scheme and introduces PDE equations, boundary and initial conditions, and provides commands for solving PDEs. The problems solved here are steady or unsteady spatial 2D. Application problems include heating a small metallic plate, drumhead vibrations, and elliptical membrane eigenvalue modes. Chapter 7 presents the PDE Modeler tool for solving 2D PDEs; here the solution steps are examined in detail along with the graphical user interface. Application problems include the momentary pressure distribution in a lubricating film between two pore-covered surfaces, unsteady thermal conductivity with a temperature-dependent material, an example of plain stress (structure mechanics), among others.

The eighth chapter describes the pdepe solver used to solve transient and spatially 1D PDEs. It is shown how various PDEs with different boundary conditions can be represented in standard forms. Applications illustrate how to solve diffusion PDE with Neumann boundaries and piecewise initial condition, Bateman-Burgers PDE, and others.

Chapter 9 covers two topics, namely those related to coupled 2D PDEs solution using the PDE Modeler, and 3D PDE solutions using the programmatic tool. Among the problems solved here are the Schnakenberg coupled PDEs for a trimolecular reaction, vibrations of a slab with elliptical hole, and the distribution of an electric potential in a plate with varying conductivity.

The final, tenth, chapter depicts life science problems, which are represented by differential equations and solved using the ODE and PDE software tools. The structure of the chapter differs from the previous one and contains application examples only. At the same time, considering the traditionally less prepared audience for programming and mathematics, problem solutions are presented with more extended explanations than in previous chapters. The applied problems addressed in this chapter include the steady-state concentration distribution in a short tube, the concentration of reagents in two reactors in series, a 1D model of the reactor, diffusing and reproducing in a bacterial culture, displacement of a homogenous membrane, and diffusion-brusselator PDEs.

The appendix provides a summary collection of over 250 variables, special characters, operators, and commands discussed in the book. In addition, a list of solved problems is provided.

The index contains about 800 alphabetical names, terms, and commands that have been explained or at least mentioned throughout this book.

1.5 MATERIAL ARRANGEMENT IN THE CHAPTER AND THE AVAILABLE PROGRAM EDITORS

The material in the chapters is presented gradually to ensure a gradual assimilation of concepts. Each chapter begins with an introduction describing the chapter content and its available features. New material, basic command forms, and their implementations are then presented. Commands are usually given in one or two of the simplest forms with possible useful extensions. Each question, if possible, is fully addressed in one subsection so that readers can attain knowledge in a focused manner. The available tables list additionally available commands, specifiers, modifiers, equations, and graphical and object geometry forms that correspond to the topics and examples included in the chapter. The chapter devoted to the differential equations has sections explaining numerical method and computation by this method, and the results are compared with the results obtained with the means of MATLAB® or PDE Toolbox. For each category of the differential equation, the general solution scheme is presented by steps with the available means to carry out each step. In the final sections of a chapter or sometimes in the middle of a chapter, application problems associated with the technological areas are solved with the commands previously presented in this and preceding chapters. Application examples from the biotechnological and chemical fields are collected in the final chapter of the book. The completed solutions are most accessible to understanding but not necessarily the shortest or most original. Composed programs and especially the form of the resulting representations vary to show the various possible solutions and thereby provide the skills to manage the solution and its visualization. Readers are encouraged to try not only the proposed solutions but also their own solutions and compare them with those in the book.

Please note that numerical values and contexts used in the various applications are not actual reference data and serve for demonstration purposes only.

To write program codes and then save them in files, MATLAB® provides two editors – the primary, which is commonly used and presented by the *Editor* window, and the more recent one, available in the *Live Editor* window, which shows the results along with the written codes in this window. These editors are described in detail in the book and are used interchangeably in order to allow the reader to master writing and running programs with each of the editors and choose the most suitable one in the future. PDE Modeler, presented by the user interface, automatically generates a program that can be launched without pre-recording in the *Editor* but can be used interactively once opened and launched in the *Live Editor*.

1.6 MATLAB® AND PDE TOOLBOX VERSIONS USED IN THIS BOOK

A new updated version of MATLAB® is released every 6 months; PDE Toolbox™ is also updated but some less often. It is generally accepted that each new version is retro compatible. Therefore, the commands applied in this book should work in any new release. The MATLAB® version used in this book is R2020b (9.7.0.1190202). For solving PDEs, version 3.5 (R2020b) of the PDE Toolbox™ is used.

Readers need to install the aforementioned or later versions of MATLAB® and PDE Toolbox on their computers in order to be able to perform all the operations presented in the book.

1.7 THE ORDER OF MASTERING THE MATERIAL

This book is designed in such a way that a reader lacking background in MATLAB® and its PDE Toolbox can progress from chapter to chapter to master PDE solutions both using programming tools and with PDE Modeler interface. Therefore, the topics are arranged sequentially in a convenient manner. The second, third, and fourth chapters introduce the basics of MATLAB®. Once the readers have mastered the basics, they can move to any subsequent chapter according to their interests – the chapters on ODE or PDE tools, or the chapter on chemical/biotechnological applications. Readers are of course not obligated to follow the order of the chapters. An experienced MATLAB® user can start with Chapter 6, which introduces the PDE Toolbox along with the applications. Those with additional interest in the ODE solutions and have experience in MATLAB® programming can start at Chapter 5.

New learners can also read the materials in a different order. For example, the Editor (Section 3.2) and Live Editor (Section 3.4) can be learned directly after the output commands (Section 2.2.7), which allows the creation of simple programs at an early stage of learning. Readers studying PDEs receive the PDE Toolbox graphical means along with the solution steps. Nevertheless, they should be familiar with the basic MATLAB® graphical tool introduced in Chapter 4. In addition, the Live Editor can be used, for example, early on to get familiar with variables and interactive calculations (prior to Section 2.2.5). This editor can be used throughout the study with examples of scripts/functions written in regular Editor, which are easily converted to a live file (Section 3.4.3).

Overall, I hope this book will help users learn the PDE Toolbox along with MATLAB® basics and use these tools effectively to solve their actual problems.

Now let's get started.

2 Basics of the Software

2.1 INTRODUCTION

More than half a century has passed since the appearance of a special computer tool and its language termed MATLAB®. The name is combined by the first three letters of two words, matrix and laboratory, and emphasizes the basic element of this language: matrix. The matrix approach unifies and facilitates computational and graphical processing. In the sequel, in a short time, MATLAB® was adapted and actively used firstly in scientific and then in engineering calculations. Today, this software and its various applications, among them the "Partial Differential Equation Toolbox", are a vital tool for scientists and technicians. To use any specialized tool, it is impossible to make this without knowledge of the MATLAB® basics. Furthermore, over time, MATLAB® and its tools have undergone significant changes in both the interface and the number and structure of its commands. The language and tools have become more diverse, efficient, and sophisticated. Therefore, it is necessary to learn the MATLAB® basics.

This chapter[1] represents the graphical user interface of the software, i.e., desktop, its menu, and the default windows; introduces a starting procedure; provides a lot of commands for simple arithmetic, algebraic, matrix, and array operations; and describes loop and logical and relational operators. For better understanding, the material is accompanied by very simple engineering examples.

2.2 RUNNING THE MATLAB®

Hereafter it is assumed that the user has obtained and installed MATLAB® along with the toolboxes required for his work. MATLAB®, like other application software for computers, is managed by a special program set termed the operating system (OS) that provides connection between the computer interface and apps. OS differs according to the computer platform; therefore, we additionally assume that MATLAB® is controlled by the OS Windows. To open MATLAB®, click the icon 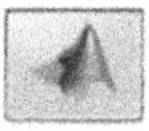 located at Start menu or at taskbar or click the 

icon on the computer Desktop; these icons show a red L-shaped membrane. If these icons were not previously placed in the indicated locations, you may start MATLAB® by typing the word MATLAB® in the Windows Search box or by clicking the

[1] Some text and table materials from Burstein, 2021a (Sections 2.1.2, 2.2, 2.43) and Burstein, 2020 (Sections 2.1.8, 2.3) are used in the chapter; with permissions from IGI Global and Elsevier respectively

DOI: 10.1201/9781003200352-2

matlab.exe file in the MATLAB bin directory. The path to this file looks like this C:\Program Files\MATLAB\R2020b\bin\matlab.exe. After that, a startup panel appears with a logo and some information about the product version, license number, and the name of the owner company. In a short time, this image disappears and the MATLAB® desktop opens.

2.2.1 Desktop, Toolstrip, and Main Windows

The main MATLAB window – desktop – includes the top toolstrip with controls and three panels, which can be undocked from the desktop and appear as separate windows: Command, Current Folder, and Workspace – see Figure 2.1.

2.2.1.1 Toolstrip

The top menu represents the Desktop strip, called toolstrip (Figure 2.2), containing three global tabs labeled HOME, PLOTS, and APPS. Each of tabs is divided into sections (e.g. FILE, VARIABLE, CODE, SIMULINK, etc. – at HOME section) combined a number of interrelated buttons, drop-down menus, and other controls (e.g., the "Open", "New", "Find Files", and other buttons of the FILE section). The HOME is the tab that is used most intensively and includes general-purpose operations such as creating new program files, importing data, and managing the workspace, as well as desktop layout options; the latter allows to restore default desktop view or/and open some non-default windows. The PLOTS tab contains buttons for generating the various charts for a variable placed in the Workspace. The APPS tab contains a gallery of applications belonging to the MATLAB® toolboxes, among them the PDE Modeler button that opens the Modeler window (see Chapter 6).

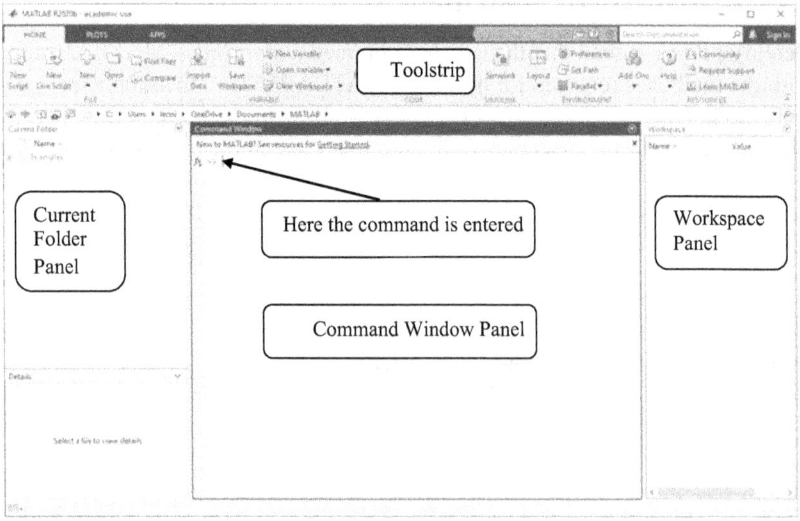

FIGURE 2.1 The default desktop view, MATLAB® R2020b.

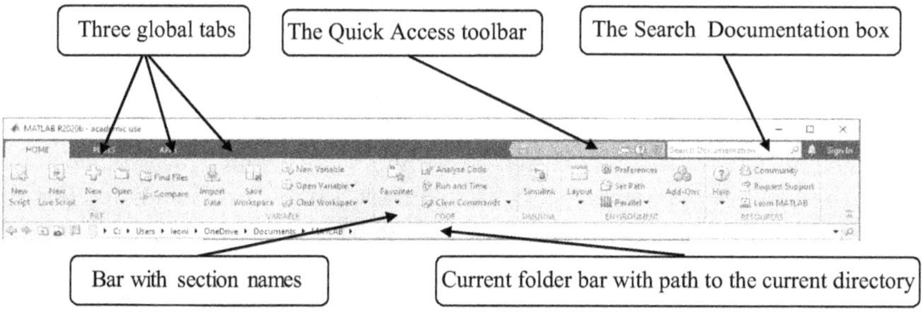

FIGURE 2.2 The desktop toolstrip.

The quick access bar locates to the right of the global toolstrip tabs. It contains some commonly used options for example Cut, Copy, and Paste; this bar can be customized with the pull-down button. In the same top line, next to the quick access bar, the Search Documentation box is accommodated which allows you to search for documentation regarding commands, windows, applications, and other topics of interest. The current folder bar is located at the bottom of the toolstrip. This bar contains file/directory management controls, shows the current working directory, and has the "Search in this folder and subfolders" button allowing you to type the searched text (e.g. *.m) in the current folder box.

2.2.1.2 Command Window

The central component of the Desktop is the Command Window. Here, the commands are entered, and after their execution, the result is displayed. Optionally, the window can be detached from the Desktop by selecting the Undock line of the "Show Command Window Actions" pull-down button locating to the right of the Command Window top bar. In the same way, it is possible to separate any other Desktop window. The separated windows can be assembled on the Desktop, by choosing the Dock line or select the Default line in the Layout option of the Desktop HOME tab.

2.2.1.3 Workspace Window

To the right from Command Window, the Workspace window is located. The window lists the variables (icon) and other objects currently located in the MATLAB® workspace. It also displays class, value, and size of each variable. The variable value can be edited here. Data values are automatically updated with calculations. Showing parameters can be edited/added/removed by clicking the right mouse button at any empty place of the menu line.

2.2.1.4 Current Folder Window

The window is located left to the Command Window. After launching MATLAB®, in the Current Folder window, we can view the files and folders located in the startup directory. The window has the Details panel located at its bottom where some details appear regarding the selected file. You can change the directory by entering the appropriate pass in the "Current folder" bar or using its folder management buttons.

There are also other windows that do not appear by default when you open the Desktop, such as Command History, Help, Editor, Live Script Editor, and Figure. They will be described later where they need to be used.

2.2.2 Simple Calculations and Math Functions

There are two main options for entering and executing the MATLAB® commands for calculations – interactive and programmatic. The first is to write commands directly in the Command Window, and the second is in usage of a program containing commands previously recorded in Editor and saved respectfully in *m*-file. In recent years, an interactive editor has appeared that combines the marked options – simultaneously write and execute commands, as well as save created program in *mlx*-file. The interactive options are briefly presented here while the programmatic and interactive editor modes will be explained later.

To execute any command, we must type and enter it in the Command Window at the place indicated by the blinking cursor (see Figure 2.1) that located immediately after the >> sign, which is called the "command prompt" or simply "prompt". Figure 2.3 represents some variables, numbers, and elementary calculating commands entered and executed in the undocked Command Window.

The  button that appears in the left frame bar is called the Function browser. It opens a drop-down list with a search box that helps to find the necessary command and obtain its syntax with a short description of its use.

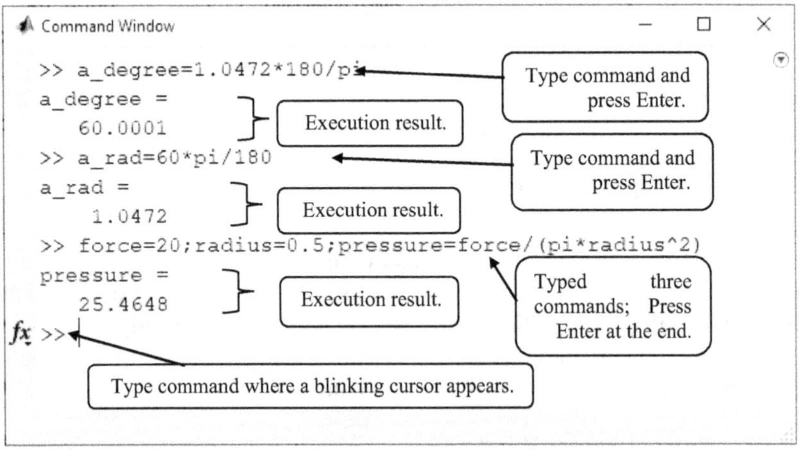

FIGURE 2.3 Examples with interactively entered and executed commands in the Command Window.

To manipulate with commands, the following rules must be mastered:

- The equal sign (=) is called the assignment operator and is used to set a value for a variable;
- Press the key Enter after the command to execute it;
- Type an ellipsis ... (three periods) to continue a long command on the next line; the total length of a single command may not exceed 4096 characters;
- Two or more commands can be written in the same line; however, the commands should be divided with semicolons (;) or commas (,);
- A semicolon, entered after the command, prevents displaying of the answer;
- A command/s located on the preceding line/s cannot be corrected/changed; to do it, press the up-arrow key ↑ and then introduce the desirable change in shown command;
- Type the % symbol (percent symbol) to introduce a comment in the line; the comments are not executed after entering; they appear in green;
- Enter the clc command to clear the Command Window;
- The result of the command execution is assigned to a variable named ans, if the result has not been previously named by another name;
- Each new entered/calculated value cancels its predecessor in the variable; for example, if the value of b was assigned early as −4.2917, thus after entering a new value by the command >> b = 94.2428, the previous value is removed and b receives the 94.2428 value.

To performed simple arithmetical and algebraic calculations, the following symbols can be used: + (addition), − (subtraction), * (multiplication), / (right division), \ (left division, mostly for matrices), and ^ (exponentiation).

In addition, there are a wide range of elementary, trigonometric, and special mathematical functions. The MATLAB® commands with these functions must be written as the name with the argument/s in parentheses, for example, tan x is written as tan(x). In trigonometric functions, the argument x is given in radians, when the command is ended with the d letter, for example, tand(x), the argument x should be given in degrees. Inverse trigonometric functions with the ending d require argument x in radians and produce the result in degrees.

A concise list of commands used in calculations with math functions is given in Table 2.1. Hereinafter, the commands/operators/statements/variables entered in the Command Window will be written after the command line prompt (>>).

When the calculations are executed by some math expression, operations in this expression are performed in the following order:

- operations in parentheses (starting with the innermost);
- exponentiation;
- multiplication and division;
- addition and subtraction;
- operations of the same priority run sequentially from left to right.

TABLE 2.1
Some Available Elementary, Trigonometric, and Specialized Mathematical Functions (Alphabetically)

No	MATLAB® Command	Name	Math Notation	Example (Input and Output)
1	abs(x)	Non-negative (absolute) x value	$\|x\|$	>> abs(−0.1728) ans = 0.1728
2	acos(x)	Inverse cosine, result in radians	$\arccos x$	>> acos(−1) ans = 3.1416
3	acosd(x)	Inverse cosine with x between −1 and 1; result in degrees between 0° and 180°	$\arccos x$	>> acosd(pi/4) ans = 38.2425
4	acot(x)	Inverse cotangent, result in radians	$\text{arccot}\, x$	>> acot(1.0000) ans = 0.7854
5	acotd(x)	Inverse cotangent; result in degrees between −90° and 90°	$\text{arccot}\, x$	>> acotd(pi/4) ans = 51.8540
6	asec(x)	Inverse secant, result in radians	$\text{asec}\, x$	>> asec(2) ans = 1.0472
7	asecd(x)	Inverse secant; results in degrees between 0° and 180°	$\text{asec}\, x$	>> asecd(2) ans = 60.0000
8	asin(x)	Inverse sine, result in radians	$\arcsin x$	>> asin(1.0000) ans = 1.5708
9	asind(x)	Inverse sine with x between −1 and 1; result in degrees between −90° and 90°	$\arcsin x$	>> asind(1.0000) ans = 90
10	atan(x)	Inverse tangent	$\arctan x$	>> atan(1) ans = 0.7854
11	atand(x)	Inverse tangent; result in degrees between −90° and 90° (asymptotically)	$\arctan x$	>> atand(pi/4) ans = 38.1460

(Continued)

TABLE 2.1 (*Continued*)
Some Available Elementary, Trigonometric, and Specialized Mathematical Functions (Alphabetically)

No	MATLAB® Command	Name	Math Notation	Example (Input and Output)
12	besselj(nu,z) nu – real constant	Bessel function of the first kind	$J_v(z)=\left(\frac{z}{2}\right)^v\sum_{(k=0)}^{\infty}\frac{\left(\frac{z^2}{4}\right)^k}{k!\,\Gamma(v+k+1)}$	>> besselj(4,5.8) ans = 0.3788
13	beta(z,w) z and w – positive integer	Beta function	$B(z,w)$ $=\int_0^1 t^{z-1}(1-t)^{w-1}\,dt$	>>beta(7,3) ans = 0.0040
14	ceil(x)	Round toward plus infinity	$\lceil x\rceil$	>> ceil(−11.99) ans = −11
15	cos(x)	Cosine	$\cos x$	>> cos(pi/3) ans = 0.5000
16	cosd(x)	Cosine with x in degrees	$\cos x$	>> cosd(60) ans = 0.5000
17	cosh(x)	Hyperbolic cosine	$\cos h\, x$	>> cosh(pi/3) ans = 1.6003
18	cot(x)	Cotangent	$\cot x$	>> cot(pi/3) ans = 0.5774
19	cotd(x)	Cotangent with *x* in degrees	$\cot x$	>> cotd(60) ans = 0.5774
20	coth(X)	Hyperbolic tangent	$\cot h\, x$	>> coth(pi/3) ans = 1.2809
21	erf(x)	Error function	$erf\ x=\frac{2}{\sqrt{\pi}}\int_0^x e^{-t^2}dt$	>> erf(2.9/ sqrt(1.98)) ans = 0.9964
22	exp(x)	Exponential	e^x	>> exp(1.98) ans = 7.2427
23	factorial(n)	Factorial; product of the integers from 1 to *n*	$n!$	>> factorial(4) ans = 24

(*Continued*)

TABLE 2.1 (*Continued*)
Some Available Elementary, Trigonometric, and Specialized Mathematical Functions (Alphabetically)

No	MATLAB® Command	Name	Math Notation	Example (Input and Output)
24	fix(x)	Round toward zero	$fix(x)$	>> fix(−1.89) ans = −1
25	floor(x)	Round toward minus infinity	$\lfloor x \rfloor$	>> floor(−11.21) ans = −12
26	gamma(x)	Gamma function	$\Gamma(x)=\int_0^{\infty} e^{-t}t^{x-1}dt$	>> gamma(6) ans = 120
27	log(x)	natural (base e) logarithm	$ln\ x$	>> log(10.0000) ans = 2.3026
28	log10(x)	Napierian (base 10) logarithm	$log_{10}\ x$	log10(10.0000) ans = 1
29	log10(x)/ log10(a)	Base *a* logarithm: $log_a x = \frac{log_{10} x}{log_{10} a}$	$log_a\ x$	>> log10(10.0000)/ log10(2.7) ans = 2.3182
30	pi	The number π (circumference-to-diameter ratio)	π -	>> 3/2*pi ans = 4.7124
31	round(x,n)	Round to the nearest decimal or integer	$[x]$	>> round(12.7252,2) ans = 12.7300
32	sec(x)	Secant	$sec\ x$	>> sec(pi/3) ans = 2.0000
33	secd(x)	Secant with x in degrees	$\sec x$	>> secd(60) ans = 2.0000
34	sin(x)	Sine	$\sin x$	>> sin(pi/2) ans = 1
35	sind(x)	Sine with x in degrees	$\sin x$	>> sind(90) ans = 1
36	sinh(x)	Hyperbolic sine	$\sin h\ x$	>> sinh(pi/2) ans = 2.3013

(*Continued*)

TABLE 2.1 (*Continued*)
Some Available Elementary, Trigonometric, and Specialized Mathematical Functions (Alphabetically)

No	MATLAB® Command	Name	Math Notation	Example (Input and Output)
37	sqrt(x)	Square root	$\sqrt{x}$	>> sqrt (1.9999/2.9999) ans = 0.8165
38	tan(x)	Tangent	tan x	>> tan(pi/3) ans = 1.7321
39	tand(x)	Tangent with x in degrees	tan x	>> tand(60) ans = 1.7321
40	tanh(x)	Hyperbolic tangent	tan$h$ x	>> tanh(pi/3) ans = 0.7807

Examples of arithmetic expressions written in the Command Window with explanations about the order of operations are provided below. Each command line is provided with explanatory text written as MATLAB® comments; for execution, expressions can naturally be typed in the Command Window without these comments (explanatory comments are used from hereinafter in this chapter).

```
>>6.28+3/5*2     % first 3/5 is executed, the result is
                 multiplied by 2, and then 6.28+
ans =
  7.4800
>>(6.28+3)/5*2      % first 6.28+3 is executed, the result is
                    divided by 5, and then *2
 ans =
   3.7120
>>2/(5*4)                 % first 2*5 is executed and then 2/
ans =
  0.1000
>>2*5/4   % first 2*5 is executed, the result is divided by 4
ans =
  2.5000
>>1.9812^4.1/3     % first 1.9812^4.1 is executed and then /3
ans =
    5.4990
>>1.9812^(4.1/3)    % first 4.1/3 is executed and then 1.9812^
ans =
  2.5457
```

```
>>1.9812^4.1/3,1.9812^(4.1/3) % two above expressions are
                              written in the same line
ans =
  5.4990
ans =
  2.5457
>>(6.9+2.1)/5.71    % first 6.9+2.1 is executed and then /5.71
ans =
  1.5762
>>8.4761+3.1\6.2    % left division: divide 6.2 by 3.1 (not
                    3.1 by 6.2) and 8.4761+ next
ans =
  10.4761
>>2.2207*10^32    % result is displayed in scientific notations
                  (Subsection 2.2.5)
ans =
  2.2207e+32
>>2.1^.12-1.72^(1.1/3.9)+log10(14.9*.1005)/... % write ellipsis,
                                                then Enter,
asin(pi/7)-sqrt(7.4)    % and continue the expression on the
                        next line
ans =
-2.4157
>>(1/(1+7.1*5))^-1.7    % innermost () are executed first, then
                        1/ and ^
ans = 452.7911
```

The numbers that have been calculated/entered are shown on the display in the default format – digits from 0.01 to 1000 with four decimal digits after a fixed point; the last decimal digits are rounded (information on the output formats is provided in Section 2.1.6).

2.2.3 About Online Help and Help Window

Thorough information about the commands, their functionality, and usage with examples can be obtained using the commands or the Help window. To access the command functionality description, type and enter help and the command name after a space. For example, entering

```
>> help format
```

provides explanations about available output formats and gives examples of its usage strictly in the Command Window.

If searching for a command concerning a topic of interest, the lookfor command may be used. For example, to obtain list MATLAB® command/s on the subject of the pressure you can type and enter

```
>>lookfor force
```

after a rather long search, a list of commands is displayed on the screen as shown below (incomplete)

`rigidode`	Euler equations of a rigid body without external forces
`snapnow`	Force a snapshot of output
`alignRows Force`	brushed rows for variables in the specified axes to have columns
`forceLayout`	Force-directed node layout
`forceLayout3 FORCELAYOUT`	Force-directed node layout
`checkInputEventMappings`	enforces the business rules that
`sf _ force _ open _ machine`	MAHCINEID = SF FORCE OPEN SELECTION (MACHINENAME)
`convforce`	Convert from force units to desired force units

...

To obtain further information, click on the selected command in the list (for example on **convforce**) or use the **help** command (e.g. `>>help convforce`). To interrupt the search process started with the **lookfor** (or also any other) command, the two keyboard keys should be pressed together - **ctrl** and **c**.

Note:

- The **lookfor** command may produce a different list of information at different computers; this is determined by the installed toolbox set. For example, for the requested `convforce`, Aerospace toolbox™ must be installed on your computer.

For opening reference page containing maximum information about the defined command, the `doc` command can be used, for example: `>>doc convforce`. In this case, the Help window (called also Help browser) will open – Figure 2.4.

The Help browser, opened with the above `doc` command, comprises the menu strip, the Search Help field, the narrow Contents pane (to the left) with Contents button , and the broad pane (to the right) with the defined specific information on the subject. Information of interest can also be obtained by typing the word/s into the Search Help field at the top of the window. In the latter case, the Contents pane contains information about product, category, or type, while the Results panel presents a list of available informational units with a brief description for each of the defined units; the description includes the toolbox and section names for which the information is relevant. For our example, the `Documentation > Aerospace Toolbox > Unit Conversions line` designates the location of the documentation page containing explanations on the requested convforce command.

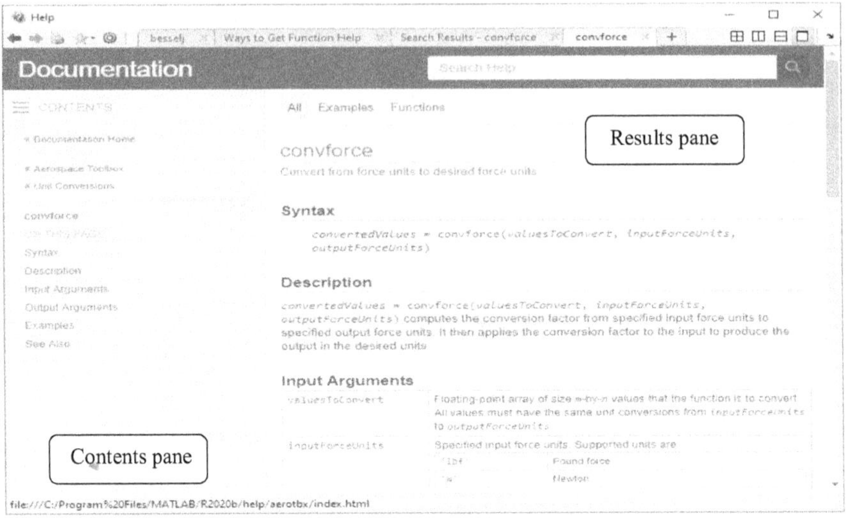

FIGURE 2.4 Help browser containing the Contents and Results panes with information about the **convforce** command, which converts force units.

The Help browser can also be accessed in various additional ways, for example, by selecting the Documentation line 

in the popup menu of the Help button 

located in the RESOURCES section of the HOME tab of the Desktop toolstrip.

2.2.4 Available Toolboxes

As mentioned above, online help provides information pertained as for the basic MATLAB® so for the installed toolboxes. The former includes info/description about mathematical functions/operations (such as sin, cos, sqrt, exp, log, etc.) used in a wide range of sciences while the latter includes info on some special commands solving the problems of some specific scientific/technical areas. A number of problem-oriented tools called toolboxes have been developed to use the software in specific scientific/technical fields. For example, commands for aerospace problems are in the Aerospace toolbox; commands related to statistics in the Statistics toolbox; and commands for neural networks in the Neural Network toolbox, to name just a few. A Partial Differential Equations Toolbox was developed to solve 2D PDE problems; later, its possibilities were expanded to solve 3D problems.

To verify which toolboxes are available on your computer, the `ver` command is used. The command has two simple forms:

`ver` and `ver` toolbox_name

After entering this first command, the header displays the product information and lists the toolbox names, versions, and releases as follows:

```
>> ver PDE
-----------------------------------------------------------------
MATLAB Version: 9.9.0.1467703 (R2020b)
MATLAB License Number: 40558492
Operating System: Microsoft Windows 10 Pro Version 10.0
(Build 19041)
Java Version: Java 1.8.0_202-b08 with Oracle Corporation
Java HotSpot(TM) 64-Bit Server VM mixed mode
-----------------------------------------------------------------
MATLAB                          Version 9.9         (R2020b)
Simulink                        Version 10.2        (R2020b)
5G Toolbox                      Version 2.1         (R2020b)
AUTOSAR Blockset                Version 2.3         (R2020b)
Aerospace Blockset              Version 4.4         (R2020b)
Aerospace Toolbox               Version 3.4         (R2020b)
...
```

Here is a shortened list of toolboxes. In reality, it can be quite long, depending on the installed toolboxes.

The second command form can be used to obtain info about some actual toolbox, for example about the PDE toolbox under study:

```
>> ver
-----------------------------------------------------------------
MATLAB Version: 9.9.0.1467703 (R2020b)
MATLAB License Number: 40558492
Operating System: Microsoft Windows 10 Pro Version 10.0
(Build 19041)
Java Version: Java 1.8.0_202-b08 with Oracle Corporation
Java HotSpot(TM) 64-Bit Server VM mixed mode
-----------------------------------------------------------------
Partial Differential Equation Toolbox Version 3.5 (R2020b)
```

Alternatively, you can get information about the available toolboxes by selecting the Manage Add-Ons 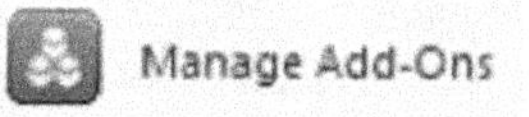

line in the Add-Ons pulldown button 

in the ENVIRONMENTS section of the MATLAB® HOME tab.

2.2.5 Variables and Managing Them

In programming, a variable is a symbol (usually a letter), namely character/s and/or number/s, to which some value is assigned. A variable contains a single number (scalar) or a table of numbers (array). MATLAB® allocates space in computer memory to store both variable name and its value/s. The variable name can contain up to 63 characters, and the first character should be a letter (not a number).

Names of the existing commands or permanently stored MATLAB® variables (e.g. sind, ans, log, pi, sqrt, etc.) are not recommended for use as variable names because they would confuse the system. Some examples of the assignment of numbers to variables and their use in algebraic calculations are:

```
>>a1=-3.142     % the value -3.142 is assigned to variable a1
a1 =
-3.1420
>>b1=pi   % the value of pi=3.1415… is assigned to variable b1
b1 =
3.1416
>>A=sqrt(a1^2+b1^2) %calculated value is assigned to variable c
A =
4.4432
>>d=log10(A)     % calculated value is assigned to variable d
d =
0.6477
```

Some specific variables, their names, and assigned values are permanently stored in MATLAB®. These variables are termed *predefined*. These variables are pi, ans, inf (infinity; is produced, for example, by dividing by zero), i or j (square root of –1), NaN (not-a-number, appears when a numerical value is moot, e.g., 0/0), and eps (smallest allowed difference between two numbers, its value being 2.2204e–16).

To manage variable, the following commands are used:

`clear` - removes all variables from memory, or `clear var1 var2 …` -removes the named variables `var1 var2 …`;
`who` -displays the names of variables;
`whos` - displays variable names, matrix sizes, variable byte sizes and variable storage classes (numeric variables pertain to the double precision class by default).

Each variable appears in the Workspace Window with the same information as in the case of whos, he marked by the icon  (this icon varies for different variable classes). As mentioned above (Section 2.2.1), possible additional information about a variable can be obtained in the Workspace window by clicking the mouse's right button when the cursor is placed on the variable header line; the pop-up menu appears with names of additional data that you want to display (namely: Bytes, Class, Min, Max, Range, etc.).

2.2.6 Screen Output Formats

The numeric result of a command execution is displayed on the screen in a certain format; by default, the four decimal digits are displayed, for example, 3.1416 (number π), last digit is rounded. This format is termed short. When the real decimal number is lesser than 0.001 or greater than 1000 (e.g. 1000.0001), the number is shown in the so-called scientific notations – shortE format. In this notations: a number between 1 and 10 multiplied by a power of 10. For example,

- the aria moment of inertia of the filled circle ($I = \pi r^4/4$) with radius $r = 0.5$ m displayed as 4.9087e–02 (m^4) should be read as $4.9087 \cdot 10^{-2}$;
- the Boltzmann constant displayed as 1.3806e–23 (in J·K^{-1}) should be read as $1.3806 \cdot 10^{-23}$,
- the modulus elasticity in tension (Young's modulus) for Brass is displayed as 1.2500e–7 (in Pa) should be read as $1.25 \cdot 10^{-7}$;

The number a = 1000.1 is displayed the same in short and shortE formats as 1.0001e+ 03 where e+ 03 is 10^3 while the whole number should be read as $1.0001 \cdot 10^3$.

Scientific notations are used in another format – engineering, shortEng; the difference is in representation exponent as multiple of three and having three-digit order. For example, the π number is displayed as 3.1416e + 000 in engineering versus 3.1416e + 00 in scientific format, and Boltzmann constant is 13.8060e–024 (note, 24 is multiple of three) in engineering versus 1.3806e–23 (23 is not multiple of three) in scientific format.

Note that scientific and engineering notations can also be used for variable input. For example, the command >> d = 5.5e–3 enters value $5.5 \cdot 10^{-3}$ into the d variable in both scientific and engineering formats.

The display format of numbers can be governed by the format command. The command takes the forms

```
format and format type_of_format
```

The first command form is used to sets the short format type. The `type_of_format` parameter in the second command form is a word that specifies the type of the displayed numbers. There are more than three described format types. So, the `long`, `longE`, or `longEng` format types should be used to show 15 decimal digits of the number. For example, when setting the longE format type and inputting the Boltzmann constant $k_B = 1.38064852 \cdot 10^{-23}$, MATLAB® yields the following results:

```
>> format longE
>> k_B=1.38064852e-23
k_B =
1.380648520000000e-23
```

Note:

- The formatting commands change the length of the numbers on the display but do not change the number representation in the computer memory, nor do they affect the type of the numbers entered.
- Once a certain format type is specified, all subsequent numbers are displayed within it.
- To restore the default (`short`) format, the `format` or `format short` command must be entered.
- The `type_of_format` can be entered with a space between `short` and `E` or `Eng` and `long` and `E` or `Eng` while the capital `E` can be written as a conventional `e`.
- To minimize the space between outputted lines the compact type of format can be used.
- The short or long formats written with the letter G provide the best representation on the display of a 5-digit or 15-digit number, respectively. For example, in the longG format, the entered Boltzmann constant appears without trailing zeros (compare with the previously inputted and outputted k_B):

```
>> format longG
k_B =
1.38064852e-23
```

The formats with `G` letter are suitable for when the outputted table contains numbers very different in size since the format picks where it is best to use the regular (e.g., `short`) and scientific (e.g., `shortE`) notation.

For more information on the described and other available formats, use the help format or doc format command.

2.2.7 Commands for Output

Apart from the formatting, two commands, disp and fprintf, are used for additional control over the output. The texts or variable values are displayed by disp without the = (equal) sign. Every new disp command yields its result in a new line. Two general forms of the command reads:

```
disp('Text string') or disp(variable_name)
```

The text between the quotes is displayed in blue.

As an example, the following lines input the value of the fuel consumption for a car Qs = 67.3 mpg and display it without and with the disp command:

```
>> Qs=67.3   % input and displaying the mu value without the
             disp command
```

```
Qs =
67.3000
>>disp('Car fuel consumption'),disp(Qs) % first disp displays
                                          the string; second
                                          - Qs value
Car fuel consumption
67.3000
```

The second of the output commands – fprintf – is used to display texts and numbers in the same line or save them in a file. Here we present the simplest form of the command, which is sufficient for displaying the calculating results:

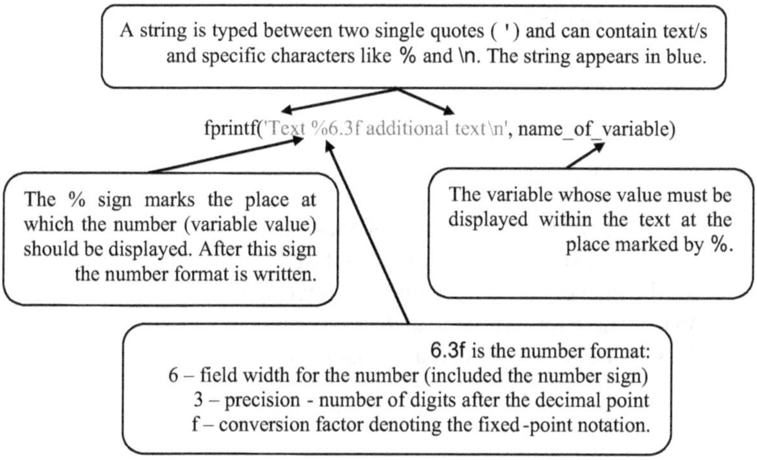

A string is typed between two single quotes (') and can contain text/s and specific characters like % and \n. The string appears in blue.
fprintf('Text %6.3f additional text\n', name_of_variable)
The % sign marks the place at which the number (variable value) should be displayed. After this sign the number format is written.
The variable whose value must be displayed within the text at the place marked by %.
6.3f is the number format:
6 – field width for the number (included the number sign)
3 – precision - number of digits after the decimal point
f – conversion factor denoting the fixed-point notation.

To use the fprintf command for output, the following rules should be used:

- Write the \n (slash n) before the word/s that you want to see on the new line, the same make for appearance of the >> prompt on the new line, after executing the fprintf command.
- The % (percent) sign and the character f (termed the conversion character) are obligatory but the field width number (represented the number of positions) number and number of digits after the point are optional. For example, if the %f characters were written instead of %6.3f (see command form above), the number will be displayed by default with six digits after the point.
- The addition of several %f units (or full formatting elements) permits the inclusion of multiple variable values in the text.
- The f character means that real number must be converted to output a value with a fixed number of digits after the decimal point.

Some additional conversion characters that can be used are:

i (or d) – integer (or decimal) notation,
e – exponential notation (e.g., 2.309123e+ 001),
g – the e or f notations whichever is more compact (for example, omitting trailing zeros).

The following example uses the fprintf command to display three lines containing the title "Fuel consumption, mpg", the word "Ford Focus" and its fuel consumption value Qs = 67.3 with one decimal digit, and the "Volkswagen Golf" word and its fuel consumption value Qs = 74.3 with one decimal digit:

```
>>Qs1=67.3;Qs2=74.3;
>>fprintf(' Fuel consumption, mpg\n Ford Focus %6.1f\n
Volkswagen Golf%6.1f\n',Qs1,Qs2)
Fuel consumption, mpg
Ford Focus 67.3
Volkswagen Golf 74.3
```

Here, exemplarily, six positions were assigned for each of the numbers; this format includes two positions for decimal digit and decimal point as well as four positions for the integer part of a number, number sign, and two preceding spaces. Note that \n used here for write title and each car model with fuel consumption on separate lines. The color of the text and characters inputted between the quotes is blue, the same as in the disp command.

In case we use the %f form (without number format), the fprintf command and its output looks like:

```
>>fprintf(' Fuel consumption, mpg\n Ford Focus %f\n
Volkswagen Golf%f\n',Qs1,Qs2)
Fuel consumption, mpg
Ford Focus 67.300000
Volkswagen Golf74.300000
```

The output commands discussed here can also be used to display tables when the variables are vectors, matrices, or arrays (Section 2.3.6).

2.2.8 Application Examples

Here are examples of using elementary commands to perform some engineering calculations in the interactive MATLAB® mode.

2.2.8.1 Voltage between Intermediate Points of the Wheatstone Bridge

Voltage between intermediate points of the measuring bridge (Wheatstone bridge) can be calculated using the following expression:

$$V_G = V_s\left(\frac{R_2}{R_1+R_2} - \frac{R_4}{R_3+R_4}\right)$$

where R_1, R_2, R_3, and R_4 are resistances of the bridge legs, and V_s and V_G are the supply and galvanometer voltages respectively.

Problem: Calculate voltage V_G when V_s = 12 V, R_1 = 110.3 Ω, R_2 = 109.4 Ω, R_3 = 112.1 Ω, and R_4 = 108.5 Ω. Display the result without and with the disp command, and in the last case add the string "Voltage, [volts]" above the calculated value.

The solution is:

```
>>Vs=12; %volts
>>R1=110.3; %ohms
>>R2=109.4; %ohms
>>R3=112.1; %ohms
>>R4=108.5; %ohms
>>V_G=Vs*(R2/(R1+R2)-R4/(R3+R4))        % output without disp
V_G =
0.0733
>>disp('Voltage, [volts]'),disp(V_G) % output using disp
Voltage, [volts]
0.0733
```

Note, two disp are written in the same line to display the text (header string) and number on the two consecutive lines. If we write these commands in two separate lines, we get the text line after the first disp input and get the line with the V_G value after the second input of the disp command. This situation arises only with interactive calculations. When commands are written as a program, each disp can be written on a separate line that leads to the same results as the above example.

2.2.8.2 Threaded Bolt: Stiffness Value Estimation

When the bolt is passed through a hole, its stiffness k (in N/m) can be estimated using the following expression (Burr & Cheatham, 1995):

$$k = \frac{\pi E d_0 \tan(30°)}{ln\dfrac{\left(d_1 + h\cdot\tan(30°) - d_0\right)\left(d_1 + d_0\right)}{\left(d_1 + h\cdot\tan(30°) + d_0\right)\left(d_1 - d_0\right)}}$$

where E is the tensile modulus (Young's modulus) of the bolt material, h is the height of the bolt, d_0 and d_1 are the hole and washer face diameter respectively.

Problem: Write commands to calculate the stiffness k of the bolt when h = 3.2 cm, d_0 = 6.4 mm, d_1 = 16 mm and E = 5.3·10^{9}N/m^2; input geometrical parameters in meters and use for numbers the scientific notations. Write commands that display the result without and with the disp command.

The solution is:

```
>>h = 3.2e-2; %scientific notations, h in m
>>d0 = 6.4e-3; %scientific notations, d0 in m
>>d1 = 16e-3; %scientific notations, d1 in m
>>E = 5.3e9; %scientific notations, E in N/m^2
>>d2 = d1+h*tand(30); %tand uses the angle in degrees
>>k = pi*E*d0*tand(30)/log((d2-d0)*(d1+d0)/(...
(d2+d0)*(d1-d0))) % ... - for continuing expr.on the next line
k =
1.3044e+ 08
>>disp('Bolt stiffness, N/m'),disp(k) %output with two DISPs
Bolt stiffness, N/m
1.3044e+ 08
```

Note the angles in trigonometric function above may be inputted in degrees because it is written with the final letter d.

2.2.8.3 Stress State for a Rectangular Plate with a Crack

The stress intensity factor for a rectangular plate with centrally located crack can be calculated by the following expression (Rooke & Cartwright, 1976)

$$K = \sigma\sqrt{\pi a}\left[\frac{1-\dfrac{a}{2b}+0.326\left(\dfrac{a}{b}\right)^2}{\sqrt{1-\dfrac{a}{b}}}\right]$$

where a is the half length of the crack, h and b are the half width and half height of the plate respectively, and σ is the uniaxial tension.

Problem: Write commands to calculate stress intensity factor K, $\text{kPa}\cdot\text{m}^{1/2}$ when $a = 0.0379\,\text{m}$, $b = 0.1009\,\text{m}$ and $\sigma = 79.9382\cdot10^3\,\text{kPa}$. Output results with and without the fprintf command. In the latter case, display in the same line:

the words "Stress intensity =";

the resulting K value with two digit after the decimal point;

the units of K.

The solution:

```
>> a=.0379;b=.1009;sigma=79.9382e3; % all assignments in one line
>> ab=a/b;
>> K=sigma*sqrt(pi*a)*(1-ab/2+0.326*ab^2)/sqrt(1-ab)
K =
2.9957e+04
>> fprintf('Stress intensity=%8.2f kPa·m^(1/2)\n',K)
Stress intensity =29957.49 kPa·m^(1/2)
```

2.2.8.4 Bravais Lattice Cell Volume

A material lattice has the unit cell of the triclinic crystal symmetry with the following parameters: $a = 3.99$ Å, $b = 4.79$ Å, $c = 2.539$ Å, $\alpha = \beta = \pi/2$ and $\gamma = 1.9$ rad. The volume of the unit cell can be calculated with the expression:

$$V = abc\sqrt{1-\cos^2\alpha-\cos^2\beta-\cos^2\gamma+2\cos\alpha\cos\beta\cos\gamma}$$

Problem: Calculate the unit cell volume with the above expression. Display the result twice: first with the disp command and then with the fprintf; in latter case display the result with three decimal points. Display the title "Volume, Angstroms^3" in each output case.

The solution is:

```
>>a=3.99;b=4.79;c=2.539;gamma=1.9; % all assignments in one line
>> alpha=pi/2;beta= alpha;
>> V=a*b*c*sqrt(1-cos(alpha)^2-... % to move to the next line
cos(beta)^2-cos(gamma)^2+2*cos(alpha)*cos(beta)*cos(gamma));
>>disp('Volume, Angstroms^3'),disp(V) % two disps in one line
Volume, Angstroms^3
45.9198
>>fprintf('Volume, Angstroms^3\n %7.3f\n',V)
Volume, Angstroms^3
45.920
```

Note: When the fprintf is used, the number can be centered with the title on the preceded line while the disp does not allow it.

2.3 VECTORS, MATRICES, AND ARRAYS

A table containing rows and columns with numerical data is essentially an array or matrix. Previously, we used scalar variables, each containing a number. In MATLAB®, each of these variables is a 1×1 matrix. To manipulate matrices and arrays, you need to use linear algebra operations for matrices and element-wise operations for arrays.

2.3.1 Managing Vectors, Matrices, and Arrays

2.3.1.1 Vector Representation

A vector is a sequence of numbers organized into rows or columns and called a row or column vector, respectively.

In MATLAB®, vector is generated with square brackets and numbers into them. When a row vector is created, the spaces or commas should be typed between the numbers. When a column vector is created, the semicolon is typed or the **Enter** key is pressed after each number.

TABLE 2.2
Thread Data for Vector Representation

Diameter, mm	1.00	5.00	10.00	20.00	30.00	45.00	60.00
Pitch, mm	0.25	0.8	1.5	2.5	3.5	4.5	5.5

As an example, generate two vectors from the two rows of Table 2.2 containing the nominal (major) thread diameter and the thread pitch data. The first row in the table represents row vector with name diam and the second row as column vector with name pitch.

```
>>diam=[1 5 10 20 30 45 60] % row vector
diam =
1 5 10 20 30 45 60
>>pitch=[0.25;0.8;1.5;2.5;3.5;4.5;5.5] % column vector
pitch =
0.2500
0.8000
1.5000
2.5000
3.5000
4.5000
5.5000
```

The generated vectors are displayed in two or more lines with a message line informing which column is presented in the line when the screen is not wide enough, for example:

```
>> diam =
 Columns 1 through 6
   1   5   10   20   30   45
Column 7
  60
```

The message states that values in columns 1...6 are shown in the next line.

The message states that the value in the 7th column is shown in the next line.

In vector analysis, the position of a point is presented by three coordinates, and for example, A-point is presented as a position vector $r_A = 3i{-}5j + 8k$ (where i, j, and k are so-called unit vectors and 3, −5, and 8 are the vector projections on the axis). In MATLAB® notations, it can be written as row vector A = [3, −5,8].

Frequently the values of the adjacent numbers within a vector differ by the same value. For example, in vector vec = [2 5 8 11 14], the spacing between two adjacent elements is 3. Such vectors can be created with one of two commands - ':' (colon) or linspace. The first of these commands has the form

```
vector=a:h:b
```

where a, b, and h are, respectively, the first and last term and the step between adjacent terms within the vector with the assigned name vector. The numbers are generated so that the last number cannot exceed the last term b. If h = 1 then the step h can be omitted, because it is assumed by default. Examples of using this command are:

```
>> p=0.015:0.16:1.9    % 1st number 0.015,last number <=1.9,
                        step 0.16
p =
Columns 1 through 6
0.0150  0.1750  0.3350  0.4950  0.6550  0.8150
Columns 7 through 12
0.9750 1.1350 1.2950 1.4550 1.6150 1.7750
>> x=-2:5% 1st number -2, last number 5, by default step 1
x =
-2-1 0 1 2 3 4 5
>>v1=0.3:0.7/4:1%1st number 0.3, last number <=1, step 0.7/4
v1 =
0.3000  0.4750  0.6500 0.8250  1.0000
>>v2=10.1:-2.1:0.45%1st number 10.1, last number >=0.45, step 2.1
v2 =
10.1000  8.0000  5.9000  3.8000  1.7000
```

The second command producing a vector with a constant step between adjacent numbers has the form

```
vector=linspace(a,b,n)
```

where a and b are the first and last numbers, respectively, and n is the required amount of numbers. The n can be omitted, in this case n is set to 100 by default.

For example:

```
>>x=linspace(0,3,4) % 4 numbers, 1st number 0, last number 3
x =
0 1 2 3
>>y=linspace(-7,7,5) % 5 numbers, 1st number -7, last number 7
y =
-7.0000  -3.5000  0 3.5000  7.0000
>> z=linspace(10.1,2.4,4)%4 numbers,1st number 10.1,last
                         number 2.4
z =
10.1000  7.5333  4.9667  2.4000
>> v=linspace(0,100)  %100 numbers(default), 1st number 0,
                       last 100
v =
Columns 1 through 6
```

```
0 1.0101  2.0202  3.0303  4.0404  5.0505
... (shortened, given only the six first numbers out of 100)
```

The position of a vector element is its address that should be a positive integer (not zero). For example, in the previously generated vector diam (with the first line numbers of Table 2.2):

- The fourth position can be addressed as diam(4), and the element located here is the number 20.
- If we enter the diam(0) command, an error message appears since the element number cannot be zero.

The last term address in a vector may be designated with the end terminator, for example, diam(end) is the last position in the vector diam and marks the number 60. Another way to address the last term is to write its position number (if you know it), in this example - diam (7).

2.3.1.2 Matrices and Arrays Representation

A two-dimensional array or matrix resembles a numerical table and contains the rows and columns with numerical data. When the number of rows is equal to the number of columns, the matrix is called square; otherwise, it is called rectangular. As with a vector, matrix elements must be entered within the square brackets with spaces or commas between the elements and using a semicolon or pressing the ENTER key between the lines. Each matrix row should have the same number of elements. The elements can be specified also using variable names or mathematical expressions.

As an example, represent some parameters of the spur gear, Table 2.3, in matrix form.

```
>>M=[10 1.062 1.564;20 1.031 1.569;30 1.021 1.570;40 1.015 1.5704]
```

A semicolon is placed before inputting the new line with numbers.

```
M =
   10.0000    1.0620    1.5640
   20.0000    1.0310    1.5690
   30.0000    1.0210    1.5700
   40.0000    1.0150    1.5704
```

TABLE 2.3
Some Parameters of Spur Gear for Matrix Representation

Number of Teeth	Teeth Height, mm	Circular Thickness, mm
10	1.062	1.564
20	1.031	1.569
30	1.021	1.570
40	1.015	1.5704

Some other matrix generation examples are:

```
>> B=[-20 41.7% here pressed Enter before inputting the next line
20-7.14]
B =
-20.0000  41.7000
20.0000   -7.1400
>>d1=-2.24; d2=pi/6;
>>D=[d1 d2; % elements are assigned with variables
sin(pi*d1/180) log(d2)] %elements are assigned using
expressions
D =
-2.2400   0.5236
-0.0391  -0.6470
```

Row and column numbers should be used to address the matrix element/s. For instance, for the M matrix with some gear data, entering M(3,2) refers to the number 1.0210 and M(2,3) –the number 1.5690. Note that row or column numbering begins with 1. Therefore, for example, the first element in matrix M designated M(1,1) is 10.0000. For addressing the several consecutive elements of a row/column, the semicolon character can be used. For example, M(1:2,3) refers to the numbers in the first and second rows of column 3 in matrix M. An address containing a single colon sign refers to an entire row or column, for example, M(:,3) refers to the elements of all rows in column 3 and M(2,:) to those of all columns in row 2.

In addition to the discussed two-number addressing (row-column), linear one-number addressing can be used. In this case, the element's address is detected sequentially, from the first element and down the first column, then continuing analogously with the second column and so forth, up to the required element in the matrix. For example, in accordance with the one-number addressing, the M(7) command refers to element M(3,2), M(8) to the M(4,2) while M(5:8) is the same as M(:,2), etc.

You can generate a new matrix by combining an existing matrix with a vector or another matrix in the square brackets. These kinds of examples are:

```
>>V=[1;2;3;4] % generates column vector V with serial
numbers
No =
1
2
3
4
>> M_V=[V M] % vector V in 1st column and matrix M in others
M
M_V =
1.0000  10.0000  1.0620  1.5640
2.0000  20.0000  1.0310  1.5690
```

```
3.0000   30.0000   1.0210   1.5700
4.0000   40.0000   1.0150   1.5704
>>M_V(3,2) % refers to the element in 3rd row and 2nd column
ans =
30
>>M_V(1:3,1)% refers to the elements in column 1 and rows 1, 2, 3
ans =
1
2
3
>>M_V(2,2:4) % refers to the elements in row 2 and columns 1…3
ans =
20.0000   1.0310   1.5690
>>M_V(3,:) % refers to all columns in row 3
ans =
3.0000 30.0000 1.0210 1.5700
>> M_V(2:4,2)=5.1% replaces the elements 2…4 in row 2 with 5.1
M_V =
1.0000   10.0000   1.0616   1.5643
2.0000   5.1000    1.0308   1.5692
3.0000   5.1000    1.0206   1.5700
4.0000   5.1000    1.0154   1.5704
```

A row/column vector can be converted into a column/row one and the matrix rows/columns can be swapped using the transpose operator ‘ (quote):

```
>>M_tr=M_V' %M_V' changes rows with columns and assigns to M_tr
M_tr =
1.0000    2.0000   3.0000   4.0000
10.0000   5.1000   5.1000   5.1000
1.0620    1.0310   1.0210   1.0150
1.5640    1.5690   1.5700   1.5704
```

Note that the rows and columns of the M_V matrix were changed, and this result was assigned to the M_tr matrix.

2.3.2 Mathematical Manipulations with Matrices

Vectors, matrices, and arrays are not always used in the same way as in various mathematical operations on individual variables. If addition/subtraction are performed in the same way as in traditional arithmetic, then multiplication/division/exponentiation is different. Moreover, there are differences between operations on arrays and matrices. The commands for manipulations with matrices are discussed below.

2.3.2.1 Addition and Subtraction

These operations are performed element by element and only matrix equality is required in size, for example, when M1 and M2 are two equal-sized matrices:

$$M1 = \begin{bmatrix} M1_{11} & M1_{12} & M1_{13} \\ M1_{21} & M1_{22} & M1_{23} \end{bmatrix} \text{ and } M2 = \begin{bmatrix} M2_{11} & M2_{12} & M2_{13} \\ M2_{21} & M2_{22} & M2_{23} \end{bmatrix}$$

the sum or subtraction of these matrices $M = M1 \pm M2$ is

$$M = \begin{bmatrix} M1_{11} \pm M2_{11} & M1_{12} \pm M2_{12} & M1_{13} \pm M2_{13} \\ M1_{21} \pm M2_{21} & M1_{21} \pm M2_{21} & M1_{23} \pm M2_{23} \end{bmatrix}$$

The commutative law is valid for these operations, namely $M1 \pm M2 = M2 \pm M1$.

2.3.2.2 Multiplication

This is a more complex operation, performed in accordance with the rules of linear algebra, and is feasible only when the number of columns in the first matrix is equal to the number of rows in the second matrix.

The first and second of the above matrices $M1$ and $M2$ have size 2×3 each and cannot be multiplied as the number of columns in the first of them is 3, and the number of rows in the second is 2. For further explanations, replace matrix $M2$ with another 3×2 matrix

$$M2 = \begin{bmatrix} M2_{11} & M2_{12} \\ M2_{21} & M2_{22} \\ M2_{31} & M2_{32} \end{bmatrix}$$

Now, the number of columns in the matrix $M1$ and the number of rows in the new matrix $M2$ (such numbers called inner size of the multiplied matrices) are equal ($2 \times \mathbf{3} * \mathbf{3} \times 2$, the inner sizes are 3, bold), and it becomes possible to multiply $M1 * M2$. The result of this multiplication by the rules of linear algebra is

$$M1 * M2 = \begin{bmatrix} M1_{11} & M1_{12} & M1_{13} \\ M1_{21} & M1_{22} & M1_{23} \end{bmatrix} * \begin{bmatrix} M2_{11} & M2_{12} \\ M2_{21} & M2_{22} \\ M2_{31} & M2_{32} \end{bmatrix}$$

$$= \begin{bmatrix} M1_{11}M2_{11} + M1_{12}M2_{21} + M1_{13}M2_{31} & M1_{11}M2_{12} + M1_{12}M2_{22} + M1_{13}M2_{32} \\ M1_{21}M2_{11} + M1_{22}M2_{21} + M1_{23}M2_{31} & M1_{21}M2_{12} + M1_{22}M2_{22} + M1_{23}M2_{32} \end{bmatrix}$$

It is not difficult to verify that the product $M1 * M2$ is not equal to the product $M2 * M1$; this means that the commutative law does not retain for matrix multiplication. The presented rule of multiplication and conclusion about the commutative law is true of course for vectors.

Some examples of addition, subtraction, and multiplication of matrices and vectors are:

```
>> A1=[0.6 2.1;1.1 0.3;3.9 2.7] % generates with; between rows
A1 =
0.6000   2.1000
1.1000   0.3000
3.9000   2.7000
>>A2=[2.2 2.3 0.1      % generates using "Enter" between rows
1.7  0.1  0.3
0.5  0.8  2.4];
>>A1*A2% A1-column number ≠ A2-row number, error message appears
Error using *
Incorrect dimensions for matrix multiplication. Check
that the number of columns in the first matrix matches
the number of rows in the second matrix. To perform
elementwise multiplication, use '.*'.
>>A=A2*A1          % A2-column and A1-row numbers are equal (=3)
A =
4.2400   5.5800
2.3000   4.4100
10.5400 7.7700
>>C1=A2(1,:),C2=A1(:,2) % gets row and column vectors from A1&A2
C1 =
2.2000   2.3000   0.1000
C2 =
2.1000
0.3000
2.7000
>> C1*C2        % multiplies column vector C1 by row vector C2
ans =
5.5800
>> C2*C1% multiplies row vector C2 by C1, not the same as C1xC2
ans =
4.6200   4.8300   0.2100
0.6600   0.6900   0.0300
5.9400   6.2100   0.2700
```

One of the possible applications of matrix multiplication to the engineering problems is presenting a set of linear equations as matrix equation $AX = B$. For example, the following set of two equations with two variables

$$a_{11}x_1 + a_{12}x_2 = b_1$$

$$a_{21}x_1 + a_{22}x_2 = b_2$$

may be rewritten in matrix form as

$$\begin{bmatrix} a_{11} & a_{11} \\ a_{11} & a_{11} \end{bmatrix}\begin{bmatrix} x_1 \\ x_2 \end{bmatrix}=\begin{bmatrix} b_1 \\ b_2 \end{bmatrix}$$

where $\begin{bmatrix} a_{11} & a_{11} \\ a_{11} & a_{11} \end{bmatrix}$ is the square matrix A and $\begin{bmatrix} x_1 \\ x_2 \end{bmatrix}$ and $\begin{bmatrix} b_1 \\ b_2 \end{bmatrix}$ are column vectors X and B, respectively.

2.3.2.3 Division

This operation is much more complicated than matrix multiplication due to the non-commutative and some other properties of matrices. A comprehensive explanation of this topic can be found in books on linear algebra. Below we briefly describe some features and MATLAB® commands for matrix division.

Identity and inverse matrices are often introduced and used in division operations.

A square matrix whose diagonal elements are equal to 1 and others equal to 0 is called the **identity** matrix I. In MATLAB®, the eye command (see Table 2.4) generates the identity matrix. When multiplying the identity matrix by any square matrix, the commutative low is retained, that is, A by I, or I by A yields the same result: $AI = IA = A$. For example

```
>>A=[80 90 27;14 10 9;12 9 94] % generates a 3x3 square matrix A
>>I=eye(3)        % eye(3) generates the 3x3 identity matrix I
I =
1  0  0
0  1  0
0  0  1
>> A*I
ans =
80  90  27
14  10   9
12   9  94
>>I*A                               % result is the same as with A*I
ans =
80  90  27
14  10   9
12   9  94
```

When AB (left multiplication) and BA (right multiplication) lead to the identity matrix: $AB = BA = I$, the matrix B is termed **inverse** to A. In MATLAB®, the inverse matrix can be written in two ways: B = A^−1 or with the inv command as B = inv(A). For example, multiplying the above matrix A by its inverse A^{-1} we get the identity matrix

```
>> A*A^-1
ans =
1.0000    0.0000   -0.0000
0.0000    1.0000    0.0000
0.0000   -0.0000    1.0000
```

Another feature that is used often when manipulating matrices is left, \, or right, /, division. For example, to solve the matrix equation $AX = B$ (see the

Multiplication above) left division should be used: $X = A\backslash B$. By contrast, to solve the same matrix equation but rewritten in form $XC = B$ the right division should be used: $X = B/C$.

For example, the set of linear equations

$$5.1x_1 - 4x_2 = 16.8$$

$$-3x_1 + 8x_2 = -7.3$$

can be represented with coefficients to the right of unknowns x_1 and x_2

$$x_1 5.1 - x_2 4 = 16.8$$

$$-x_1 3 + x_2 8 = -7.3$$

These two possible forms of the equations representation correspond to the two matrix forms:

$$AX = B \text{ where } A = \begin{bmatrix} 5.1 & -4 \\ -3 & 8 \end{bmatrix}, B = \begin{bmatrix} 16.8 \\ -7.3 \end{bmatrix}, \text{ and } X = \begin{bmatrix} x_1 \\ x_2 \end{bmatrix}$$

and

$XC = D$ with coefficient written as $C = A' = \begin{bmatrix} 5.1 & -3 \\ -4 & 8 \end{bmatrix}$, $D = B' = \begin{bmatrix} 16.8 & -7.3 \end{bmatrix}$, and $X = \begin{bmatrix} x_1 & x_2 \end{bmatrix}$.

To obtain the same solution for the two discussed forms of the set of equations, two forms of matrix division are required – left and right division, the commands being:

```
>>A=[5.1-4;-3 8];               % A is the matrix for form AX=B
>>B=[16.8;-7.3];    % B is the column vector,form AX=B
>> X_left=A\B  % defines X with left division
X_left =
3.6528
0.4573
>> C=A'; % C is the transposed matrix A; form XC=B
>> D=B'; % D is the row vector (transposed B); form AX=B
>> X_right=D/C % defines X with right division
X_right =
3.6528   0.4573
```

Further, in Section 2.5.2, an application example with matrix division is presented.

2.3.3 Elementwise Operations

As stated above, for matrix operations the rules of linear algebra are applied. However, in various engineering calculations, many operations must be performed strictly element-by-element. In such cases, we use the term *array* instead of *matrix*. Just as with addition or subtraction, multiplication, division, and exponentiation are performed element by element, and the equality of the number of elements in the arrays is required. To provide such operations, the *, /, \, and ^ operators are fronted by the period sign, namely

```
.* (for elementwise multiplication);
./ (for elementwise right division);
.\ (for elementwise left division);
.^ (for elementwise exponentiation).
```

For example, if we use these operators for two row vectors of three elements $v = [\,v_1\; v_2\; v_3]$ and $c = [\,c_1\; c_2\; c_3]$, then the multiplication, division, and exponentiation are performed as follows:

$v.{*}c = [\,v_1{*}c_1\; v_2{*}c_2\; v_3{*}c_3]$, $v./c = [\,v_1/c_1\; v_2/c_2\; v_3/c_3]$, $v.\backslash c = [\,v_1\backslash c_1\; v_2\backslash c_2\; v_3\backslash c_3]$, and $v.\hat{}\; c = \left[\,v_1^{c_1}\, v_2^{c_2}\, v_3^{c_3}\,\right]$.

The same operations applied for two arrays $a = \begin{bmatrix} a_{11} & a_{12} & a_{13} \\ a_{21} & a_{22} & a_{23} \end{bmatrix}$ and $b = \begin{bmatrix} b_{11} & b_{12} & b_{13} \\ b_{21} & b_{22} & b_{23} \end{bmatrix}$ are performed as follows:

$$a.*b = \begin{bmatrix} a_{11}b_{11} & a_{12}b_{12} & a_{13}b_{13} \\ a_{21}b_{21} & a_{22}b_{22} & a_{23}b_{23} \end{bmatrix},\; a1./a2 = \begin{bmatrix} a_{11}/b_{11} & a_{12}/b_{12} & a_{13}/b_{13} \\ a_{21}/b_{21} & a_{22}/b_{22} & a_{23}/b_{23} \end{bmatrix},$$

$$a.\backslash b = \begin{bmatrix} a_{11}\backslash b_{11} & a_{12}\backslash b_{12} & a_{13}\backslash b_{13} \\ a_{21}\backslash b_{21} & a_{22}\backslash b_{22} & a_{23}\backslash b_{23} \end{bmatrix},\text{ and } a.\hat{}\,b = \begin{bmatrix} a_{11}^{\;b_{11}} & a_{12}^{\;b_{12}} & a_{13}^{\;b_{13}} \\ a_{21}^{\;b_{21}} & a_{22}^{\;b_{22}} & a_{23}^{\;b_{23}} \end{bmatrix}.$$

Element by element operators are frequently used to calculate a function with an argument specified as a series of values (e.g., x^2 for $x = 1, 2, \ldots, 10$). Array manipulations are demonstrated in the following examples:

```
>>a=[3.81 1.63;-3.37 0.05;-4.02 0.86] % generates a 3x2 array a
a =
3.8100   1.6300
-3.3700  0.0500
-4.0200  0.8600
>>b=[-0.32-2.4;0.73-7.33;0.07 0.95] % generates a 3x2 array b
b =
-0.3200  -2.4000
0.7300   -7.3300
0.0700    0.9500
>>a.*b % Element-by-element multiplication
-1.2192-  3.9120
-2.4601  -0.3665
```

```
-0.2814    0.8170
>>a*b%matrix operation(* without preceded point)does not work here
Error using *
Incorrect dimensions for matrix multiplication. Check
that the number of columns in the first matrix matches
the number of rows in the second matrix. To perform
elementwise multiplication, use '.*'.
>>a./b % Element-by-element right division
ans =
-11.9063   -0.6792
-4.6164    -0.0068
-57.4286    0.9053
>> a.\b % Element-by-element left division of A by B
ans =
-0.0840   -1.4724
-0.2166   -146.6000
-0.0174    1.1047
>>b.^2            % each term in b is raised to the power of 2
ans =
0.1024    5.7600
0.5329    53.7289
0.0049    0.9025
>>x=0:3% generates four-element vector for the two next commands
x =
0  1  2  3
>> y=3.8+exp(-1.2*x.^0.9)%element by element exponentiation: x^0.9
y =
4.8000  4.1012  3.9065  3.8397
>>y=sqrt(x)./(x.^2.2+1) % elementwise division and exponentiation
y =
0   0.5000   0.2528   0.1418
```

2.3.4 Supplementary Commands for Matrix/Array Manipulations

Vectors, matrices, or arrays can contain certain identical or random values generated by special commands. The following commands

```
ones(m,n) and zeros(m,n)
```

generate matrices of m rows and n columns all terms of which are equal to 1 and 0, respectively.

Many scientific and engineering problems, including those described by differential equations, are associated with random numbers, for which the following pseudo-random number generators should be used:

```
rand(m,n) or randn(m,n)
```

The former command yields an m by n matrix with uniformly distributed numbers from 0 to 1, while the latter produces the same size matrix with normally distributed numbers with a mean of 0 and a standard deviation of 1. These commands can be abbreviated to rand(n) and randn(n) to generate a square n by n matrix. Examples are:

```
>>ones(3,2) %generates a 3x2 matrix in which all elements are 1s
ans =
1  1
1  1
1  1
>>zeros(2,3)%generates a 2x3 matrix in which all elements are 0s
ans =
0  0  0
0  0  0
>>a=rand(3,2)%Uniform distr.,3x2 matrix, numbers between 0
and 1
ans =
0.8147  0.9134
0.9058  0.6324
0.1270  0.0975
>>v=rand(1,3)%Uniform distr.,row vector, numbers between 0
and 1
v =
0.8147  0.9058  0.1270
>>b=randn(2,4) % Normal distr., 2x4 matrix with random numbers
b =
0.5377  -2.2588   0.3188  -0.4336
1.8339   0.8622  -1.3077   0.3426
>>w=randn(4,1)% Normal distr., column vector with random numbers
w =
0.5377
1.8339
-2.2588
0.8622
>>d=randn(4) % Normal distr., 4x4 matrix with random numbers
d =
0.5377    0.3188   3.5784   0.7254
1.8339   -1.3077   2.7694  -0.0631
-2.2588  -0.4336  -1.3499   0.7147
0.8622    0.3426   3.0349  -0.2050
```

The randi command (see Table 2.4) generates the integer random numbers.

Note: New random number values are generated every time we reuse the rand, randn or randi command. To restore the random number generator settings to retrieve the same random number values as when you **restart** MATLAB®, you must enter the rng default command as shown below:

TABLE 2.4
Some Commands for Matrix Manipulations and Numerical Analysis (Alphabetically)

Command	Parameter/s	Description	Example (Inputs and Outputs)

char(s1,s2,s3,...)	s1, s2, s3, ... are the strings; may not be the same length	Creates a matrix of string rows, each row contains one string with a length equal to the longest row; missing characters in shorter lines are added with spaces	`>>char('Plant','Gear');`
			`ans =`
			`2×5 char array`
			`'Plant'`
			`'Gear '`

cross(a,b)	*a* and *b* are three-element vectors each	Calculates the cross product of 3D-vectors $\mathbf{a} = a_1\hat{i} + a_2\hat{j} + a_3\hat{k}$, $\mathbf{b} = b_1\hat{i} + b\hat{j} + b\hat{k}$:
		$\mathbf{c} = \mathbf{a} \times \mathbf{b} = (a_2b_3 - a_3b_2)\ \hat{i}$ +
		$(a_3b_1 - a_1b_3)\ \hat{j}$ +
		$(a_1b_2 - a_2b_1)\ \hat{k}$
			`>>a=[2-4 7];`
			`>>b=[1 3-2];`
			`>>c=cross(a,b)`
			`c =`
			`-13 11 10`

det(A)	A – square matrix	Calculates the determinant of A	`>> A=[2 16 8.1;4-9 0;8.1 7 3];`
			`>> det(A)`
			`ans =`
			`571.2900`

diag(V)	V – vector	Generates a matrix with elements of V placed diagonally	`>> V=1.1:2:5.1;diag(V)`
			`ans =`
			`1.1000 0 0`
			`0 3.1000 0`
			`0 0 5.1000`

(Continued)

TABLE 2.4 (*Continued*)
Some Commands for Matrix Manipulations and Numerical Analysis (Alphabetically)

Command	Parameter/s	Description	Example (Inputs and Outputs)

diff(V)	V – vector	Calculates differences between adjacent elements and can be used to approximate derivatives	`>>x=linspace(0,pi/4,4);`
			`>>V=sin(x)`
			`V=`
			`0 0.2588 0.5000 0.7071`
			`>>diff(V)`
			`ans =`
			`0.2588 0.2412 0.2071`
			`>>dVdx=diff(V)./diff(x)`
			`dVdx=`
			`0.9886 0.9212 0.7911`
			`>>diff(V)`
			`>>dVdx=diff(V)./diff(x)`

dot(a,b)	*a* and *b* are the *n*-element vectors each	Calculates the scalar product of two vectors: a·b = $a_1b_{1+}a_2b_2+\ldots+a_nb_n$; the result is a scalar	`>>a=[2.9 0.9 5.1];`
			`>>b=[3 1 6.1];`
			`>>c=dot(a,b)`
			`c =`
			`40.7100`

a = full(S)	S is a sparse matrix;
	A – is the matrix converted to the full
		Converts the sparse S matrix to the full matrix a	`>> % S see below (in sparse command)`
			`>> a=full(S)`
			`a =`
			`1 0 0 0`
			`0 1 0 0`
			`0 0 1 0`
			`0 0 0 1`

(*Continued*)

TABLE 2.4 (*Continued*)
Some Commands for Matrix Manipulations and Numerical Analysis (Alphabetically)

Command	Parameter/s	Description	Example (Inputs and Outputs)

vi=interp1(x,v,xi,' m_name','extrap')	x and v are the vectors with *y* values for *x* points;
	xi is the input point/s for which the vi interpolated output values are searched;
	m_name is a string with the method name;
	'extrap'- is used for interpolation outside the *x* interval
		Inter/extrapolates y(x) data to the xi points using the following methods:
		'linear'
		'pchip'
		'spline'
			`>>x=linspace(0,pi,10);`
			`v=sin(x);`
			`>>xi=linspace(pi/4,3/2*pi,5);`
			`>>vi=interp1(x,v,xi,'spline','extrap')'`
			`vi =`
			`0.7071`
			`0.9808`
			`0.3827`
			`-0.5645`
			`-1.0282`

length(V)	V – vector	Calculates the amount (length) of the elements in vector V	`>>V=[-2.3 6 7.26 3];`
			`>>length(V)`
			`ans =`
			`4`

max(a)	a – vector or matrix	Returns a row vector with maximal numbers of each column in the matrix a.
		When a is a vector, it returns the maximal number in a
			`>>a=[5.9 6;2 3;7 1];`
			`>>b=max(a)`
			`b =`
			`7 6`
			`>>v=[2.1-9 2 1 8 2];`
			`>>b=max(v)`
			`b =`
			`8`

(*Continued*)

TABLE 2.4 (*Continued*)
Some Commands for Matrix Manipulations and Numerical Analysis (Alphabetically)

Command	Parameter/s	Description	Example (Inputs and Outputs)

mean(a)	a – vector or matrix	Analogous to max but for arithmetical mean/s	`>>a=[5.9 6;2 3;7 1];`
			`>>b=mean(a)`
			`b =`
			`4.9667 3.3333`
			`>>v=[2.1-9 2 1 8 2];`
			`>>b=mean(v)`
			`b =`
			`1.0167`

median(a)	a – vector or matrix	Analogous to max but for median/s	`>>a=[5.9 6;2 3;7 1];`
			`>>b=median(a)`
			`b =`
			`5.9000 3.0000`
			`>>v=[2.1-9 2 1 8 2];`
			`>>b=median(v)`
			`b=`
			`2`

min(a)	a – vector or matrix	Analogous to max but for minimum value/s	`>>a=[5.9 6;2 3;7 1];`
			`>> b=min(a)`
			`b =`
			`2 1`
			`>>v=[2.1-9 2 1 8 2];`
			`>> v=min(a)`
			`b =`
			`1`

(*Continued*)

TABLE 2.4 (*Continued*)
Some Commands for Matrix Manipulations and Numerical Analysis (Alphabetically)

Command	Parameter/s	Description	Example (Inputs and Outputs)

num2str(a)	a – vector or matrix	Converts a single number or numerical vector/matrix elements into a string representation	`>>a=[-27.63746 6.1];`
			`>>v=['a=' num2str(a)]`
			`v =`
			`'a=-27.63746 6.1'`

P=polyfit(x,y,n)	x and y are the vectors with the *y*(*x*) data;
	P – defined fit coefficients of a polynomial *p*(x) of degree n
		Fits the *y*(*x*) curve/data;
		Polynomial p(x) has form:
		$p=P_1x^n+P_2x^{n-1}+\ldots+P_nx+P_{n-1}$
			`>>x=linspace(0,pi,5)`
			`>>y=sin(x);`
			`>>P=polyfit(x,y,4)`
			`P=`
			`0.0376-0.2361 0.0583 0.9820-0.0000`

y=polyval(P,x)	x – vector of the points;
	y value of the polynomial (see polyfit); P – vector of the coefficients of the polynomial
		Calculates values of the p(x) polynomial at x points; polynomial is given by its P coefficients	`% p from the polyfit example`
			`>>yp=polyval(p,x);`
			`>>[y' yp']`
			`ans =`
			`0-0.0000`
			`0.7071 0.7071`
			`1.0000 1.0000`
			`0.7071 0.7071`
			`0.0000-0.0000`

randi(imax,m,n)	imax – maximal integer value,
	m - number of rows,
	n – number of columns
		Returns an m by n matrix of integer random numbers from value 1 up to imax –maximal integer value	`>>randi(9,2,3)`
			`ans =`
			`8 2 6`
			`9 9 1`

(*Continued*)

TABLE 2.4 (*Continued*)
Some Commands for Matrix Manipulations and Numerical Analysis (Alphabetically)

Command	Parameter/s	Description	Example (Inputs and Outputs)

repmat(a,m,n)	a - matrix,
	m – number of rows,
	n – number of columns
		Generates the large matrix containing m × n copies of a	`>>a=[4 3;5 2]`
			`A=`
			`4 3`
			`5 2`
			`>>b =repmat(a,1,2)`
			`b =`
			`4 3 4 3`
			`5 2 5 2`

reshape(a,m,n)	a - matrix,
	m – number of rows,
	n – number of columns
		Returns an m-by-n matrix whose elements are taken column-wise from a. Matrix a must have m*n elements	`>>a=[6 7;2 3;8 1]`
			`a =`
			`6 7`
			`2 3`
			`8 1`
			`>>reshape(a,1,6)`
			`ans =`
			`6 2 8 7 3 1`

sign(x)	x – vector or single number	signum or sign function;
		returns:
		1 if the element of x >0;
		0 if x=0; and
		–1 if x<0.
		x ’/l x ’l, when x is complex
			`>>x=[51.2-3.2 0 5.7];`
			`>>sign(x)`
			`ans =`
			`1-1 0 1`

(*Continued*)

TABLE 2.4 (*Continued*)
Some Commands for Matrix Manipulations and Numerical Analysis (Alphabetically)

Command	Parameter/s	Description	Example (Inputs and Outputs)

size(a)	a – matrix or vector	Returns a two-element row vector; the first element is the number of rows in matrix a while the second is the number of columns	`>>a=[6 7;2 3;8 1];`
			`>>size(a)`
			`ans=`
			`3 2`

[y,ind]= sort(a,dim,mode)	a –vector/matrix;
	dim – dimension;
	mode are the 'ascend' or 'descend' sorting order;
	y – ordered vector/matrix;
	ind – indices a for each y
		Sorts elements of a along columns (dim=1) or along rows (dim=2) and in 'ascend' (default) or 'descend' mode order. It returns ordered y and indices of the a for each y	`>>a=[6 7;2 3;8 1]`
			`a =`
			`6 7`
			`2 3`
			`8 1`
			`>> [y,ind]=sort(a,2,'descend')`
			`y =`
			`7 6`
			`3 2`
			`8 1`
			`ind =`
			`2 1`
			`2 1`
			`1 2`

S= sparse(a)	a – full matrix	Converts full matrix a to the sparse form S by squeezing out any zero terms	`>>a=eye(4)`
			`a =`
			`1 0 0 0`
			`0 1 0 0`
			`0 0 1 0`
			`0 0 0 1`

(*Continued*)

TABLE 2.4 (*Continued*)
Some Commands for Matrix Manipulations and Numerical Analysis (Alphabetically)

Command	Parameter/s	Description	Example (Inputs and Outputs)

			`>>S=sparse(a)`
			`S=`
			`(1,1) 1`
			`(2,2) 1`
			`(3,3) 1`
			`(4,4) 1`

std(a)	a – vector or matrix	Calculates standard deviation/s for each column as $\left[\frac{1}{n-1}\sum_{i=1}^{n}(a_i-\mu)^2\right]^{\frac{1}{2}}$, with μ as a mean of *n* elements for each column of matrix a.
		If a is a vector, it returns the standard deviation value of vector
			`>>a=[5.9 6;2 3;7 1];`
			`>>std(a)`
			`ans =`
			`2.6274 2.5166`
			`>>v=[2.1-9 2 1 8 2];`
			`>> std(v)`
			`ans =`
			`5.5174`

strvcat(t1,t2,t3,...)	t1, t2, t3, ... - strings	Generates the matrix containing the t1, t2, t3... strings as rows	`>>t1 = 'Electricity';`
			`>>t2 = 'Wear';`
			`>>t3 = 'Lubrication';`
			`>>strvcat(t1,t2,t3)`
			`ans =`
			`3x11 Char array`
			`'Electricity'`
			`'Wear '`
			`'Lubrication'`

(*Continued*)

TABLE 2.4 (*Continued*)
Some Commands for Matrix Manipulations and Numerical Analysis (Alphabetically)

Command	Parameter/s	Description	Example (Inputs and Outputs)

sum(a)	a – vector or matrix	Analogous to max but for column (matrix) or row (vector) sums of elements	`>> a=[5.9 6;2 3;7 1];`
			`>> sum(a)`
			`ans =`
			`14.9000 10.0000`

trapz(V)	V – vector of data	Integrates V(x) data by the trapezoidal rule	`>> x=1:5;`
			`>> V=x.^2`
			`V=`
			`1 4 9 16 25`
			`>> I=trapz(V)`
			`I=`
			`42`

Note:
- In this table, commands are presented in their simplest forms; use help or doc commands for more information.
- The sparse matrix representation is used to reduce storage space for large matrices containing many zero elements.

```
rng default                  % for a starting generator settings
>> a=rand(2,4) % 1st use of the rand, seed random number values
a =
0.8147  0.1270  0.6324  0.2785
0.9058  0.9134  0.0975  0.5469
>> a=rand(2,4)    % reuse rand yields new random number values
a =
0.9575  0.1576  0.9572  0.8003
0.9649  0.9706  0.4854  0.1419
rng default             % restoring a starting generator settings
>> a=rand(2,4)%reuse rand yields the initial random number values
a =
0.8147  0.1270  0.6324  0.2785
0.9058  0.9134  0.0975  0.5469
```

In addition to the above commands, there are a lot of others that can be used for manipulation with vectors, matrices, and arrays, some of which are listed in Table 2.4. A small number of numerical analysis commands using vectors/ matrices have also been added to the table.

2.3.5 Strings as Variable and Strings as Matrix Elements

In all calculations discussed so far, except for some of those in Table 2.4, we used numbers. However, single variable or matrix elements can also contain strings. A string is an array of characters and numbers, enclosed in single quotes:

```
>>str1='Engineering is the design and manufacturing things
using science and mathematics' % generates string vector str1
str1 =
'Engineering is the design and manufacturing things using
science and mathematics'
>>str2='Bearing' % generates the 7-element string vector str2
str2 =
'Bearing'
>>length(str2)% the str2 length is 7, the ' signs are not counted
ans =
7
```

Each character of the entered string is treated and stored as a coded number (thus the sequence of characters represents a vector) and can be addressed like an element of a vector, for example, str2(4) in the string Bearing is the letter 'r'. Single quotes are not counted in the string length. Some examples with string manipulations are:

```
>>str2(6) % 'n' is the 6th element of str2
ans =
'n'
>>str2(3:5) % 'a', 'r' and 'i' are the 3rd ...5th elements of str2
```

```
ans =
'ari'
>>str2([7 2:4])%'g','e','a' and 'n' are the 7th, 2nd…5th elements
ans =
'gear'
```

The string is displayed in single quotes, to display string without quotes, the disp command can be used, e.g.

```
>>disp(str1)
Engineering is design and manufacturing things using science
and mathematics
>>disp(str2([7 2:4]))
gear
```

Like numerical arrays, the strings within a row are divided by a space or comma, and strings between the rows are by a semicolon (;). Each row of strings must be the same length as the longest row of strings. To achieve equality of characters in rows of strings, the spaces should be added to shorter strings; for example,

```
>> List_of_names=['Gear';'Piston';'Bolt';'Crankshaft']
Error using vertcat
```

Dimensions of arrays being concatenated are not consistent.

This error message indicates unequal length of the strings in rows. The number of characters in the words Gear, Piston, and Bolt is 4, 6, and 4 while in Crackshaft – 10. To correct the array, the six spaces should be added after Gear and after Bolt, and four spaces after Piston; causing all the strings to be of the same length. This gives the same length to all strings. Thus, the List_of_names array can be successfully entered and displayed.

```
>>List_of_Names=['Gear      ';'Piston    ';'Bolt      ';'Crankshaft'];
>> disp(List_of_Names)
Gear
Piston
Bolt
Crankshaft
```

Determining a longer string requires counting the number of spaces added to each row of a column, which is a tedious procedure for the user. To avoid this, you can use the char or strvcat command as shown in Table 2.4.

2.3.6 About Displaying a Table

Earlier (Section 2.2.6) it was shown how display single numbers as well as vectors, matrices, and captions with the disp and fprintf commands. The same commands can be used to output data in tabular form.

2.3.6.1 The disp Command

Display for example, the gear teeth height and thickness data (Table 2.3) together with caption "Table" and column header:

```
>> M=[10  1.062  1.564;20  1.031  1.569
30 1.021 1.570;40 1.015 1.5704] %matrix M with data from Table 2.3
M =
10.0000  1.0620  1.5640
20.0000  1.0310  1.5690
30.0000  1.0210  1.5700
40.0000  1.0150  1.5704
>> disp('     Table'),disp('   N     H, mm      S, mm'),disp(M)
Table
N H, mm S, mm
10.0000  1.0620  1.5640
20.0000  1.0310  1.5690
30.0000  1.0210  1.5700
40.0000  1.0150  1.5704
```

The first two disp commands show the table captions and the third – outputs the M array.

Note that all disp commands must be written on the same command line for the table to be displayed as presented.

2.3.6.2 The fprintf Command

The command allows you to format the output numbers, so in the above example, the N values can be displayed without decimal digits while H and S with three decimal digits. The two following commands can do this

```
>> fprintf('        Table\n N H,mm S,mm\n'),fprintf('%3.0f
%6.3f %6.3f\n', M')
Table
N H,mm S,mm
10  1.062  1.564
20  1.031  1.569
30  1.021  1.570
40  1.015  1.570
```

The first fprintf command displays two lines of the table header, and the second displays numbers of the M array. Similar to the previous case, two fprintf commands must be written on the same command line to display the table in the presented view.

Note:

- the **fprintf command prints array/matrix rows as table columns.** Thus the above 4 × 3 M matrix was transposed to the 3 × 4 matrix (written as M') and the number format for the outputting values was provided for three columns;
- two symbols \ and n (back slash and letter n without space) must be typed after the last of the column numbers formats.

2.3.7 Application Examples

2.3.7.1 Cuboid Lattice Cell Volume

To define volume via vectors for a face-centered cuboid lattice cell, the following expression can be used

$$V = \left| \vec{a_1} \cdot \left(\vec{a_2} \times \vec{a_3} \right) \right|$$

where $\vec{a_1}$, $\vec{a_2}$, and $\vec{a_3}$ are tree-element vectors, each a is the lattice constant, and the $|\cdot|$ and $\times$ signs designate the dot and cross products, respectively.

Problem: Consider the unit cell volume V when $\vec{a}_1 = (0.5\mathbf{j}+0.5\mathbf{k})$a, $\vec{a}_2 = (0.5\mathbf{i}+0.5\mathbf{k})$a $\vec{a}_3 = (0.5\mathbf{j}+0.5\mathbf{k})$a while the a value is 3.567 Å (diamond). Use the dot and cross commands as per Table 2.4. Display the defined values of the vectors and volume with the fprintf commands; get each number with three decimal digits.

The commands solving the problem are

```
>>a=3.567;                                   % assigns 3.567 to a
>>a1=[0 0.5 0.5]*a;           % generates tree-element vector a1
>>a2=[0.5 0 0.5]*a;           % generates tree-element vector a2
>>a3=[0.5 0.5 0]*a;           % generates tree-element vector a3
>>v=dot(a1,cross(a2,a3));                   % calculates volume
>>a123=[a1;a2;a3];     % generates matrix with a1, a2 a3 rows
>>A=[(1:3)' a123];                 % adds row with numbers 1 2 3
>>fprintf('a%1d=[%6.3f %6.3f %6.3f]\n',A')% outputs a1,a2,a3
a1=[ 0.000 1.784 1.784]
a2=[ 1.784 0.000 1.784]
a3=[ 1.784 1.784 0.000]
>>fprintf('Volume=%8.4f Angstrom^3\n',v) % outputs volume
Volume= 11.3462 Angstrom^3
```

These commands perform the following:

- The four first commands input a value and generate three three-element vectors named a1, a2, and a3;
- The fifth command calculates V with the dot and cross commands;
- The last four commands display results with the two fprintf commands, for this:
    - matrix a123 is generated with the a1, a2, and a3 vectors as matrix rows;
    - matrix A is generated to combine the serial numbers 1, 2, and 3 (for further displaying character a with trailing number) and matrix a123;
    - in the first fprintf command, the first row values of A are defined as an integer and a1, a2, a3 values – as a fixed point with three decimal digits; matrix A transposed since the rows are displayed as columns with the fprintf command;
    - the second fprintf command displays text together with the V value, i.e.: Volume = 11.3462 Å^3.

2.3.7.2 Table Containing Strings and Numbers

The available linear temperature expansion coefficients α data for some technological/industrial materials are shown in the table:

Material	$\alpha \cdot 10^{-6}$, m/(m K)
Aluminum	23
Copper	16
Polyethylene	108
Titanium	0.5
Polypropylene	54
Magnesium	26
Steel	12
Silicon Carbide	450

Problem: Generate and display the above table sorted by the material names, use for this purpose the char and sort commands (Table 2.4); use the fprintf command to display the table header and disp for output the table as two-column arrays.

The commands solving the problem:

```
>>Name=char('Aluminum','Copper','Polyethylene','Titanium',...
'Polypropylene','Magnesium','Steel','Silicon Carbide');
>>alpha=[23 16 108 0.5 54 26 12 450]';
>>[Name_sort,ind]=sort(Name(:,1),1); % sorts 1st Name column
>>Name_sort=[Name_sort Name(ind,2:end)];% joins 1st&other
                                          columns
>>alpha_sort=alpha(ind)          % puts in order alpha by ind
>>Table=[Name_sort num2str(alpha_sort)]; % table for fprintf
>>fprintf('Material   Expan sion·10^-6\n               m/(m
K)\n'),disp(Table)          % in one line for interactive mode
Material Expansion·10^-6
m/(m K)

Aluminum           23
Copper             16
Magnesium          26
Polyethylene      108
Polypropylene      54
Steel              12
Silicon Carbide   450
Titanium          0.5
```

The above commands perform the following actions:

- The char command contains strings with material names and generates the Name matrix, each row of which is a string with name;
- The second command generates a column vector alpha with numbers representing the linear temperature expansion values taken from the second table column;
- In the next line, the sort command orders the first characters of the material names and outputs two column vectors with ordered characters (Name_sort) and the addresses (ind) of its previous places;
- The fourth line command concatenates the ordered first characters of the names with the second, third, and until the last characters to generate a complete column with the material names;
- The command in the fifth line generates a column of thermal expansion values alpha sorted by ind;
- The sixth line command combines two previously produced separate columns Name_sort and alpha_sort into the two-column matrix table. As we remember, all matrix elements must be of the same type, for this the num2str command (Table 2.4), which converts the numeric α data into strings;
- The last line contains two output commands – fprintf and disp; the first one displays the table header in two lines with the column names and thermal expansion units, and the second displays a two-column array with the material names and the corresponding expansion values. Note: spaces must be added in the fprintf command in the table header string to center the table captions over the displayed columns.

2.3.7.3 Voltage and Current in an RC-type Circuit

In some circuits with resistance R, capacitor C, switch, and source supplying the input voltage V_0, the time-dependent voltage V (in V) and current i (in Amp) can be calculated by:

$$V = V_0\left(1 - e^{\frac{-t}{RC}}\right)$$

$$i = \frac{V_0}{R} e^{\frac{-t}{RC}}$$

where the RC product is the time constant while t is the time, both in s; R in Ω, and C in F

Problem: Calculate voltage and current after the switch is closed at times $t = 0, 3, \ldots, 24$ s when $V_0 = 24$ V, $R = 3000\ \Omega$, $C = 3900 \times 10^{-6}$ F. Display results as three-column table with captions, use the fprintf commands which output time as integer, voltage with one decimal digit, and current with six decimal digits.

The commands and their result are as follows:

```
R=3000;C=3900e-6;V0=24;                    % inputs the constants
t0=R*C;                                    % calculates RC constant
t=0:3:24;                                  % generates row vector t
d=exp(-t./t0);                   % calculates exp for further using
V=V0*(1-d);                                    % calculates voltage
i=V0/R*d;                                      % calculates current
Table=[t' V' i']; % generates three-column matrix for fprintf
fprintf(' Time,s Voltage,V Current,Ohm\n');fprintf(' %3d
%5.1f    %10.6f\n',Table') % in one line (interactive mode)

Time,s  Voltage,V   Current,Ohm
 0        0.0       0.008000
 3        5.4       0.006191
 6        9.6       0.004790
 9       12.9       0.003707
12       15.4       0.002869
15       17.3       0.002220
18       18.8       0.001718
21       20.0       0.001329
24       20.9       0.001029
```

The commands act as follows:

- The first command line contains three commands assigning the R, C, and V_0 values to the variables R, C, and V0;
- the second line command calculates the time constant t0;
- The main MATLAB window – desktop – includes the top toolstrip with controls and three panels, which can be undocked from the desktop and appear as separate windows: Command, Current Folder, and Workspace – see Figure 2.1.
- the three further commands calculate voltage V and current i using the exponent d that appears in the expressions for V and i;
- the last two lines (a) generate the three-column matrix table assembling the t, V, and i vectors transposed from row to column each; and (b) displays the resulting table using two fprintf commands to output the table header and numbers of the table matrix in the required format; note that the fprintf command outputs rows of the matrix as columns.

2.3.7.4 Momentary Position of the Piston Pin

The position x of the engine piston pin is changed each moment and can be determined with respect to the angle of rotation θ of the crankshaft by the expression:

$$x = r\cos\theta + \sqrt{l^2 - r^2\sin^2\theta}$$

where r and l are the lengths of the crank arm and connecting rod, respectively.

Problem: Determine the piston pin instantaneous position at ten rotation angles in range 0…2π for the crank and rod lengths 0.12 and 0.25 m, respectively. Display the results as a two-column table with header. Calculated θ and x values should be represented in the first and second table columns respectively. Use the disp command to display the table header, and the fprintf command to display the θ and x values, formatted to two and four decimal digits respectively.

The commands are:

```
>>r=0.12;l=0.25;                                          % in m
>>Thet=linspace(0,2*pi,10);        % generates 1X10 vector Thet
>>x=r*cos(Thet)+sqrt(l^2-r^2*sin(Thet).^2);% calculates position
>>Table=[Thet;x]; % collects two row vectors in two-row matrix
>>disp(' Angle, rad Position, m'),fprintf(' %7.2f %7.4f \n',...
Table)   % disp and fprintf in one line for interactive mode

Angle, rad Position, m
0.00    0.3700
0.70    0.3297
1.40    0.2411
2.09    0.1674
2.79    0.1338
3.49    0.1338
4.19    0.1674
4.89    0.2411
5.59    0.3297
6.28    0.3700
```

The commands operate as follows:

- In the first two lines, the commands assign r and l values, and generate a ten-element row vector with θ values from the range 0…2π, the latter is realized using the linspace command;
- The next command calculates vector x using the elementwise operation for $\sin^2\theta$;
- The fourth line command collects two separate row vectors θ and x into a two-row matrix Table;
- The commands in the last line contain the disp command with the table header and the fprintf command with the required format of the θ and x numbers; note that the fprintf command outputs rows of the matrix as columns.

2.4 FLOW CONTROL

Commands written sequentially one after the other represent a computational program. Until now, all inputted commands have been executed in the order in which they were written. Frequently, the order of execution of the commands must be changed depending on some conditions. For example, when calculations must be repeated each time with new parameters or when it is necessary to select one of several calculating expressions justified within a different area each. Other possible situation: a result is with certain accuracy is required that can be ensured by repeating the commands several times until the error in the calculating value diminishes to the required size. To implement such sort of calculations, flow control is applied. MATLAB® has various means that provide a selection of commands for next execution, which use some relational, logical, and conditional operators. The most common flow control commands are described below.

2.4.1 Relational and Logical Commands

As stated above, flow control manipulations are carried out using relational and logical commands. These commands compare pairs of values or statements; in the first case, the commands operate mainly with numeric values, and the second – with logical (Boolean) values.

2.4.1.1 Relational Operators

The commands that compare values pairwise are called relational or comparison operators. The result of their action is 1 (true value) or 0 (false value), for example, the expression x>6.21 results in 1 if x greater than 6.21 and 0 otherwise.

The relational operators are < (less than), > (greater than), <= (less than or equal to), >= (greater than or equal to), == (equal to), and ~= (not equal to).

Note: Two-sign operators are written without spaces between the signs, for example, == and not = =.

When a relational operator compares elementwise matrices or arrays of the same size, the result is an array of the same size, containing 1's where the relation is true and 0's where it is not. In case one of the compared operands is a scalar while the other is an array, the scalar is compared with each element of the array, and result is an array of equal size. Ones and zeros are presented in MATLAB® as logical data types, not numerical; nevertheless, they can be used in arithmetical operations in the same way as numerical data.

Here are some illustrative examples:

```
>>3+4==14/2          % the answer is 1 since 3+4=7 and 14/2=7
ans =
logical
1
>>cos(pi/4)~=0% the answer is 1 as cos(pi/4)≃0.7 and it is not 0
ans =
logical
1
>>M=[-8.21 109.14-11.3;-5 4.1 0.9]  % produces a 2x3 matrix M
M =
-8.2100  109.1400  -11.3000
-5.0000  4.1000      0.9000
>>B=M<=0            % assigns 1s and Os to B as results of M<=0
```

```
B =
2×3 logical array
1  0  1
1  0  0
>>M(B) % Selects negative M-elements, using logical B as
       numerical
ans =
-8.2100
-5.0000
-11.3000
>>M(M<0)=0%Assigns 0s to the elements of M that are less than 0
M =
0  109.1400        0
0    4.1000   0.9000
```

2.4.1.2 Logical Operators

These operators work with logical expressions, say $(x > 6)$ & $x(<2\pi)$, and operate on logical values true (1) or false (0). Like relational operators, they can be used as addresses in vectors, matrices, or arrays; see, for instance, the last two commands in the example above.

There are three logical operators:

& or command AND – logical *and*, e.g. A&B is equal to and(A,B),
| or command OR – logical *or*, e.g. A|B is equal to or(A,B),
~ or command NOT – logical *not*, e.g. A~B is equal to not(A,B).

In these operators, A and B denote the compared operands. Similar to relational operators, they perform an elementwise comparison that yields logical 1s or 0s. The result of operations with logical operators depends on the order in which logical operations are performed. The order is determined by the following precedence rules (from highest to lowest): parentheses, exponentiation, NOT (~), multiplication/division, addition/subtraction, relational operators, AND (&), OR (|). In actual expression, the desired order of execution of operators can be set/changed using parentheses.

Some examples are:

```
>>x=-5.3;          % assigns -5.3 to x
>>-3<x<-4 %runs from left to right:-3<-5.3 gives 0 and 0<-3 is 0
ans =
logical
0
>>x>-6&x<2 %  first -5.3>-6 and -5.3<2 both are 1 then 1&1 is 1
ans =
logical
1
>>~(x<-5)    % -5.3<-5 runs first and is true (1), then ~1 is 0
ans =
```

```
logical
0
>>~x<5 %~(-5.3) runs 1st and is 0 as x is nonzero, then 0<5
is true
ans =
logical
1
```

Additional information about logical functions can be obtained by entering the >> `doc 'logical operations'` command.

One of the frequently used MATLAB® logical functions is the find command. Its simplest forms are read as

```
ind=find(x) or ind=find(A>c)
```

where ind is a vector of the addresses (indices) at which the nonzero elements of x or where A is greater than c are located. In the second form of this command, any of the relational operators can be used, namely, <, >=, etc. These forms of the find command use the one-number linear addressing (Section 2.3.1). For example,

```
>>Vec=[8.14 9.06-1.27 0 6.32-0.9754];    % assigns vector Vec
>>i1=find(Vec)% returns the addresses of the nonzero Vec-elements
i1 =
1 2 3 5 6
>>i2=find(Vec<3)% returns the addresses of the V with elements <3
i2 =
3 4 6
>>M=[0.1 -3;7.2 -7.43;4.2 1];          % assigns 3x2 matrix M
>>i=find(M<2)%returns the LINEAR addresses of M where
elements <2
i =
1
4
5
6
```

2.4.1.3 Application Example: Determining Outlying Results in a Sample

Test results for one of linear dimensions *LD* of a sample containing 15 measured mechanical parts are: 11.0, 10.4, 9.6, 10.4, 10.1, 10.2, 9.8, 14.9, 10.1, 6.3, 10.5, 10.5, 10.0, 10.3, and 10.5 mm. These observations contain values that are sharply different from the rest (e.g., 14.9, 6.3). Can we discard the outplayed data for proper data analysis? For solving this problem, we can use the following criterion:

If

$$\frac{|LD_i - \mu|}{\sigma} > t_q$$

thus the *i*-element/s of the *LD* can be discarded. In this expression, μ and σ are the sample mean and standard deviation respectively, and t_q is the critical value that is equal to 2.409 for our case (corresponds to the 15th processed values at a significant level of 5%).

Problem: Define outlier/s and generate two vectors with LD data: a) without outliers, b) outliers only; for (a) use the relational command/s, and for (b) use the find function. Display results with captions using the disp command.

The commands that solved this problem are:

```
LD=[11.0 10.4 9.6 10.4 10.1 10.2 9.8 14.9...
10.1 6.3 10.5 10.5 10.0 10.3 10.5];           % assigns LD data
>>tq=2.409;                          % assigns criterion value to tq
>>sigm=std(LD),mu=mean(LD); % calculates the st. dev. and mean
>>crit=abs(LD-mu)/sigm;   % calculates value of the criterion
>>out_logic=crit>tq; %lesser(0) and greater(1) tg,logical values
>>outliers= find(out_logic); % finds addresses of the outliers
>>normal= LD(out_logic==0); % finds normal(non outlier) values
>>disp('Normal LD, mm'),disp(normal') % two disp-s in one line
Normal LD, mm
11.0000
10.4000
 9.6000
10.4000
10.1000
10.2000
 9.8000
10.1000
10.5000
10.5000
10.0000
10.3000
10.5000
>>disp(' Outliers, mm'),disp(LD(outliers)')% displays outliers
Outliers, mm
14.9000
6.3000
```

These commands perform the following operations:

- The first two commands assign *LD* and t_q values to variables of the same names;
- The following command line contains two commands calculating the mean average and standard deviation sigm;
- The command in the fourth line calculates abs(LD-mu)/sigm and assigns the results to the crit vector for further checking the inequality of the criterion;
- The following command determines the locations of the outliers using the crit >tq relational operator and saves the results in the out_logic vector with logical 1's and 0's at the places where the outlying and non-outlying (normal) LD values are located, respectively.

- The find command returns the addresses of nonzero elements in the out_logic vector, in which logical 1's indicate the locations of the outliers.
- The next logical command generates the normal vector containing only non-outliers;
- The last two command lines contain two **disp** commands each, and they display the obtained results together with headers above the normal and outlier numbers.

2.4.2 The If Statements

The order of the execution of commands can be managed with so-termed conditional statements. The if statement is one of them. It has three basic forms: **if … end**, **if … else … end** and **if … elseif … else … end**. Each **if** statement form should be terminated with the word end. The words of the statements appear on the screen in blue. Design of the **if** statement forms with its description is shown in Table 2.5:

TABLE 2.5
Forms of the If Statement

Design	Description

if conditional expression
{
… MATLAB® command/s
}
end
	Executes the inner MATLAB® command/s (indicated by ellipsis) when the conditional expression is true

if conditional expression
{
… MATLAB® command/s
}
else
{
… MATLAB® command/s
}
end
	Executes the command/s that is/are located between the words **if** and **else** in case the conditional expression is true; otherwise, the command/s placed between **else** and **end** are executed

if conditional expression
{
… MATLAB® command/s
}
elseif conditional expression2
{
… MATLAB® command/s
}
else
{
… MATLAB® command/s
}
end
	Executes the command/s that is/are located between the words **if** and **elseif** in case the first conditional expression is true; otherwise, when the second conditional expression, conditional expression2, is true the command/s located between **elseif** and **else** is/are executed, and if this expression is false the commands between **else** and **end** is/are executed

In this table, the conditional expression term denotes any relational or logical operator/s; for example, a<c1&a> = c2 or b == v.

When the if word is typed in the Command Window next to the prompt >>, the new line appears each time after pressing enter without the prompt until the final end is typed and entered.

An application example showing the if statement usage is presented in Section 2.5.2.

2.4.3 Loop Commands

An important group of commands used in flow control is the commands for loop constructions. These commands allow you to repeat circularly one or more commands. Each command/s re-execution is termed a pass. There are two main commands to the loop design: for … end and while … end. The loop you commands are displayed in blue. Like the if statement, each for or while construction must terminate with the word end.

The general view of each statements is presented in Table 2.6:

The commands that can be included between for and end are repeated k times; k is changed in every loop pass in accordance to the next number in the brackets (Table 2.6) or by the addition of the step-value to the initial/preceded number. This process continues until k reaches the final number or exceeds the final value.

Note:

- In the expression for k (Table 2.6), the square brackets mean that k can be specified as a sequence of numbers, for example in such way k = [32.051–0.32 1:6], where the first k is 32.05, then –0.32, and then 1, 3, and 5.
- If the expression for k contains only the colon operators, then the brackets can be omitted, e.g., k =10: –3:0.

TABLE 2.6
Command Forms for Loop Construction

Design	Description

for k = [initial : step : final]
{
… command/s
}
end
	Executes k-times the internal command/s (signed by ellipsis); k can be specified as a sequence of numbers with the colon command or simply by numbers written in the square brackets

while conditional expression
{
… command/s
}
end
	Executes the internal command/s (signed by ellipsis) repeatedly while the conditional expression is true

Sometimes, the for... end loop can be replaced with the matrix operations, which are actually superior as for...end loops work slowly. This advantage is negligible for short loops with a small number of commands but appreciable for large loops with numerous commands.

After the loop is ended, in other words, immediately after the last pass, the command following the loop is executed.

The while ... end construction is applied when the number of passes is not known in advance. In this case, the commands between the while and end are executed, and such passes are repeated until the conditional expression is true. An incorrectly written while loop may continue indefinitely, for example

```
>>a=3.2;
>>while a >0 %a is always >0, so the loop continues indefinitely
b=0.1+sqrt(a)
end
b =
1.8889
...
```

This result (b = 1.8889) appears repeatedly on the screen. To interrupt the loop, press the Ctrl and C keys simultaneously.

To demonstrate the work of both described loop functions, consider an example the exponential function e^x via the Taylor series at $\sum_{n=0}^{\infty}\frac{x^n}{n!}$ for x 1.0472 using the for ... end and while ... end commands. Assign $n = 0, 1, \ldots, 7$ in case of for ... end, and assume that the conditional expression requires $\frac{x^n}{n!}$ should be greater than 0.0001 for the while ... end. Assign $x = 1.8$.

The solutions are:

```
>>x=1.8;     % assigns x for the both loop-statement examples
>>%                                           for ... end loop
>>s=0; % sets s = 0 for further summation of the series terms
>>n=10;      % assigns 10 to determine 11 as last term number
>>for k=0:n %                                  for... end loop
s=s+x^k/factorial(k);     %   s increases by x^k/k! each pass
end                            % the end of the for... loop
>>s                  % displays the resulting s that is the e^x
s=
6.0496
>>%                                         while ... end loop
>>term=1;                 % it is the 1st series' term at k=0
>>k=1;                     % sets k=1 for the next while loop
>>s=term; % sets s = 0 for further summation of the series terms
>>while term>=.0001                         % while... end loop
term=x^k/factorial(k);               % calculates new term value
s=s+term;                        % s increases by term each pass
```

```
k=k+1;                                    % new number of k
end                                 % the end of the while... loop
>>fprintf('exp(%3.1f)=%7.4f n=%i\n',x,s,k)%displays x, e^x and n
exp(1.8)= 6.0496 n=11
```

This example in the first part of commands examines the for ... end loop. In the beginning of the first loop pass, the s value is zero. During this pass, the first term (its k number is 0) is calculated and added to the s. In the second pass, the second series term ($k = k + 1 = 0 + 1 = 1$) is calculated and added to the previous s value. This procedure is repeated up to $k = n = 10$. Then the loop ends and the resulting value is displayed by entering the variable s. In the case of the for ... end loop, the number of passes must be set in advance.

The further group of the example commands examines the while ... end loop and implements a bit more complicated solution. After the word while, we must indicate the condition under which the loop should be terminated. According to our problem, we can assume that the kth term should not be greater than 0.0001 (this number represents the accuracy of the calculated value). Before the start of the loop, the first term is assigned equal to 1 at $k = 0$ (as $x^k/k! = x^0/0! = 1$). Then the s and k (k is sometimes called a counter) values are equal to zero and 0, respectively. In the first pass, the first term of the series is checked if it is greater than 0.0001. If this condition is true, the new term is calculated and added to the sum s; now the k is increased by 1. Then the new term is checked if it is greater than 0.0001. If this condition is true, the next pass is started for the next term calculation. If it is false, the loop ends and the fprintf command displays the x value; the obtained s value and the number of terms n of the series are defined as the number of the loop passes. As the latter number is an integer, the conversion character i is used for its displaying.

Note:

- The while loop does not summarize a value that does not meet the required conditional statement. As a result, the last value of the counter k is increased by 1 if the starting k-value was 1; in the example above the initial $k = 0$ (Taylor series expression), therefore decreasing the last k by 1 is not required when we want to know the correct number of summed terms;
- The for ... end and while ... end loops and if statements can be incorporated (nested) within other loop and/or if-statement. The order and quantity of these statements are not limited and are predetermined only by the purposes of the calculations.

2.5 APPLICATION EXAMPLES

2.5.1 Currents in an Electrical Circuit

An engineer analyzes some electrical circuit containing resistors and voltage sources. He defines that using the mesh current method based on Kirchhoff's voltage law, and the following set of linear equations should be solved to define the mesh currents:

$$V_1 - R_1 i_1 - R_3(i_1 - i_3) - R_2(i_1 - i_2) = 0$$

$$-R_5 i_2 - R_2(i_2 - i_1) - R_4(i_2 - i_3) = 0$$

$$-V_2 - R_6 i_3 - R_4(i_3 - i_2) - R_3(i_3 - i_1) = 0$$

After some reorganizing, these equations can be presented in two possible matrix forms $AX = B$ or $XA = B$ as:

$$\begin{bmatrix} -(R_1 + R_2 + R_3) & R_2 & R_3 \\ R_2 & -(R_2 + R_4 + R_5) & R_4 \\ R_3 & R_4 & -(R_3 + R_4 + R_6) \end{bmatrix} \begin{bmatrix} i_1 \\ i_2 \\ i_3 \end{bmatrix} = \begin{bmatrix} -V_1 \\ 0 \\ -V_3 \end{bmatrix}$$

or

$$\begin{bmatrix} i_1 & i_2 & i_3 \end{bmatrix} \begin{bmatrix} -(R_1 + R_2 + R_3) & R_2 & R_3 \\ R_2 & -(R_2 + R_4 + R_5) & R_4 \\ R_3 & R_4 & -(R_3 + R_4 + R_6) \end{bmatrix}$$

$$= \begin{bmatrix} -V_1 & 0 & -V_1 \end{bmatrix}$$

Note: In general case, to solve the equations written in the second matrix form XA = B, the matrix A and the row vector B should be transposed (using the 'operator); but for A in the above set of three linear equations, this is not necessary since A = A'.

Problem: Define the i_1, i_2, and i_3 currents with left- and right-division when V_1 and V_2 are 22 and 13 V, respectively; R_1, R_2, R_3, R_4, R_5, and R_6 are 17, 9, 15, 6, 16, and 8 Ω, respectively; display the resulting currents on three lines using the fprintf command with each line being "Current i(X) = X.XXXX A", where X designates the place of the actual digit.

The commands solving the problem are:

```
>>V1=22;V2=13;    % assigns values to the variables V1 and V2
>>R1=17;R2=9;R3=15;R4=5;R5=14;R6=7;%assigns values to the R1...
>>A=[-(R1+R2+R3) R2 R3                    % generates 3x3 matrix A
R2-(R2+R4+R5) R4
R3 R4-(R3+R4+R6)];
>>B=[-V1;0;V2];                              % generates vector B
>>i=A\B;                      % solves form A\B with left division
>>for j=1:3 %for…end loop for further displaying current number
```

```
fprintf('\n Current i(%i)=%7.4f A\n',j,i(j))%displays results
end

Current i(1)= 0.4985 A
Current i(2)= 0.1279 A
Current i(3)=-0.1808 A

>>i=B'/A;                % solves form A\B with left division
>>for j=1:3 % for...end loop for further displaying current number
fprintf('\n Current i(%i)=%7.4f A\n',j,i(j))%displays results
end

Current i(1)= 0.4985 A
Current i(2)= 0.1279 A
Current i(3)=-0.1808 A
```

The above commands act as follows

- The commands of the two first lines generate the V1, V2 and R1, …R6 row variables with initial data;
- The third line command generates a 3 × 3 matrix A of the first matrix form, in which the rows are the row coefficients of the above equation for the first matrix form;
- The following command generates a column vector B according to the right side of the first form of the matrix equation;
- The command of the fifth line calculates the *i*-currents using the left division A\B;
- The following for …end loop is used to display the numbers 1, 2, and 3 next to the character i using the fprintf command; the latter displays three lines with the results, and each line looks like this: "Current i(X)= X.XXXX A", where X designates the actual digit;
- The following command calculates currents using the right-hand division B/A, for which the column vector B is transposed into the row vector; this solution should verify the previous solution;
- The next for …end loop and the fprintf command inside it, as in previous case, are used to display the numbers 1, 2, and 3 next to the character i and to display results as "Current i(X) = X.XXXX A", where X designates the actual digit.

2.5.2 Resistance of Volume of a Material–Bulk Modulus

The bulk modulus k (in Pa) of a material was measured at different applied pressures, P, in range 0 … 8 GPa. The observed results were described by the following linear equation:

$$K = 550.8 + 10.1P$$

Problem: Calculate and display k values if the pressure is in allowed pressure range and displays message "This pressure is out of allowed range"; otherwise, use the loop command to obtain results at the following P values – 0, 4, 8, and 12 Gpa.

The commands solved this problem are:

```
>>P=0:4:12;                % assigns pressure, GPa, to vector P
>>n=length(P);                  % calculate number of pressures
>>for i=1:n  % for... end loop to calculate k at all P values
>>if P(i)>=0&P(i)<=8 % checks if k is in the available range
k=550.8+10.1*P(i);                           % k in available range
fprintf('k= %6.1f at P= %4.1f\n',k,P(i))% displays k, P
else                         % otherwise, if k is out of range
fprintf('P is %4.1f, This pressure is out of allowed range\n'...
,P(i))% displays P and message that P is out of range
end                               % terminates the if statement
end                                  % terminates the for … loop

k= 550.8 at P= 0.0
k= 591.2 at P= 4.0
k= 631.6 at P= 8.0
P is 12.0, This pressure is out of allowed range
```

The commands operate as follows:

- The first command assigns pressures to the 1×4 vector P;
- The second command determines the number of pressures to further usage in the for ... end loop;
- The next command is the for... end loop in which every i-pass the new P value is defined by its index (address) as P(i);
- In the following lines, the if... else... end statement is inside the loop; here the current value of P is checked with the P(i)> = 0&P(i)< = 8 conditional expression, if this expression is true, the value of k is calculated and displayed with the P(i) value, otherwise the message "P is XX.X, This pressure is out of allowed range" is displayed. The calculated k, P, and messages are displayed with the fprintf commands, in which the P format is given with one decimal digit.

2.5.3 Compression Piston Ring: Radial Thickness

The designed radial thickness t_r (in mm) of the compression piston rings can be calculated as:

$$t_r = D\sqrt{\frac{3P_w}{\sigma}}$$

where D is the cylinder bore diameter, mm; P_w is the gas pressure, N/mm^2 and σ – allowable bending stress, MPa (Sobhy M., n.d.).

Problem: Consider the thickness of the piston ring when cylinder bore diameter is 130 mm and gas pressure and bending stress ranges are $P_w = 0.025 \ldots 0.042$ N/mm^2 and $\sigma = 85...110$ MPa respectively; take the four values of P_w and five values of σ. Solve the problem in two ways: with and without the for... end loops (using the vectors only). Display the results in a tabular form in which each column shows the thickness at a constant σ-value while each row corresponds to a constant pressure; use the fprintf command.

The commands solves this problem are:

```
>>D=130;n_Pw=4;n_sigma=5;                                % assignments
>>Pw=linspace(0.025,0.042,n_Pw);%generates row vector Pw, N/mm^2
>>sigma=linspace(85,110,5); % generates row vector sigma, MPa
>>tr=zeros(n_Pw,n_sigma);                  % preallocated tr matrix
>>for i=1:n_Pw                                 % the external loop
for j=1:n_sigma                                % the internal loop
tr(i,j)=D*sqrt(3*Pw(i)/sigma(j));                 % calculates tr
end                                      % ends the internal loop
end                                      % ends the external loop
>>fprintf(' Thickness using loops\n'),fprintf(...
'%4.1f %4.1f %4.1f %4.1f %4.1f\n',tr') %displays title and tr
Thickness using loops
3.9  3.7  3.6  3.5  3.4
4.3  4.1  4.0  3.9  3.8
4.7  4.5  4.3  4.2  4.1
5.0  4.8  4.7  4.5  4.4

>>tr=D*sqrt(3*Pw'*(1./sigma)); %   1./sigma first then 3*Pw'*
>>fprintf(' Thickness without loops\n'), fprintf(...
'%4.1f %4.1f %4.1f %4.1f %4.1f\n',tr') %displays title and tr
Thickness without loops
3.9  3.7  3.6  3.5  3.4
4.3  4.1  4.0  3.9  3.8
4.7  4.5  4.3  4.2  4.1
5.0  4.8  4.7  4.5  4.4
```

The above commands realize both required solution ways and act as follows:

- The commands in the first three lines assign values to the variables D, n_Pw, and n_sigma and generate the Pw and sigma row vectors with the corresponding specified values;
- The following commands calculate the thickness t_r using a command located inside two for... end loops: external for *Pw* and internal for σ; before these loops, the null t_r matrix is generated to reduce computer computation by preventing matrix resize during each pass of the loop.
- The two fprintf commands on the next line display the heading "Thickness using loops" and the calculated Pw values with one digit after the decimal point.

- The following commands implement the second required solution – no loops. To do this, the term under the square root in the t_r expression is written as 3Pw(1/ σ), so division in brackets comes first and generates (using element-wise division) a 1 × 5 row vector. To make the inner dimensions of the 1./sigma and Pw row vectors equal, we transpose Pw (using the ' sign) into a 4 × 1 column vector. The next multiplication by three does not change the vector size. Finally, the product of the [4 × 1] * [1 × 5] matrices is a 4 × 5 matrix that is identical with the obtained in the solution of the first required way.
- The last line with the two fprintf commands displays the heading "Thickness without loops" and the calculated *Pw* values with one decimal digit.

3 Program Managing
Editor and Live Editor

3.1 INTRODUCTION

The commands and its managing were performed interactively in the preceding chapter. With this way of working, the commands are not saved and should be re-entered in the Command Window each time you need to repeat the calculations. This is very inconvenient and is an essential disadvantage of the interactive mode; another inconvenience is that corrections of any command already entered are possible only in a new command line. If multiple commands were entered to obtain a result, to correct any of these commands, all predecessors together with the corrected and subsequent commands must be repeated to obtain the correct result.

The situations described are uncomfortable, and the reader has undoubtedly experienced this. The solution to this problem is to write all the required commands into a list (termed a program), save them to a file, and run when needed. MATLAB® has two types of such files, called script and function files. Their creation and storage using the available editors, launching the created program, as well as examples of applications are discussed below[1].

3.2 SCRIPTS AND SCRIPT FILES

3.2.1 Editor and Creating, Saving, and Running a Script File

The list of commands written as they should be executed is called a script and is a program. The special window called Editor is used to type a script and save it in a file, after which this file can be run. Corrections or new commands can be performed directly into the Editor window. Saved files get the extension '.m' and termed m-files.

To launch the Editor enter the edit command in the Command Window or click the New Script icon located on the toolstrip in the File section of the MATLAB® desktop Home tab. After this, the Editor window appears. The screenshot of the window separated from the desktop is shown in Figure 3.1.

The window contains a menu on the top toolstrip and a large blank field in the rest; the program should be typed in this field. The toolstrip includes the EDITOR, PUBLISH, and VIEW tabs. The EDITOR tab (default appearance) is

[1] Some text and table materials from Burstein, 2021a (Section 3.4) and Burstein, 2020 (Sections 4.4, 4.5) are used in the chapter; with permissions from IGI Global and Elsevier respectively.

DOI: 10.1201/9781003200352-3

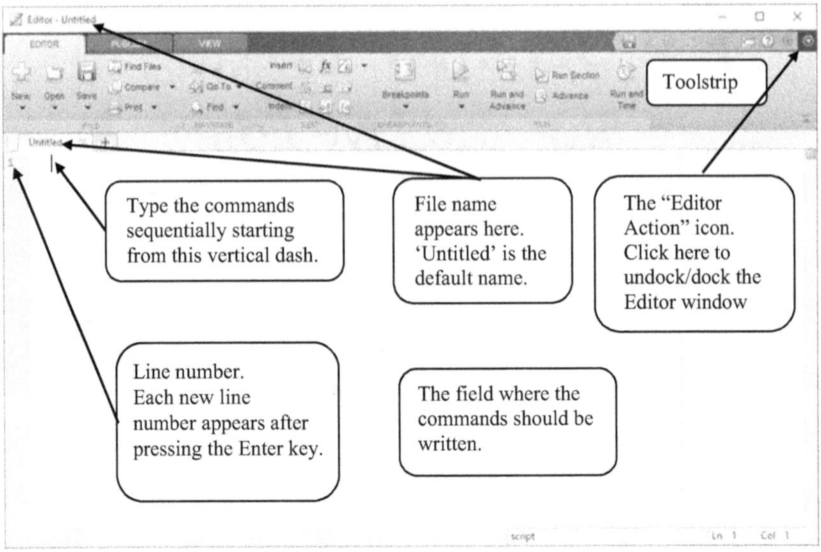

FIGURE 3.1 The Editor window.

commonly used for writing/editing/debugging programs as well as for saving/opening and running.

Starting from the place marked by the vertical dash in Figure 3.1, the script commands should be typed sequentially line by line. Two or more commands can be placed on the same line; these commands should be separated by commas or semicolons. A new line is available after pressing the Enter key. Each line with command gets a serial number that appears automatically when the line is opened. A typical script file created in the Editor window is shown in Figure 3.2. The file is named EditorView_ScriptExample and calculates spur gear chordal thickness. The first four lines of this file are the explanatory comments; they are preceded by the % comments sign and are displayed in green. Comments are not a part of the execution and do not get a dash against the number. Two further lines contain commands for calculating and displaying the tooth thickness s, in mm, according to a given modulus m and the number of teeth Z. For better legibility, the commands in the Editor appear in black.

The right vertical frame bar of the Editor Window is a bar on which the message markers are displayed – horizontal colored strokes. These are messages of the code analyzer (also called the M-Lint analyzer). This analyzer detects the script, identifies possible errors, comments on them, and recommends corrections to improve program performance. The small square at the top of the message bar indicates the presence or absence of the errors and/or makes some warnings. When the indicator is green, it means no errors in the script; a red indicator means that syntax errors were detected; an orange indicator warns of the possibility of improving the script (but no errors). A place/command that contains an error

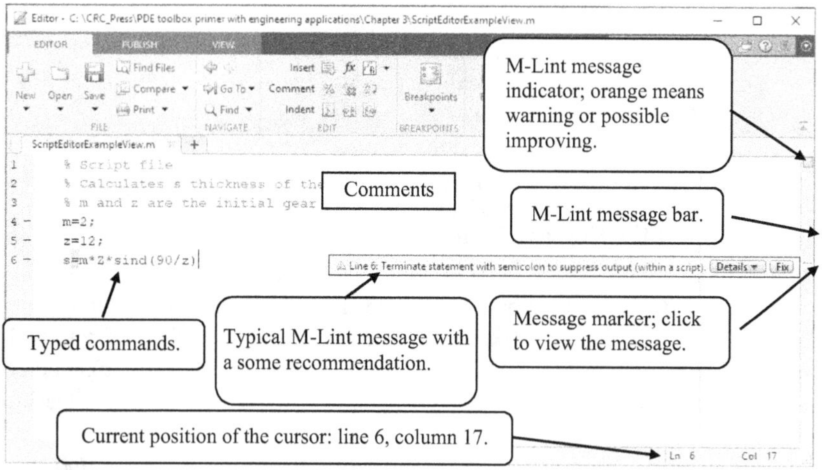

FIGURE 3.2 Editor window with script file and *M*-Lint message.

or needs improvement is underlined/highlighted and a horizontal colored dash appears in the message bar. By moving cursor to this dash or to the marked place in the command, we can obtain the error/warning message; an example of such a message can be seen in Figure 3.2. Not every warning recommendation should be followed; for example, the recommendation shown in this figure (adding a semicolon) should not be followed because we want to display the resulting value of the cell volume (the semicolon suppresses the display of the resulting value). The code analyzer appears and operates by default, nevertheless, it can be disabled by unmarking the "Enable integrated warning and error messages". For this, check the box in the "MATLAB Code Analyzer Preferences" panel that appears after selecting the "Code Analyzer" line of the "Preferences" window, which should be pre-opened by clicking the Preferences line in the Environment section of the desktop Home tab. The "MATLAB Code Analyzer Preferences" panel contains many other default/active settings that you can adjust.

3.2.1.1 Saving the Script File

After preparing it in the Editor, the program must be saved. To perform this, select the "Save As …" option of the "Save" button of the Editor window (located at the toolstrip in the FILE section of the EDITOR tab). Then, in the appeared "Select File for Save As" window, select the desired file location in the directory/file tree on the left panel, and also type a name in the "File name:" field. The default save file location is the "MATLAB" folder in "My documents" directory.

To select name a file, follow these guidelines:

- It is desirable that the name begins with a letter and not with a number;
- The name cannot be longer than 63 characters;

- The file name should not repeat the name of another existing file or the name of a MATLAB® command/function/predefined variable;
- The signs of the mathematical operations (e.g. +, –, /, \, *, ^) should not be used within the file name;
- It is not recommended to use dots, commas, spaces, or other punctuation marks in the file name.

3.2.1.2 About the Current Folder

You can see the path to the current folder in the first top bar of the Editor window (Figure 3.3). Note: This path may differ from the one that appeared on the MATLAB® desktop. In this desktop, the directories and files of the present folder are displayed in the Current Folder panel (to the left of the desktop). The full path to the user file can be seen in the current folder bar just under the Desktop toolstrip.

To set a non-default folder containing an m-file previously created and saved with Editor, the following operations should be performed:

- Open the "Select a new folder" dialog box by clicking the "Browse for folder" icon located on the current directory bar;
- In the left panel, select the line with the desired directory name and then click the "Select Folder" button at the bottom of the dialog box; the

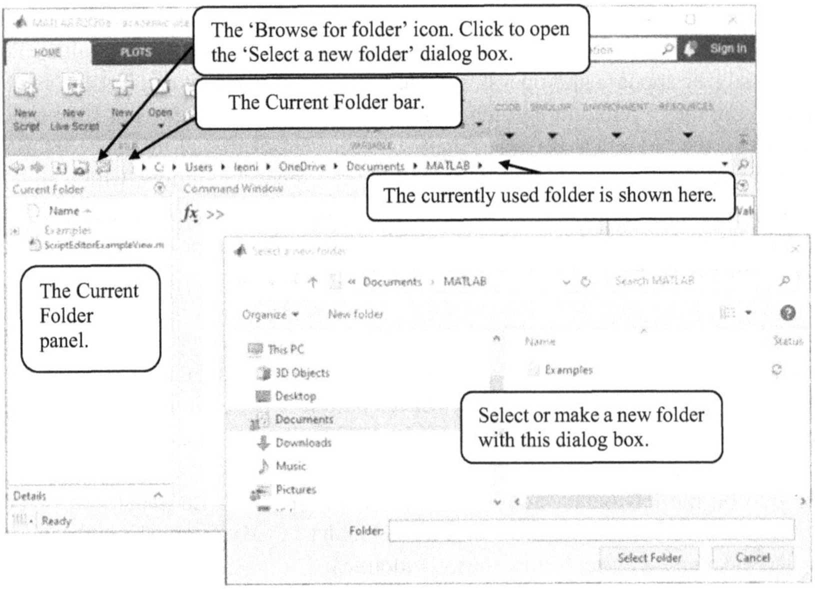

FIGURE 3.3 Desktop with opened "Select a new folder" dialog box and explanatory frames.

selected directory appears in the current directory field at the top of the box. As mentioned above, the selected folder should be the same as in the Editor window.

For example, Figure 3.2 shows the separate file ScriptEditorExampleView.m stored next to the Examples subfolder; the path to this file is C:\Users\leoni\OneDrive\Documents\MATLAB.

3.2.1.3 Running Created Script File

The developed script file should be run to execute it and display the calculation results. To execute a script file:

- Check first if the required file is in the installed current folder; otherwise, set the appropriate folder;
- Enter the file name (without m-extension) in the command line of the Command Window or click the "Run" icon on the Editor window toolstrip.

Following this, to run the ScriptEditortExampleView script we need to check/install the C:\Users\leoni\OneDrive\Documents\MATLAB path, and then to type and enter the file name in the Command Window. The command and displayed result:

```
>> ScriptEditorExampleView    % enter to run the script file
s =
3.1326
```

3.2.2 Input Values to the Program Variables from the Command Window

Whenever a script file is to be used for recalculations with new parameters, the new values should be typed into the script, then the corrected script must be saved, and run. To prevent the alterations of the script file, you can use the input command, this command has the following forms:

```
Numeric_Variable = input('Explanatory text')
Character_Variable =input('Explanatory text','s')
```

where 'Explanatory text' is a message displayed in the Command Window to help you understand what to input; the inputted number is assigned to the Numeric_Variable, and the string is assigned to the Character_Variable; the 's' signs means that the inputting text is a string that can be written without single quotes. The Explanatory text can contain more than one line, and the \n specifier should be entered at the end of each line.

When the input command is met and initiated in the running script file, the text written in single quotes is displayed on the screen and the user should type and enter a number or string depending on the user command form.

As an example, create a script file that converts international system SI standard density units, kg/m^3, into the US (Imperial) density units, lb$_m$/ft^3- pound mass per cubic foot; use this expression $d_{US} = 0.062428 \cdot d_{SI}$:

To perform this, we should type in the Editor the following commands:

```
% density convertor: kg/m^3 to lb/ft^3
d_SI=input('Enter density in kg/m^3, d = ');
d_US=0.062428*d_SI;
fprintf('\n Density in lb/ft^3 is %10.4f \n',d_US)
```

Before running, save these commands in the m-file with the name Density_SI2US. After entering this name in the Command Window, the input command prompt 'Enter density in kg/m^3, d = ' appears on the screen. Type now a density value (in kg/m^3) and press enter; the inputted density value is assigned to the d_SI variable and then converted to lb/ft^3 with the next command; the results display as the string 'Density in lb/ft^3 is' and the obtained value.

Running command, appeared prompt of the input command, inputted pressure in kilogram per cubic meter, displayed string of the fprintf command together with and defined density in lb/ft^3 are:

```
>> Density_SI2US
Enter density in kg/m^3, d = 2755
Density in lb/ft^3 is 171.9891
```

The input command allows you to enter a series of numbers in view of the vectors or matrices that can be done by typing numbers in squared brackets. Use, for example, the Density_SI2US script file to input just two densities and convert them:

```
>> Density_SI2US
Enter density in kg/m^3, d = [2755 1023]
Density in lb/ft^3 is 171.9891
Density in lb/ft^3 is 63.8638
```

3.3 USER-DEFINED FUNCTIONS AND FUNCTION FILES

3.3.1 Function Creation

The simplest form of a function f in classical mathematics is $y = f(x)$, and it connects the independent x and the dependent y parameters (called also arguments or variables). The right and left parts of the function can contain sets of variables, for example, $\boldsymbol{l} = f(x_1,x_2,x_3)$ with $\boldsymbol{l}$ as a three-element vector. In terms used in programming, the independent parameters on the right side of the function are input parameters, and on the left side are output parameters. Many of the commands discussed in the preceding chapter have the function form, such as exp(x), log(x),

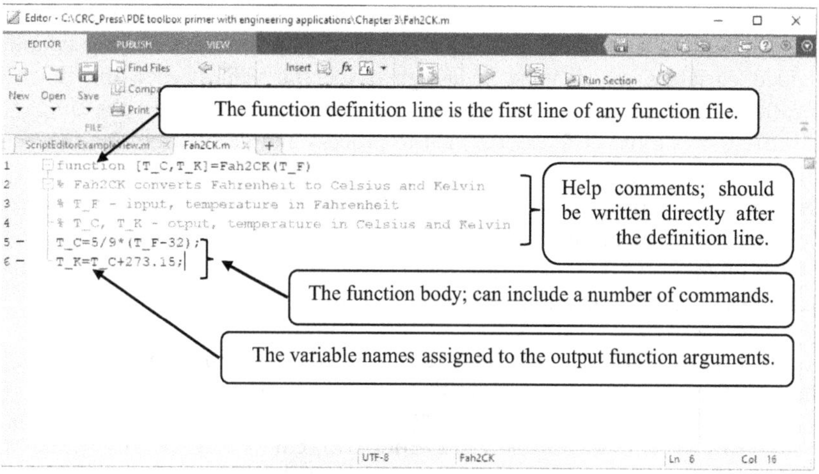

FIGURE 3.4 Editor window with the function file and a user-defined function.

sqrt(x), sin(x), cos(x), etc.; such form enables to use commands in interactive calculations or in programs by entering the function name with the appropriate input argument. In MATLAB®, in addition to the available functions, the user can create any new function and then reuse it in the same way as a regular one; such functions are termed "user-defined". An individual expression and a complete program with many commands can be defined as a function, saved as a function file, and then used in any required calculations.

In general, the developed user-defined function should comprise the following parts: the definition of the function, explanatory help lines, and the body of the function with program commands. The named components should be written sequentially in the Editor window, as shown in the example in Figure 3.4.

The program shown in this figure is the user-defined function Fah2CK saved in a file with the same name; the function converts the temperatures specified in Fahrenheit to temperatures in Celsius and Kelvin (SI units). In accordance with the definition line, the function has: one input parameter – the vector T_F with temperature in Fahrenheit that should be converted – and two output parameters – the converted temperatures in Celsius and in Kelvin.

The required forms for each part of a user-defined function are as follows.

3.3.1.1 Function Definition

The first line of a function contains its definition and reads

```
function [output _parameters]=function_name(input_
parameters)
```

The word function is the first word of the user-defined function, and after typing, it is highlighted in blue. The function_name is the name that is placed to the right of the " = " assignment symbol, and it must be specified according to the variable naming rules (Section 2.2.4). The input_parameters denotes comma-divided variables which must be assigned specific values when the function is called. These variables can be scalars or the vectors/matrices; the output _ parameters is a list of those we want to process and derive from the function. The input parameters must be written between the parentheses and the output parameters – between square brackets. When there is a single output parameter, the brackets can be dispensed with.

The input and/or output parameters can be completely or partially omitted. Examples of various function definition lines are:

```
function [A,B,D] = fnc1(a,b,c,d) - full record, function
named 'fnc1' with four input (a, b, c, and d) and three
output (A, B and D) parameters;
function A= fnc2(a,b) - full record, function named 'fnc2'
with two input (a and b) and one output (A) parameters;
function [A,B]= fnc3 - partial record, function named
'fnc3' with two output parameters (A and B), no input
parameters;
function fnc4(a,b,c) - partial record, function named
'fnc4' with three input arguments (a, b, and c), no output
parameters;
function fnc5 - partial record, function named 'fnc5'
without input and output parameters.
```

Note:

- the word "function" should be written in lowercase letters;
- the amounts and names of the input and output function arguments may differ from those shown in the examples above.

3.3.1.2 Help Lines

These lines contain the comments about the function; they should be placed just after the function definition line. The first line of the function help should contain the function name and a short info about the function. The lookfor command searches the first lines of the regular and user-defined functions and displays it, for example, typing lookfor Fah2CK (this function is shown in Figure 3.4) in the Command window yields:

```
>>lookfor Fah2CK
Fah2CK - converts Fahrenheit to Celsius and Kelvin
```

All help lines written by your user-defined function are displayed when the help command with a name of the function is entered, for example:

```
>> help Fah2CK
Fah2CK converts Fahrenheit to Celsius and Kelvin
```

```
T_F - input, temperature in Fahrenheit
T_C, T_K - otput, temperature in Celsius and Kelvin
```

Note: The function help lines are optional and can be omitted.

Hereafter, to be short, we will write the function help part with only two lines: the first line with function info, and then a line with an example of the run command.

3.3.1.3 Function Body, Local and Global Variables

The commands written after the function help are the body of a function that should comprise one or more commands for actual calculations. The obtained values should be assigned to the output parameters; for example, in Figure 3.4 the output parameters are T_C and T_K; thus the two commands of this function calculate and assign defined values to these parameters. Note: The user-defined function can only be executed if the actual values were assigned to the input arguments.

The body of a function can include other user-defined functions; such functions are called sub-functions.

Variables inside a user-defined function are local and relevant only within this function. Such variables are not saved and do not remain in the workspace after the function finishes executing. Nevertheless, some or all of the function variables can be shared with other function/s. For this we have to make them available outside of this function, which can be performed using the **global** command:

```
global var1 var2 ...
```

The **global** command with the space-divided list of variables (designated here as **var1 var2** ...) must be written within the function before the variables are first used, and this command should be repeated in other functions that intend to use these variables.

3.3.2 About Function File

Before running, user-defined function created in the Editor window should be saved in a file. This can be realized exactly as for the script file: select "Save As" line of the "Save" option (located in the "File section" of the "Editor" tab) and then in the appeared dialog box "Select file for save as ..." enter the desired location and name of the saving file. It is highly desirable to name the file with the name of the function. For example, the **Fah2CK** function should be saved in a file named **Fah2CK.m**.

Here are some examples of function definition lines and corresponding function file names:

```
function [i1,i2]=current(V1,R1,R2,R3) - the function file
should be named and saved as current.m;
function piston_ring(m,n) - the function file should be
named and saved as piston_ring.m;
```

```
function [sigmax,sigmay]=stiffness - the function file
should be named and saved as stiffness.m;
function PDEmodel - the function file should be named and
saved as PDEmodel.m.
```

If a function comprises several sub-functions, the function file should match the name of the main function (from which the program is started).

3.3.3 Running a User-defined Function

Any user-defined function stored in the corresponding file can be launched from another function/script program or from the Command Window as follows: the function file definition line should be typed without the word 'function' and to the input parameters should be assigned their values. The latter can be performed before the function call by pre-assigning actual values to the input parameters or strictly within the function launch command by replacing the variables with their values. For example, the Fah2CK function file (Figure 3.4) can be run from the Command Window with the following command:

```
>>[T_C,T_K]=Fah2CK([80 85])       % T_F replaced by its values
T_C =
26.6667 29.4444
T_K =
299.8167 302.5944
```

Alternatively, with pre-assignment of the input variables:

```
>> T_F=[80 85];                 % pre-assigns T_F with its values
>> [T_C,T_K]=Fah2CK(T_F)          % launches the Fah2CK function
T_C =
26.6667 29.4444
T_K =
299.8167 302.5944
```

A user-defined function can be used to calculate some expression containing one of the function output parameters. For example, to calculate the water dynamic viscosity by the expression $\mu = 2.414–10^{-2}\cdot 10^{247.8/(T-140)}$ (where μ in mPa·s and T in K) at the temperature $T = 80$ F, we must convert Fahrenheit to Kelvin using the Fah2CK function and then calculate μ; the following commands solve the problem interactively:

```
>>T_F = 80;    % assigns temperature in K to the T_F variable
>>[T_C,T_K] = Fah2CK(T_F);              % converts F to C and K
>>mu_water = 2.414e-2*10^(247.8/(T_K-140))%usesT_K output
parameter
mu_water =
0.8576
```

3.3.3.1 Comparison of Script and Function Files

A beginner often does not see the difference between script and function files. Indeed, many actual problems can be solved using an ordinal script-type file. For a better understanding of the above two file forms, their similarities and differences are outlined below.

- Both types of files are saved with the extension *m*, as *m*-files.
- The user-defined function that forms the function file has the definition line as the first line; this feature is absent in the script file.
- The function file name must be the same as the function name; this requirement does not make sense for a script file, since the latter does not have a definition line.
- Function files can adopt and return data through the input/output parameters of the user-defined function; script files do not have this capability, and the values of their variables must be assigned directly within the file or entered, for example, with the input commands.
- The script files use all the pre-defined/calculated variables that are in the workspace, while the function files only use those that are in the body of the function.
- Only function files can be used as functions in other calculating programs, or simply in the Command Window.

Eventually, the user must decide to himself which file is preferable to develop to solve his problem.

3.3.4 Anonymous Function

When a user-defined function has one line and in addition should use some constant/s, it is convenient to use a so-called anonymous function that has the following form:

```
name_fun=@(arg1,arg2,…)expression
```

where name_fun is the name of the anonymous function; arg1,arg2,... - independent variables that should be passed to the function; expression is a mathematical expression that can include independent variables, constants, other ordinary or anonymous functions; the @ symbol is used to denote an anonymous function.

An anonymous function can be written and used directly in Command Window, in a script file, or in a user-defined function. Once this function has been inputted, it can be used by typing its name with the argument value in parentheses.

For example, BTU, British thermal units, can be converted to kW, kilowatt, by the formulae $kW = 2.93 \cdot 10^{-4} \cdot BTU$, this expression can be inputted in the Command Window as the following anonymous function with one argument:

```
>>kW=@(BTU)2.93e-4*BTU
kW =
@(BTU)2.93e-4*BTU
After that we can use this function to convert, say, 2600
BTU to kW:
>>kW(2600)
ans =
0.7618
```

To convert multiple BTU values, for example, 3000, 3100,…3500, the vector with these values can be inputted when calling the kW anonymous function:

```
>> kW(3000:100:3500)
ans =
0.0879 0.0908 0.0938 0.0967 0.0996 0.1026
```

Anonymous functions can be used as parameters in other command/s and will be used further in chapters on differential equation solvers.

3.4 INTERACTIVE SCRIPT AND FUNCTION PROGRAMS: LIVE EDITOR

To perform a program written in the Editor window and saved as script or function file, it should be launched in the Command Window. Thereby, two different windows should be used to operate with these files: Editor window and Command Window. Moreover, when these programs generate any graph, they use an additional graphical window (see Chapter 4). This way of working is inconvenient, especially for non-programmers; it would be more suitable to use the same window to record programs, run them, and display the results. Therefore, in 2016 MATLAB® announced a new developed tool called the Live Editor. This editor allows you to generate interactive script programs that include commands, explanatory text, and images along with the numerical and graphical output of the program. In 2018, the Live Editor has been enhanced and now the user-defined functions can also be created with this editor. Here we bring a brief description of the Live Editor and show how to create a live script file containing an engineering example.

3.4.1 Opening the Live Editor

To open the Live Editor window – click the New Live Script button  located in the FILE section of the MATLAB® Desktop toolstrip. The Live Editor window appears - Figure 3.5.

This window toolstrip comprises three tabs – LIVE EDITOR, INSERT, and VIEW. The LIVE EDITOR tab contains the sections with buttons that are used to edit the scripts. The INSERT tab contains functions to insert texts, images, equations, code examples, section breaks, controls, and others to inserting items into

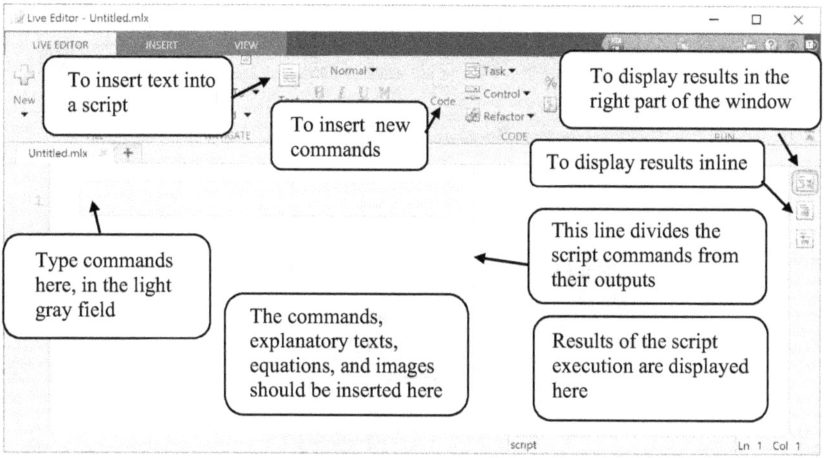

FIGURE 3.5 The Live Editor window.

the script. The VIEW tab includes various icons to control the live script output. In the following, we describe those of the tab buttons that are commonly used to create an actual live script.

The Live Editor window contains a vertical dividing line (by default) that separates the window into two parts: the left one for entering commands, explanatory texts, equations, and images; and the right one for displaying numerical and graphical results. To show the outputs inline, instead of on the right, the "Output inline" icon in the right frame bar should be clicked. After that, the dividing line disappears and the results (numbers, strings, graphics) are displayed directly under the appropriate command/s.

3.4.2 Creating the Life Script

To demonstrate the Live Editor use, write, for example, the program presented in Figure 3.2 for the chordal tooth thickness of the spur gear.

Problem: Write the live script with the text and code parts that calculate and output inline the gear tooth thickness for the modulus m = 2 and teeth number z = 17. Display result with one decimal place using the fprintf command.

The program LiveScriptEx1 that solves this problem is presented in Figure 3.6.

The following steps were performed to build the above live script.

Step 1. Inserting the explanatory text.

The Text button was clicked in the TEXT section in the LIVE EDITOR tab to insert the text that you can see before image and after the equation. The text was typed into the white field that appears above the light gray command field immediately after clicking the Text button.

Live Editor - C:\My_Folder\Leonid Burstein\Articles\CRC_Press\PDE tool... — □ ✕

LIVE EDITOR INSERT VIEW

New Open Save Find Files NAVIGATE TEXT CODE SECTION RUN
 Compare
 Print
 FILE

LiveScriptEx1.mlx ✕ +

Tooth thichness s, mm, via the modulus m and the number of teeth z

$$s = m \cdot z \cdot \cos\frac{90}{z}$$

where 90/z in degrees.

```
1     m=2;
2     z=17;
3     s=m*z*sind(90/z);
4     fprintf('s=%4.1f mm\n',s)

      s= 3.1 mm
```

UTF-8 script

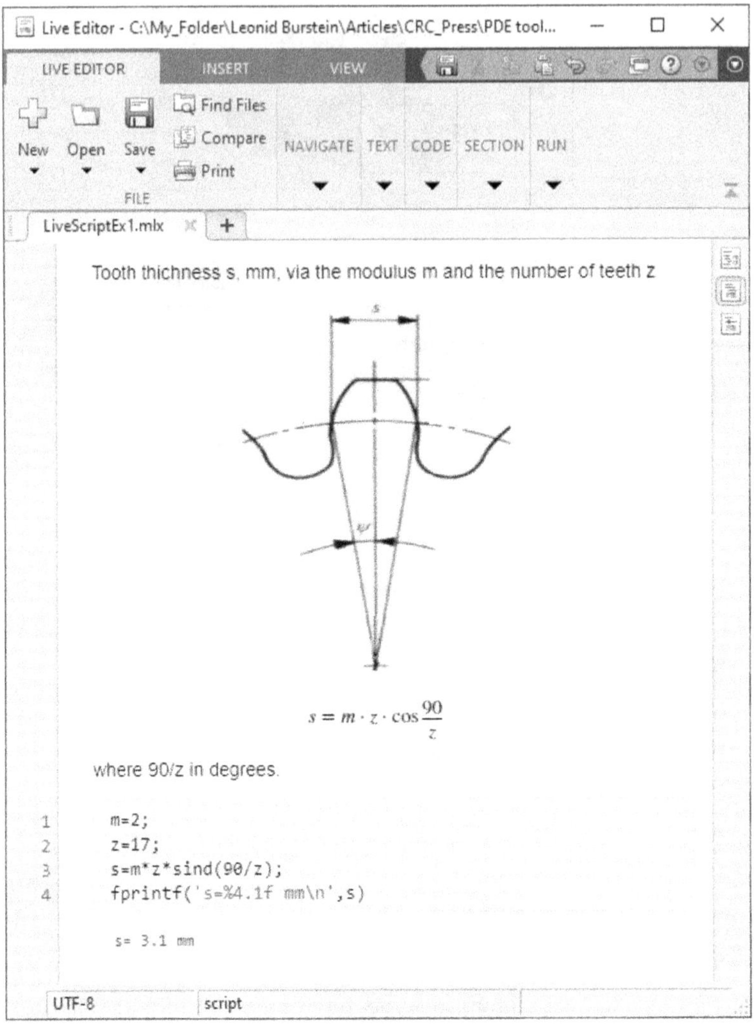

FIGURE 3.6 View of the Live Editor with a live script program.

Step 2. Inserting the image that illustrates the gear tooth.

The gear tooth image was previously created and saved as a jpg-file (also allowed bmp-, gif-, png-, or wbmp-file). Then the Image button was clicked on the INSERT tab. In the appeared "Load Image" window, the image file was selected and then the "Open" button was clicked. After this, the image appeared in the Live Editor text field.

Step 3. Inserting the equation in the text field.

The "Equation" line of the "Equation" button located on the INSERT tab was clicked to insert the solving equation as text below the image. The equation was typed in the appeared "Enter your equation" framed field. After selecting the "Equation" option, the additional EQUATION tab appears on the toolstrip; the tab allows you to type an equation using the necessary symbols, structures, and matrices.

Step 4. Entering commands for thickness calculations.

The commands were typed within the light gray field to calculate and display the result with one decimal place. For the latter, the fprintf command was used. The commands are as follows:

```
m = 2;
z = 17;
s = m*z*sind(90/z);
fprintf('s = %4.1f mm\n',s)
```

Step 5. Selecting the "Output inline" option.

The "Output on right" button displays the results to the right of the dividing line (default). To output directly under the command, the "Output Inline" button was clicked on the VIEW section of the VIEW tab to display the thickness, as shown in Figure 3.6. Alternatively, you can click the button on the editor right bar to set the inline mode.

Step 6. Running the created live script.

To run the live script, the Run button located in the RUN section of the LIVE EDITOR tub should be clicked. After some time that maybe slightly greater than in the case of the simple script running, the results are output immediately after the command that does not have the ending semicolon, or/and after the output command (in our case, this is the fprintf command).

Step 7. Saving created live script.

The developed script was saved to a file by selecting the "Save as …" line in the popup menu that appears after clicking the "Save" button on the toolstrip. In the appeared "Select file for Save as" dialog box the file location was selected and the file name was typed; the saved script file gets the *mlx* extension. The life script file shown in Figure 3.6 named LiveScriptEx1.mlx.

If necessary, the script can be converted into a pdf file by selecting the "Export to pdf …" line in the popup menu of the "Save" button. Another possibility provided by this button is the "Export to Word …" option that allows you to save the script in Word format.

3.4.3 Additional Information for Using the Live Editor

3.4.3.1 Separate Sections in the Live Program

When the live program is long, it is better to divide the commands into separate sections. For this, before writing the text and commands, the "Section Break" button (located in the SECTION section of the INSERT tab) should be clicked.

Where you want to finish this section and start a new one, you need to click the "Section Break" button again. To run the script section, click the blue bar (vertical margin of the Live Editor frame) to the left of the section or alternatively click the "Run Section" button 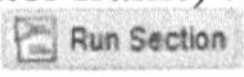

located in the SECTION section of the LIVE EDITOR tab.

3.4.3.2 Creating Live Function

Steps described in the receded section produce a live script. To create a live function, open the Live Editor and select the Live Function line in the popup menu of the "New" button located in the LIVE EDITOR tab of the toolstrip. The Live Editor opens with the untitled function (default) – Figure 3.7.

Note: The live function should have the end as its last line.

Two ways can be used to run the live function:

a. enter in the Command Window the function name with the values assigned to the input parameters of the function; the results of function execution are displayed in this window;
b. type out the function in the Live Editor window or in another program, the function name with the values assigned to the input parameters; the results are displayed in the Live Editor window before the function definition line (see Section 3.5.4).

3.4.3.3 Converting a Script/Function to a Live Script/Function

Any previously developed *m*-file containing script can be opened and saved as a live script *mlx*-file. For this, when the file opened in the Editor, click on the Save button of the Editor tab, select Save As line in the popup menu, and in the "Save

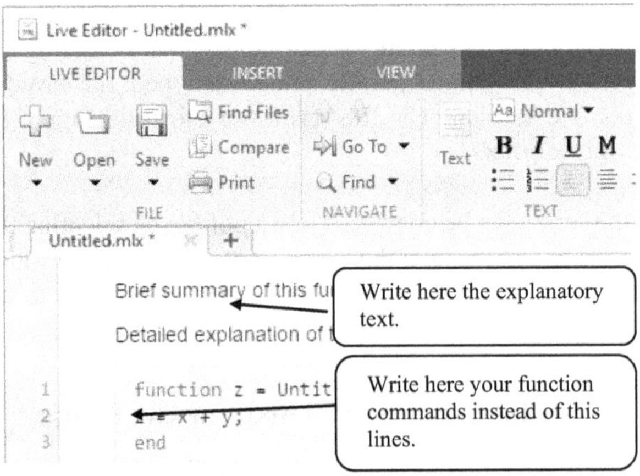

FIGURE 3.7 View of the Live Editor window that opens to create a live function. The Untitled.mlx is the default file with an example of function calculating output variable **z** as sum of the two input parameters x and y.

as type" field of the opened window, select the "to MATLAB Live Code Files (*.mlx)" option and click Save. Opening/saving a script as a live script creates a copy of the original script. The comments of the script are converted into the texts of the generated live script.

The existing m-file containing function can be open as a live function mlx-file in the same way as for a script file – selecting the "to MATLAB Live Code Files (*.mlx)" option (see above).

3.4.3.4 Text Formatting Options

To format your typed text use the following TEXT section options of the LIVE EDITOR tab:

- Text Style: Normal (default), Heading 1, or 2, or 3, Title;
- Alignment: Left, Center, Right;
- Lists: Numbered list, Bulleted list;
- Standard Formatting: Bold, Italic, Underline, Monospace;
- Change Case (select the text and right click): uppercase, lowercase.

3.4.3.5 About the Interactive Controls

For interactive control of variables, the Live Editor has the following means: Numeric Slider, Drop Down, Check box, Edit Field, and Button. These controls can be added to live script or function. For example, to insert a slider in the above LiveScriptEx1 script – click the Control button of the CODE section of the INSERT tab and select the Numeric Slider line; then specify the minimum, maximum, and step values of the variable in the "Values" section of the opened dialog box. Thereby, the command using the liner for modulus *m* looks like . The value to the left of the slider line is the current value of the variable m.

If the Drop Down control was selected, the command for the number of teeth Z looks like – . The highlighted value is the current Z value. The numbers in the drop-down list are entered in the dialog box that opens after selecting the Drop Down line in the list appears after the Control button is clicked, as described above for the slider.

Note: Live Editor is a newer and more suitable tool for interactive programming than the regular MATLAB® Editor, but many life scripts and programs that contain/use a large amount of commands or data run slower than regular scripts or functions. Moreover, this editor has restrictions on the use of certain classes of variables. Thus, the user should decide which editor he should use to solve his particular problem.

3.5 APPLICATION EXAMPLES

Below are simple practical examples with script and function programs as well as with live script and live function.

3.5.1 Converting Brinell Hardness to Vickers and Rockwell Hardness

The hardness H of a material is measured in different units, Brinell, HB, Vickers, HV, Rockwell B, HRB, and others. To compare different hardness data, the values should be converted to the same hardness scale. There are no exact conversion expressions, thus approximate expressions have been taken in this example:

- For Brinell to Vickers

$$HV = \frac{HB}{0.95}$$

- For Brinell to Rockwell B

$$HRB = 134 - \frac{6700}{HB}$$

Problem (Based on Burstein, 2020)
Develop using the Editor a script program that converts from Brinell to both Vickers and Rockwell B hardness. The program should request and retrieve the Brinell hardness value/s and display the calculated HB and HV hardness together with the inputted HB hardness with one decimal digit. Save program in the ApExample_3_1.m file

The program that solves this problem is:

```
%             the program converts hardness in HB to HV and HRB
%                                        To run: >> ApExample_3_1
HB=input('Enter hardness in the Brinell's units ');
HV=HB/0.95;                                            % HB to HV
HRB=134-6.7e3./HB;                                     % NB to HRB
Results=[HB HV HRB]';                 % three-row matrix for output
fprintf('\n %5.1f HB is %5.1f in HV and %5.1f in
HRB\n',Results)
```

These commands were entered in the Editor window and saved in a file named ApExample_3_1. When the script file name, ApExample_3_1, entered into the Command Window, the input command displays the input command prompt for data entry, after which the value/s should be typed and entered. If we need to convert several values of hardness, they should be entered as vector – in square brackets with space-divided values; after we press the Enter key, the HB, HV, and HRB are calculated and displayed using the fprintf command with one digital number after the decimal point.

Below is the Command window after the file has been launched and the hardness values were entered:

```
>> AApExample_3_1
Enter hardness in the Brinell's units [302 479 130]
302.0 HB is 479.0 in HV and 130.0 in HRB
317.9 HB is 504.2 in HV and 136.8 in HRB
111.8 HB is 120.0 in HV and 82.5 in HRB
```

3.5.2 Bending Shaft Stress

Stress concentration factor K_t for a shaft of the diameter D with a transverse hole of diameter d can be calculated by the expression (Norton, 2010).

$$K_t = 1.58990 - 0.63550 \log \frac{d}{D}$$

Problem: Create a program in the form of a user-defined function to calculate the stress concentration factor for the ratios $d/D = 0.01$, 0.05, 0.1, 0.15, 0.2, and 0.25. The program should comprise a main function named ApExample_3_2 and a sub-function named Stress. The main function must be specified without input/output parameters and should contain: the d/D data, calls to the sub-function, and displays results in two d/D and K_t columns. The sub-function should be specified with the d/D and K_t as its input and output parameters respectively, and it must calculate the shaft concentration factor and pass it to the main function.

The program solving the problem is written below.

```
function ApExample_3_2        % main function, definition line
% calculates the stress factor of the shift with hole in
bending
%                                       To run: >> ApExample_3_2
dD=[.01.05:.05:.3];                          % define the d/D values
Kt=Stress(dD);                         % runs the subfunction Stress
Out=[dD;Kt];              % organize the output as two-row matrix
fprintf('d/D=%4.2f Kt=%4.1f\n',Out)        % outputs d/D and Kt
function Kt=Stress(dD)              % subfunction, definition line
Kt=1.58990-0.63550*log10(dD);                       % calculates Kt
```

This program contains two user-defined functions: the main function ApExample_3_2 that has not input and output parameters in the definition line, and the Stress subfunction, which has one input parameter dD and one output parameter Kt.

Within the main function, the d/D values are assigned to the dD variable, the K_t values are obtained using the Stress subfunction, and the desired values are displayed using the fprintf function, which respectfully displays d/D and K_t values with two and one decimal digits. The Stress subfunction inputs dD values and outputs Kt values which are calculated using the above expression.

The run command and the displayed results are:

```
>> ApExample_3_2
d/D=0.01 Kt= 2.9
d/D=0.05 Kt= 2.4
d/D=0.10 Kt= 2.2
d/D=0.15 Kt= 2.1
d/D=0.20 Kt= 2.0
d/D=0.25 Kt= 2.0
```

3.5.3 Dew Point

The dew point temperature is the air temperature T_d when water vapor begins to condense out of the air at actual relative humidity RH. The T_d value (in °C) can be approximately calculated as:

$$T_d = \frac{243.5 ln\frac{e}{6.112}}{17.67 - ln\frac{e}{6.112}}$$

where $e = 6.112e^{\frac{17.67T_w}{T_w+243.5}} - p(T - T_w)0.00066(1+0.00115T_w)$, and $RH = 100 \frac{e}{6.112e^{\frac{17.67T}{T+243.5}}}$

In these equations T is the actual, dry-bulb, air temperature in °C, T_w is the wet-bulb temperature (thermometer covered by water-soaked cloth) in °C, p is barometric pressure in mbar, and RH is relative humidity in %.

Problem

Write and save a live script file that calculates and displays T_d and RH values for the selected (with the Numeric Slider controls) T, T_w and p. Specify range T as 5 … 50 °C, range T_w as 8 … 35 °C, and p –900 … 1100 mbar. Add appropriate explanatory text to the live script that includes calculating expressions. Present results, for example, for the T = 25 and T_w =19 °C at p = 985.1 mbar; use the `fprintf` command that displays T and T_d with 1 decimal digit, and RH as integer. Use "Output inline" options of the Live Editor.

Created live script is presented in Figure 3.8.

Steps to produce the above `ApExample_3_3` live script:

- After opening the Live Editor window for the further inline output, click the "Output Inline" button located within the VIEW tab at the LAYOUT section;
- Click the Text button in the TEXT section of the LIVE EDITOR tab and type the explanatory text in the appeared white field;
- Select the Equation line in the popup menu of the "Equation" button located in the EQUATION section of the INSERT tab, and type the above equations into the "Enter your equation" box (you can perform this three times for each of the three equations); when entering equations, use the STRUCTURES button or the Brackets part of the SYMBOLS section of the appeared EQUATION tab;
- Click the "Code" button in the CODE section of the LIVE EDITOR tab and type the T = characters in the light gray field; then select the Numeric Slider line in the popup menu of the "Control" button located into the CODE section of the INSRERT tab. A dialog box appears with three fields: Min, Max, and Step; type the T-values 5, 50, and 1 in the corresponding fields;

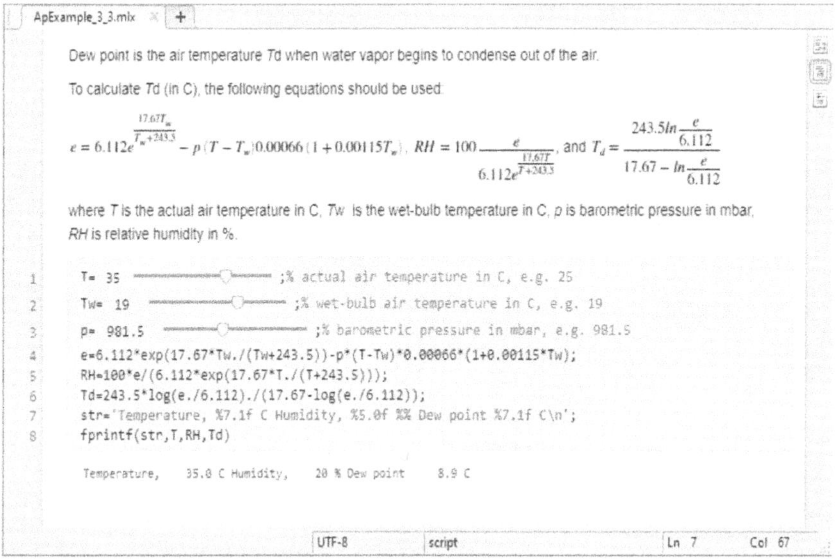
ApExample_3_3.mlx
Dew point is the air temperature Td when water vapor begins to condense out of the air.
To calculate Td (in C), the following equations should be used:
$e = 6.112e^{\frac{17.67T_w}{T_w+243.5}} - p(T-T_w)0.00066(1+0.00115T_w)$, $RH = 100\frac{e}{6.112e^{\frac{17.67T}{T+243.5}}}$, and $T_d = \frac{243.5ln\frac{e}{6.112}}{17.67 - ln\frac{e}{6.112}}$
where T is the actual air temperature in C, Tw is the wet-bulb temperature in C, p is barometric pressure in mbar, RH is relative humidity in %.
T= 35 ;% actual air temperature in C, e.g. 25
Tw= 19 ;% wet-bulb air temperature in C, e.g. 19
p= 981.5 ;% barometric pressure in mbar, e.g. 981.5
e=6.112*exp(17.67*Tw./(Tw+243.5))-p*(T-Tw)*0.00066*(1+0.00115*Tw);
RH=100*e/(6.112*exp(17.67*T./(T+243.5)));
Td=243.5*log(e./6.112)./(17.67-log(e./6.112));
str='Temperature, %7.1f C Humidity, %5.0f %% Dew point %7.1f C\n';
fprintf(str,T,RH,Td)
Temperature, 35.0 C Humidity, 20 % Dew point 8.9 C
UTF-8
script
Ln 7
Col 67

FIGURE 3.8 The Live Editor window with the life script program ApExample_3_3.mlx designed to calculate the dew point temperature.

- In the next code line, type the *Tw* = and select the Numeric Slider line again from the popup menu of the "Control" button located in the CODE section of the INSERT tab; type T_w-values 8, 35, and 1 in the Min, Max, and Step fields, respectively; do the same for the p = with the new Numeric Slider control by entering 900, 1100, and 0.5 in the Min, Max, and Step fields;
- Then enter the commands to calculate and output the dew point T_d together with the inputted T, and calculated HW using the `fprintf` command; note that the string line for this command, together with the required number formats, was written separately as string `str`;
- Save the program in the ApExample_3_3.mlx file by selecting the line "Save as…" as explained earlier – Subsection 3.4.2, step seven.
- Set the T, Tw, and p slider indicators by moving each of them to the desired *T*, T_w, and *p* values; the inputted *T* and calculated *RH* and T_d values appear immediately below the last command; any changes in the indicator/s location/s lead to the recalculation and the appearance of a new *T*, *RH*, and T_d values.

3.5.4 Live Function for Two-Stage Gear Train Calculations

Image in Figure 3.9 represents a two-stage, compound reverted gear train that should be designed (Budinas & Nisbett, 2011). The expressions that should be

Typical two-stage gear reduction

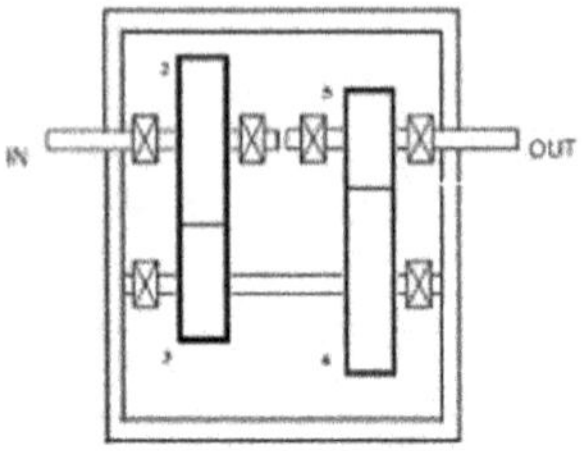


```
Gear(1750,85,3,20,1)
```

```
Teeth, unit
 N2 N3 N4 N5
 16 72 16 72

Speeds, rpm
omega_2 omega_3 omega_4 omega_5
1750.0  388.9   388.9    86.4

Torques, N.m
torque_2 torque_3 torque_5
  81.4     366.2    1648.0

The output gear speed is acceptable
```

```
function Gear(om_in,om_out,om_del,H,k)
% Gear - calculates the teeth numbers, speeds, and torques
om2=om_in;om5=om_out;e=om5/om2;m=1/sqrt(e);phi=20*pi/180;
% Teeth calculations
fprintf('Teeth, unit\n')
N2=2*k/((1+2*m)*sin(phi)^2)*(m+sqrt(m^2+(1+2*m)*sin(phi)^2));
N2=ceil(N2);                               % teeth, rounded to inf
N3=fix(m*N2);                            % teeth, rounded down to 0
N4=N2; N5=N3;
fprintf(' N2 N3 N4 N5\n %i %i %i %i\n',N2,N3,N4,N5)
% Speed calculations
fprintf('\nSpeeds, rpm\n')
om5_cal=(N2/N3)*(N4/N5)*om2;om3=round(N2/N3*om_in,1);om4=om3;om5=om5_cal;
fprintf('omega_2 omega_3 omega_4 omega_5\n%5.1f  %5.1f   %5.1f
%5.1f\n',om2,om3,om4,om5)
% Torque calculations
fprintf('\nTorques, N.m\n')
T2=H/om2*745.7/(2*pi)*60;T3=T2*om2/om3;T5=T2*om2/om5;
fprintf('torque_2 torque_3 torque_5\n %5.1f     %5.1f     %5.1f\n',T2,T3,T5)
% Output speed check
om5_up=om_out+om_del;om5_low=om_out-om_del;
    if om5_cal<=om5_up&om5_cal>=om5_low
        fprintf('\nThe output gear speed is acceptable\n')
    else
        fprintf(['\nThe output gear speed is unacceptable,';' try another
output speed\n'])
    end
end
```

FIGURE 3.9 The user-defined live function **Gear** together with the function call command (first command line), and the outputted results.

used for gear speeds ω (rev/min), teeth number, N (units), torques T (N·m), and gear ratios (units) are:

$e = \frac{\omega_5}{\omega_2}$ – train value, $\omega_2 = \omega_{in}$ and $\omega_5 = \omega_{out}$ are the input and output rotation speeds respectively that should be specified,

$\frac{N_2}{N_3} = \frac{N_4}{N_5} = \sqrt{e}$ – teeth ratio for equal reduction on both gear train stages,

$m = \frac{1}{\sqrt{e}}$ – gear to pinion teeth ratio,

$N_2 = N_4 = \frac{2k}{(1+2m)\sin^2\varphi}\left(m + \sqrt{m^2 + (1+2m)\sin^2\varphi}\right)$ –number of teeth on the second and fourth pinions, for 20 degrees of the pressure angle, φ,

$N_3 = N_5 = mN_2$ – number of teeth of the third and fifth pinion,

$\omega_5 = \frac{N_2}{N_3}\frac{N_4}{N_5}\omega_{in}$– output speed, if this value is not within the specified limits, $\omega_5 \pm \Delta$, try again the calculation with a new initial ω_5 selected within the required limits,

$T_2 = \frac{60 \cdot 745.7H}{2\pi\omega_2}$ – torque on the input shaft,

$T_3 = T_2\frac{\omega_2}{\omega_3}$ – torque on the median shaft,

$T_5 = T_2\frac{\omega_2}{\omega_5}$ – torque on the output shaft.

In these expressions, H is the power to be delivered, hp; m is the teeth ratio, units; $k = 1$ for studied case of the full-depth teeth; Δ is the possible error in the output rotation speed; lower indices "in" and "out" designate the input and output speeds respectively; the numbers 2, 3, 4, and 5 indicate the number of gear unit.

Problem (Based on Burstein 2021):
Write a live program named ApExample_3_4 to calculate the number of teeth, speeds, and torques, for the above gear train parts. The program should include the command to run the life function and the live function itself named Gear, which has only the following input parameters: ω_{in}, ω_{out}, Δ, H, and k. The input values are: $\omega_{in} = 1750$ rpm, $\omega_{out} = 85$ rpm, $\Delta = 3$ rpm, H = 20 hp, and k = 1. The live function should include an image to show the solved gear train and display the calculated value with the appropriate header using the fprintf command.

The created Live Editor program with the obtained results is shown in Figure 3.9; since the relatively large size of the program, the Live Editor window is not copied here with its menu and borders. To produce the program copy, the "Export to Word…" option of the Save button in the LIVE EDITOR tab was used.

The steps performed to create, save, and run the above live program are:

- After opening the Live Editor window, the "Output Inline" button was clicked in the VIEW tab at the VIEW section for inline output;

- Then the "Text" button was clicked (TEXT section of the LIVE EDITOR tab) to type an explanatory text in the white field;
- The command called the function Gear was written before the function commands; the function call is the name of function with parameter values to be transmitted to the function;
- The "Image" button of the INSERT tab was clicked to insert an image into the program; in the appeared "Load image" dialog box, a file containing the gear train image was selected; Note: you should prepare this file in advance;
- After that, the function definition line was written together with the input parameters necessary for solving the problem (the Live Editor automatically adds the final end to the entered word function), and then the help line with a brief explanation of the purpose of the function was entered;
- Then all calculating commands were written, after each part of the calculations – teeth, speeds, and torques – the appropriate `fprintf` commands was added to display the captions and results;
- In the final part, the `if` … `else` … `end` statement was introduced to compare the calculated and specified values of the output shaft speed and display a string containing the conclusion about the acceptability or unacceptability of the specified output speed.

4 Basics of Graphics

4.1 INTRODUCTION

The presentation of the results of observations, tests, theoretical and engineering calculations in the form of graphs or diagrams is a widespread practice in science and technology. MATLAB® provides many commands for this purpose. The available commands allow you to create 2D (sometimes called XY), 3D (XYZ), and some other specialized graphics. Many toolboxes have the additional ability to plot graphs using commands specifically designed for the purposes of that toolbox. For example, the PDE Toolbox has special commands that allow you to draw mechanical parts and geometrical shapes, and represent in graphs the results of PDE solutions; these commands will be described further, in the places where they will need to be applied.

Here we describe the basic MATLAB® commands that generate 2D and 3D graphs[1]. In between, there are 2D graphics commands and formatters that allow you to generate various linear, nonlinear, and semi- or logarithmic graphs, bar plots, pie charts, and others with two or more curves on one graph, as well as allow you to build multiple graphs in a separate Figure window. The commands for 3D graphics described in this chapter allow you to build plots of spatial lines, meshes, and surfaces, as well as various mathematical shapes and images.

The following description assumes that the reader has thoroughly studied the previous chapters of the book. Therefore, in most cases, the commands are written without explanatory comments (%), and the necessary explanations are given directly in the text of the chapter.

4.2 2D PLOTS GENERATING AND FORMATTING

4.2.1 One or More Curves on the 2D Plot

The basic command used to build 2D graph is the plot command, for one curve per graph the simplest forms of which are:

```
plot(y1) or  plot(x1,y1)
```

and for two, three, or n curves in the same graph:

```
plot(x1,y1,x2,y2,…,xn,yn)
```

where x1 and y1, x2 and y2, …, xn and yn are pairs of equal-length vectors containing the x and y-coordinates of the plotting points.

[1] Some text and table materials from Burstein, 2021a (Section 4.1) are used in the chapter; with IGI Global permission

DOI: 10.1201/9781003200352-4

The first command form, single y1, draws a line by the points at which the y1-vector values are the y-coordinates of the points to be plotted while the x-coordinates are the location numbers (indices) of the value in the y1-vector. The second form of the plot command, (x1, y1)-pair, requires both vectors –x1 and y1 with the x, and y-coordinates for each of the points. The third form of the command requires x1 and y1, x2 and y2, ..., xn and yn pairs of vectors for each x,y series of points.

After inputting the appropriate plot command, one or more $y(x)$ curves are created in the MATLAB® Figure window.

Note:

- By default, a graph with linear scale axes is generated with a blue solid line between the unmarked points;
- To open additional Figure window, the command figure should be entered before the plot command.

For example, let's generate two separate graphs with (a) one curve and (b) two curves; use for that the friction coefficient – normal force data for two tested materials (e.g., rubber and aluminum). The normal force F_N are 1, 3, 10, 32, and 40 N. Friction coefficient μ = 1.2, 0.75, 0.5, 0.37, 0.32 for one of the two substances, and μ = 1.55, 1.1, 0.81, 0.69, 0.65 (dimensionless). The one-curve graph we build by the first substance data, while the two-curve plot by the data for both substances. To present this data in two required plots, we must type the following commands in the Command Window:

```
>>FN = [1 3 10 32 40];    % creates vector with normal forces
>>mu1 = [1.2 0.75 0.5 0.37 0.32];  % creates vector with mu1
>>mu2 = [1.55 1.1 0.81 0.69 0.65]; % creates vector with mu2
>>plot(FN,mu1)                 % creates single curve in the plot
>>figure                       % to open additional Figure window
>>plot(FN,mu1,FN,mu2)            % creates two curves in one plot
```

After entering these commands, the two Figure windows are opened with the $\mu(F_N)$ plots as shown in Figure 4.1a and 4.1b.

To generate three or more curves in the same plot, the x- and y-vectors should be introduced in the plot command for each additional curve.

Note: When the y-data is not a vector, but a n-column matrix, the plot(x,y) command draws n curves; in this case, each curve corresponds to a column of the y-matrix; for example, the two curves shown in Figure 4.1b can be generated by entering the following commands

```
>>mu = [mu1;mu2]'    % combines two vectors into a 2-column
                     matrix
>>plot(FN,mu)       % draws two columns as two Fn(mu) curves
```

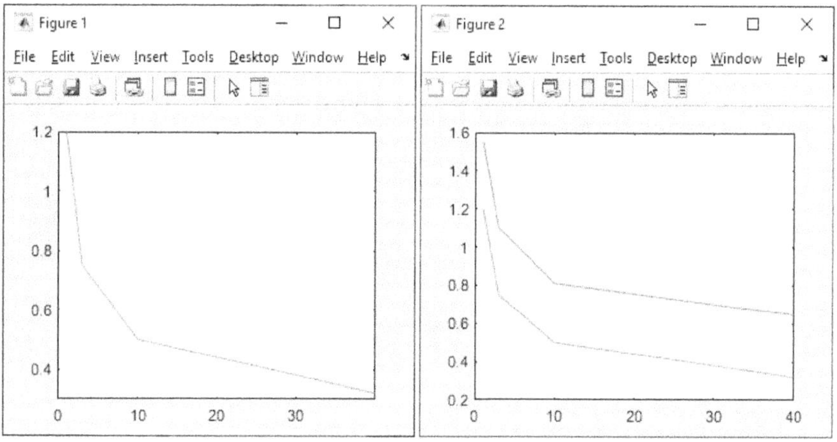

(a) (b)

FIGURE 4.1 Two Figure windows one with single $\mu(F_N)$ curve (a) and the other with two $\mu(F_N)$ curves (b); default settings.

The line style, marker type, thickness, color, and other parameters of the graph were performed by default. To change the default setting, these parameters may be optionally added to the plot command just after each pair of the x and y identifiers:

```
plot(x1,y1,'Line_Specifiers1', x2,y2,'Line_Specifiers2',…,
'Property_Name', Property_Value,…)
```

where the 'Line_Specifiers1', 'Line_Specifiers2', …are the parameters specified the line type, marker symbol, and color of the plotted lines (see Table 4.1–4.3). The pair 'Property_Name', Property_Value represents one of the possible curve properties: the 'Property_Name' is a string containing a name of the desired property; and the Property_Value is a number or a quoted string that specifies the property itself (see Table 4.4). There may be more than one property pair.

In Tables 4.2, 4.3, and 4.4, the y label command was used to label the *y*-axis; the command is explained in Section 4.2.4.1. Line specifiers, as well as property names and their values (when the value is a string), should be typed in inverted commas. The symbols within the specifiers and property name-value pairs can be written in any order, with the option of omitting any of them. Omitted properties will be assigned by default.

TABLE 4.1
The 'Line_Specifiers' of the Marker Symbol

Marker Symbol Specifier	Graphical Representation	Marker Symbol Specifier	Graphical Representation
* (asterisk)	`>>y=1;` `>>plot(y,'*','MarkerSize',10)` 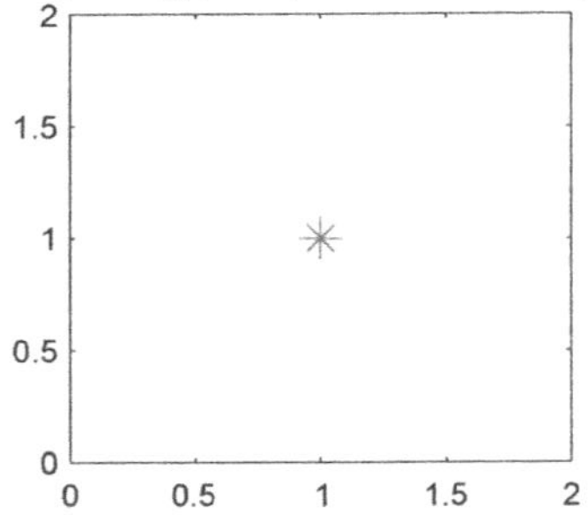	o (circle)	`>>y=1;` `>>plot(y,'o', 'MarkerSize',10)` 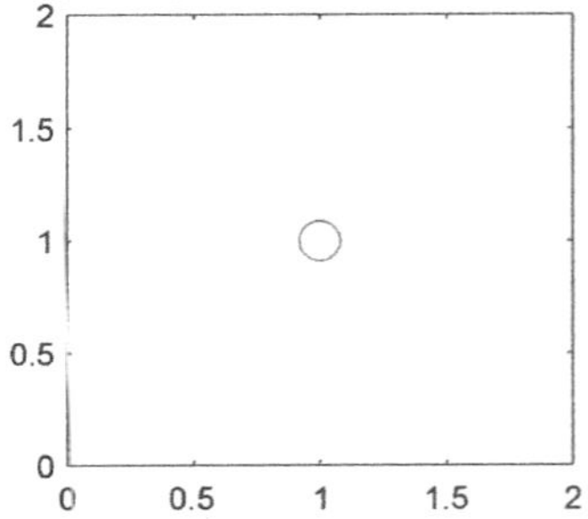
x (cross)	`>>y=1;` `>>plot(y,'x','MarkerSize',10)` 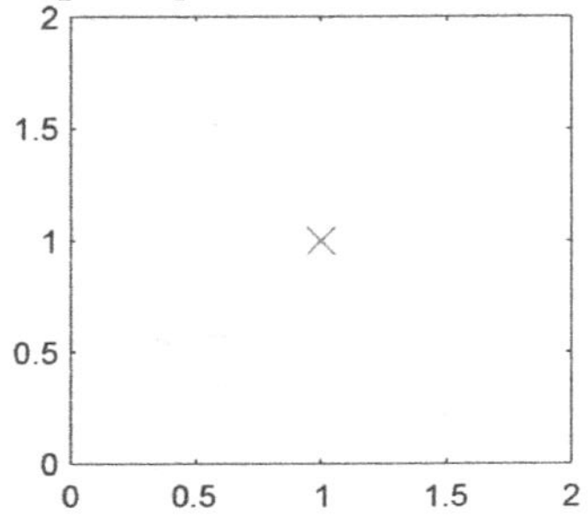	d (diamond)	`>>y=1;` `>>plot(y,'d','MarkerSize',10)` 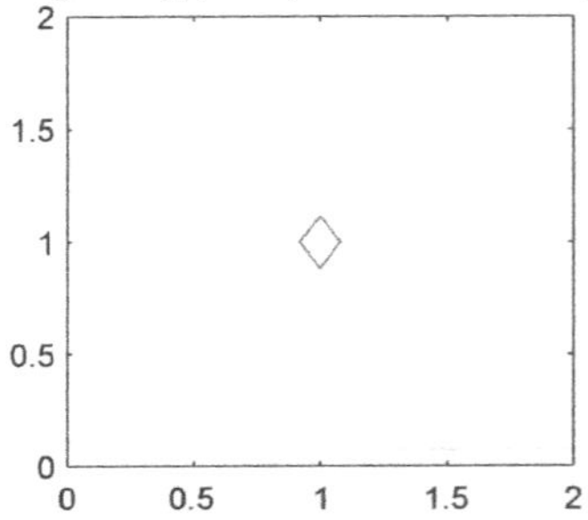

(Continued)

TABLE 4.1 ***(Continued)***

The 'Line_Specifiers' of the Marker Symbol

Marker Symbol Specifier	Graphical Representation	Marker Symbol Specifier	Graphical Representation

p
(five-pointed asterisk)
	`>>y=1;`
	`>>plot(y,'p','MarkerSize',10)`
	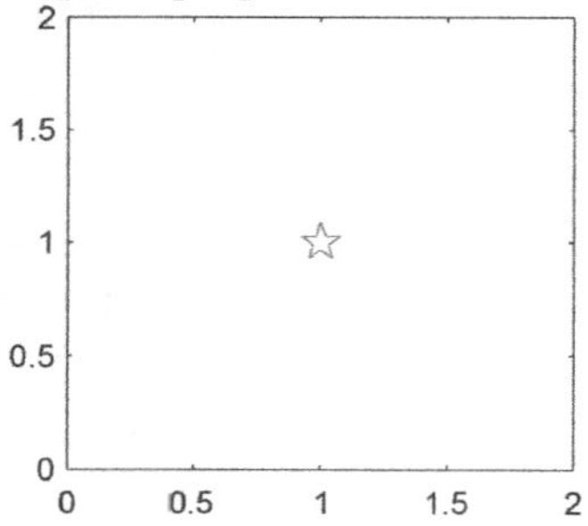
		.
		(point)
			`>>y=1;`
			`>>plot(y,'.','MarkerSize',10)`
			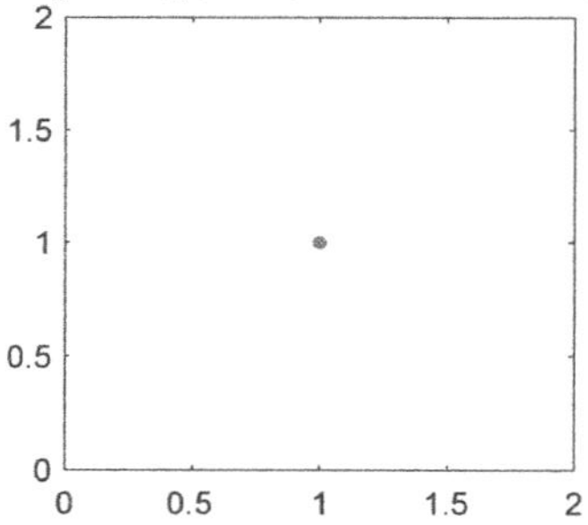

+
(plus)
	`>>y=1;`
	`>>plot(y,'+','MarkerSize',10)`
	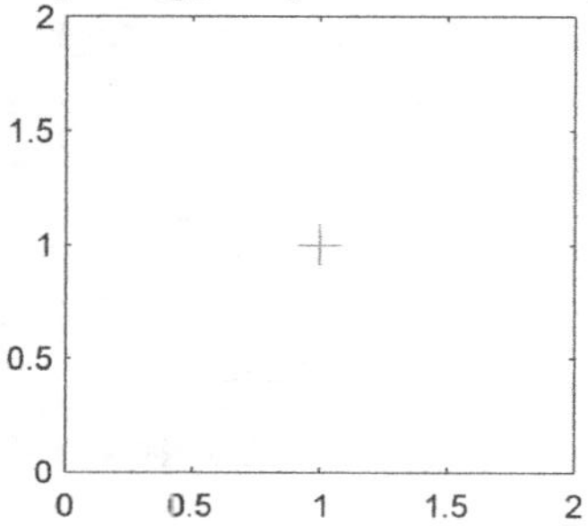
		s
		(square)
			`>>y=1;`
			`>>plot(y,'s','MarkerSize',10)`
			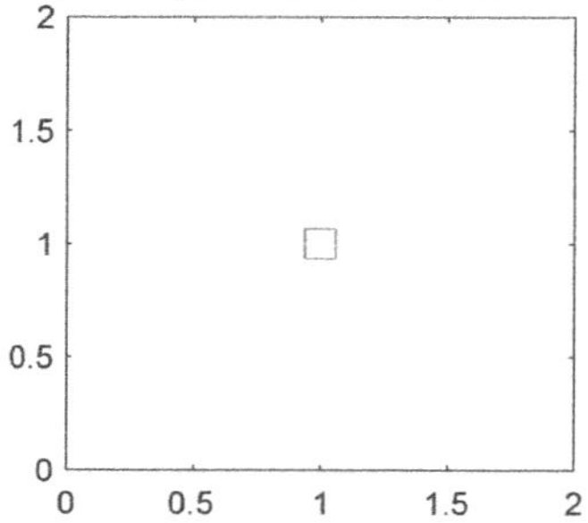

(Continued)

TABLE 4.1 (*Continued*)

The 'Line_Specifiers' of the Marker Symbol

Marker Symbol Specifier	Graphical Representation	Marker Symbol Specifier	Graphical Representation

h
(six pointed asterisk)
	`>>y=1;`
	`>>plot(y,'h','MarkerSize',10)`
	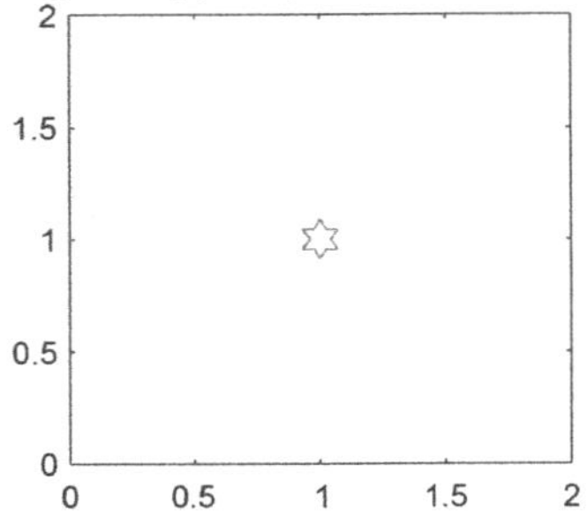
		V
		(inverted triangle; v-key)
			`>>y=1;`
			`>>plot(y,'v','MarkerSize',10)`
			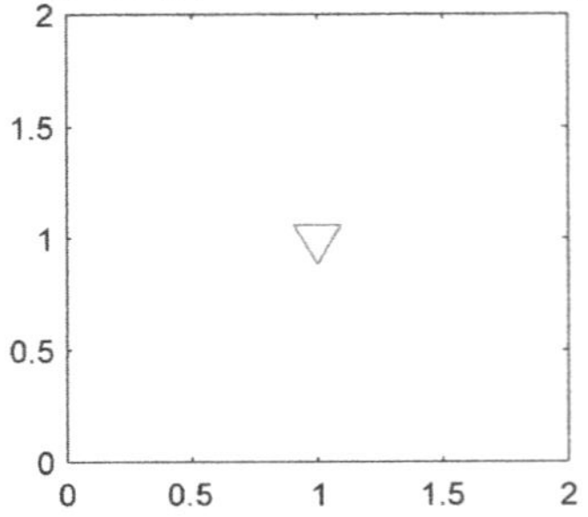

^
(triangle; upright-key)
	`>>y=1;`
	`>>plot(y,'^','MarkerSize',10)`
	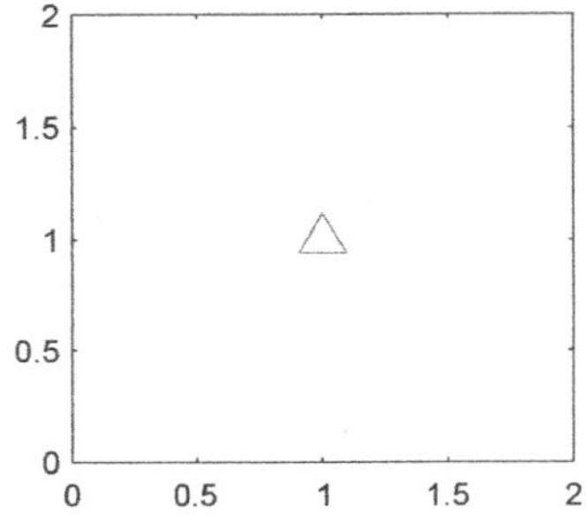		

TABLE 4.2
The 'Line_Specifiers' of the Line Style

Line Style Specifier	Graphical Representation	Line Style Specifier	Graphical Representation

-.
(dash-dot)
	`>> y=[24 38 38 64];`
	`>> plot(y,'-.', 'linewidth',3)`
	`>> ylabel('w')`
	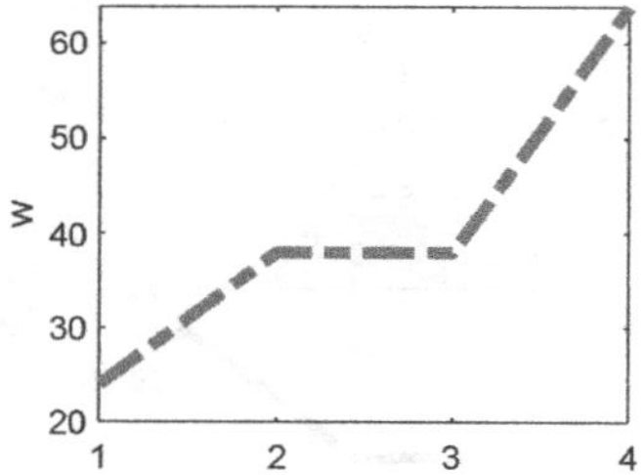
		--
		(dashed; two minuses without space between them)
			`>> y=[24 38 38 64];`
			`>> plot(y,'--')`
			`>> ylabel('w')`
			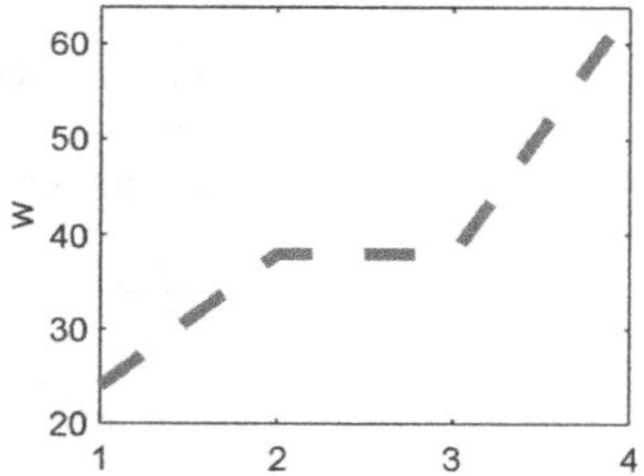

:
(dotted)
	`>> y=[24 38 38 64];`
	`>> plot(y,':')`
	`>> ylabel('w')`
	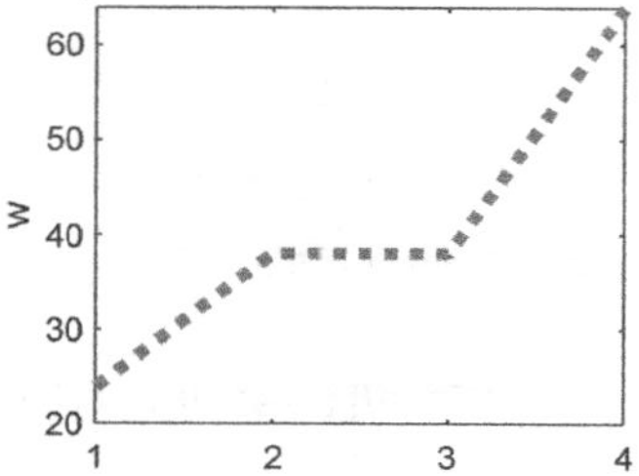
		-
		(solid; default)
			`>> y=[24 38 38 64];`
			`>> plot(y,'-')`
			`>> ylabel('w')`
			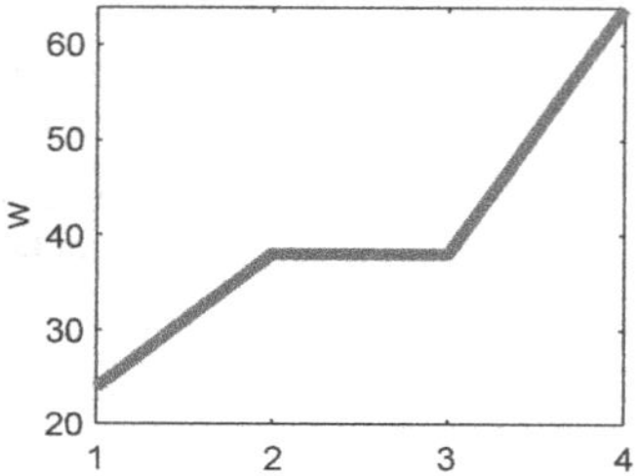

TABLE 4.3
The 'Line_Specifiers' of the Line Color

Color Specifier	Graphical Representation	Color Specifier	Graphical Representation

k (black)	`>>y=[24 38 38 64];`
	`>>plot(y,'-k', 'linewidth',3)`
	`>>ylabel('w')`
	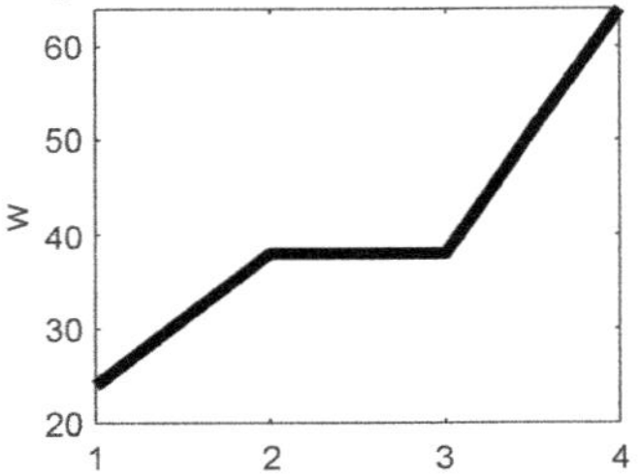
		b (blue; default for single line)	`>>y=[24 38 38 64];`
			`>>plot(y,'-b', 'linewidth',3)`
			`>>ylabel('w')`
			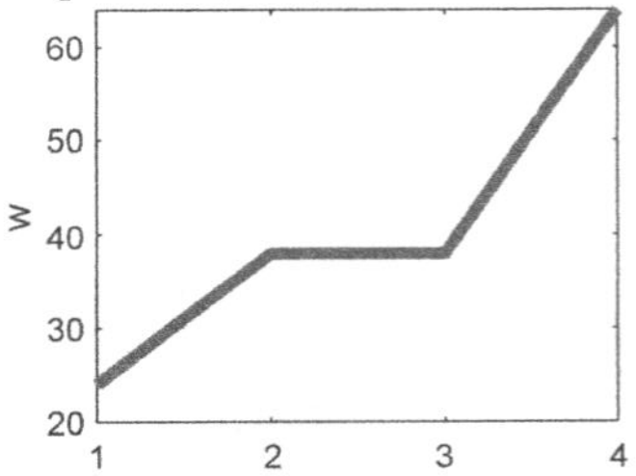

c (cyan)	`>>y=[24 38 38 64];`
	`>>plot(y, '-c', 'linewidth',3)`
	`>>ylabel('w')`
	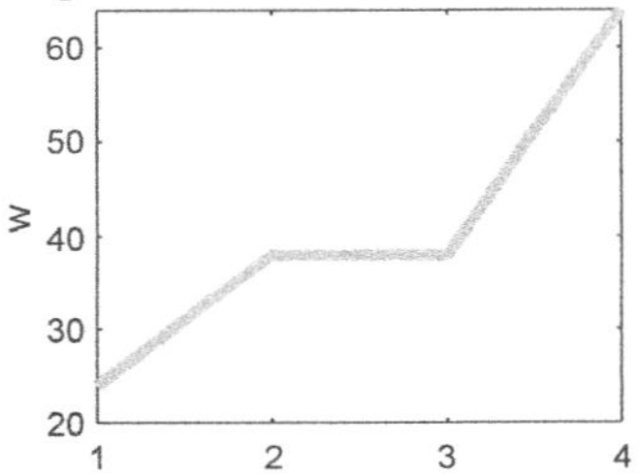
		g (green)	`>>y=[24 38 38 64];`
			`>>plot(y,'-g', 'linewidth',3)`
			`>>ylabel('w')`
			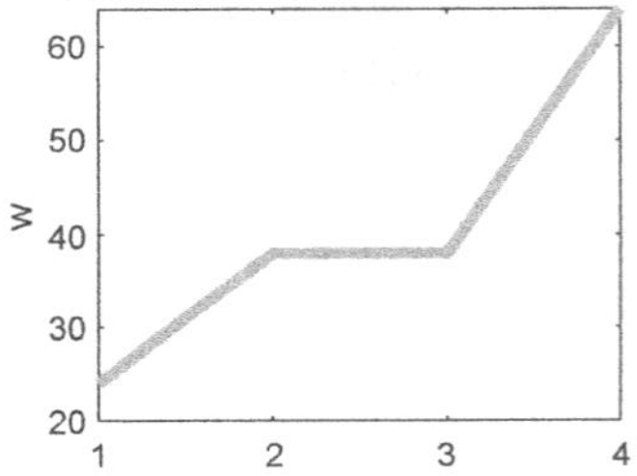

(Continued)

TABLE 4.3 (*Continued*)

The 'Line_Specifiers' of the Line Color

Color Specifier	Graphical Representation	Color Specifier	Graphical Representation

r
(red)
	`>>y=[24 38 38 64];`
	`>>plot(y,'-r', 'linewidth',3)`
	`>>ylabel('w')`
	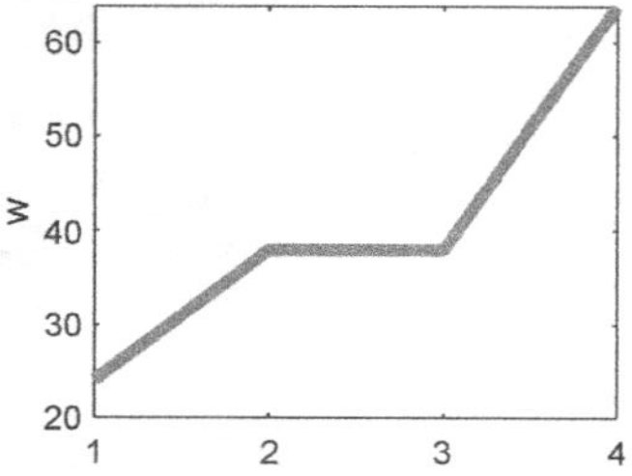
		m
		(magenta)
			`>>y=[24 38 38 64];`
			`>>plot(y,'-m', 'linewidth',3)`
			`>>ylabel('w')`
			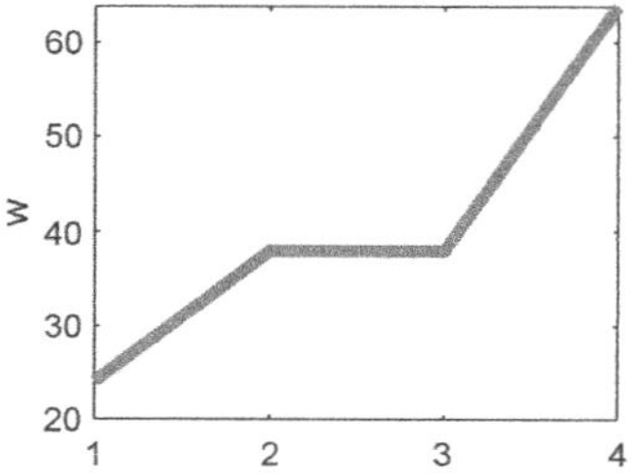

y
(yellow)
	`>> y=[24 38 38 64];`
	`>> plot(y,'-y', 'linewidth',3)`
	`>> ylabel('w')`
	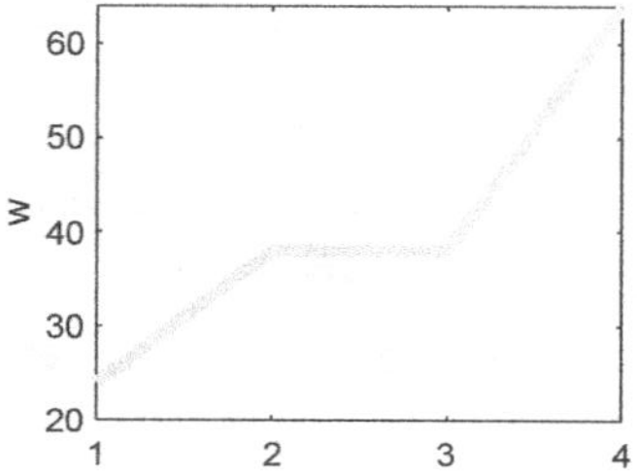		

TABLE 4.4
Some Commonly Used 'Property_Name' - Property_Value Pairs

Property Name	What Means	Property Value	Graphical Representation

LineWidth
or
linewidth
	The width of the drawn line	A decimal number that represents the line width measured in the points (1 point is 1/72 inch or approximately 0.35 mm).
		The default line width is 0.5 points.
			`>>y=[24 38 38 64];`
			`>>plot(y,'-b', 'linewidth',5)`
			`>>ylabel('w')`
			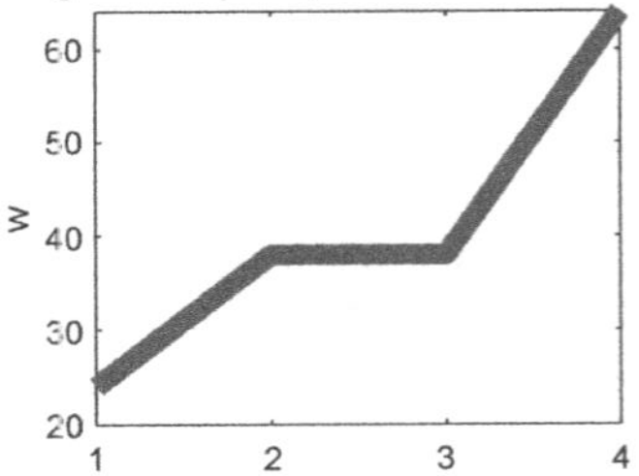

MarkerSize
or
markersize
	The size of the marker	A decimal number in points.
		Default value is 6.
		In case of the '.' marker is 1/3 of specified size
			`>>y=1;`
			`>>plot(y,'o','markersize',20)`
			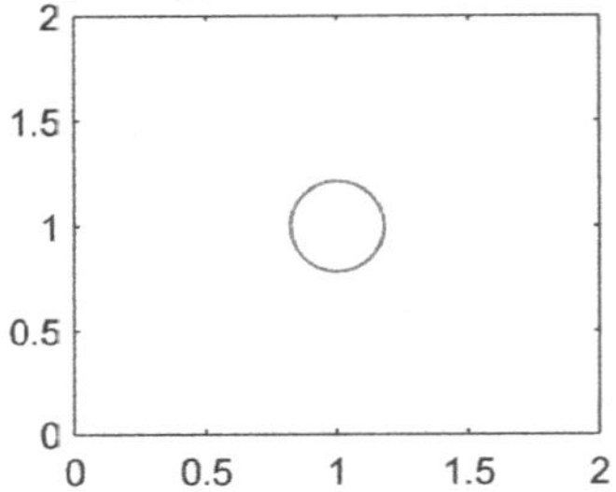

(Continued)

TABLE 4.4 (*Continued*)
Some Commonly Used 'Property_Name' - Property_Value Pairs

Property Name	What Means	Property Value	Graphical Representation
MarkerEdgeColor or markeredgecolor	Marker color for empty markers, or edge line color for filled markers	A color character in accordance with the color_specifiers (Table 4.3)	`>>y=1;` `>>plot(y,'o','markeredgecolor','r','markersize',20)` 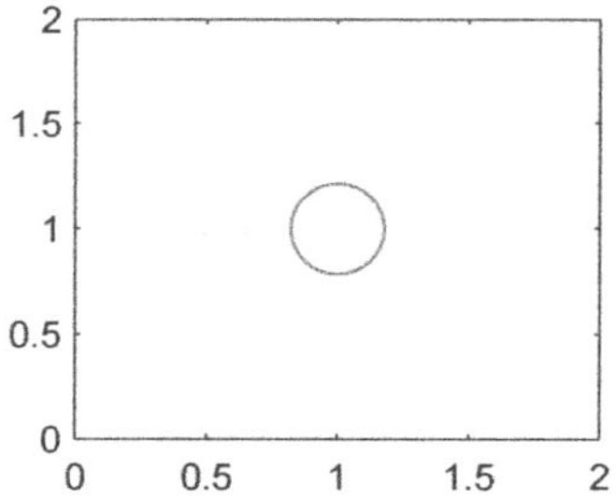
MarkerFaceColor or markerfacecolor	Fills with color the markers that have closed area (e.g., circle or square)	Color character in accordance with the color_specifiers (Table 4.3)	`>>y=1;` `>>plot(1,'o','markerfacecolor','g','markeredgecolor','r','markersize',20)` 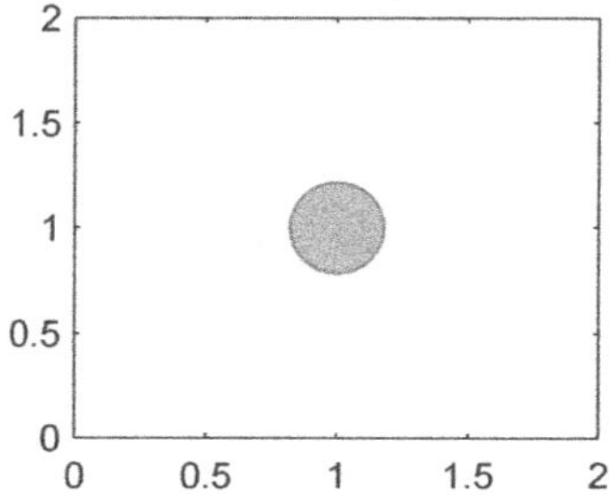

Some examples of the plot commands with various Line Specifiers and Name-Value properties:

`plot(y,'-m')` generates the magenta solid line between the unmarked points, with y - as *y*-coordinates and *x* as addresses of the elements of the vector y.
`plot(x,y, 'o')` generates the points (not connected by a line) with *x,y* -coordinates marked by the circle.
`plot(x,y,'y')` generates the yellow solid (default) line that connects by the points marked by default with point '.'.
`plot(x,y,'-h')` generates the blue (default) solid line that connects the points marked with the six-pointed asterisks.
`plot(x,y,'k:d')` generates the black dotted line with points marked with the diamond.
`plot(x,y,'-gx','markersize',9,'markeredgecolor','g')` generates a green solid line (by width 0.5 points - default) with 9-point size marker represented by the green crosses (x).
`plot(FN,mu, '-ms','LineWidth',3,'MarkerSize',14,'MarkerEdgeColor','k', 'MarkerFaceColor','y')` plots as magenta 3-points solid line with FN, mu -values marked with the 14 points black-edged yellow squares.

Formatting the graph in Figure 4.1b using the previously entered force FN and friction coefficient mu1 and mu2 values and entering the following plot command

```
>>plot(FN,mu1, '-ms',FN,mu2,'--r^','LineWidth',3,'MarkerSize'
,14, 'MarkerEdgeColor','k', 'MarkerFaceColor','y')
```

we can obtain a new view of the FN(mu) graph – Figure 4.2a.

The above *x*,*y*-values were specified before plotting (which is possible with measured or tabular data); however, if necessary, the coordinates can be determined

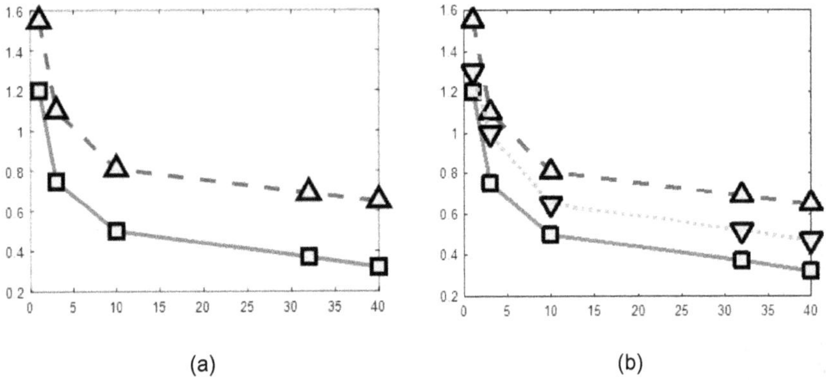

FIGURE 4.2 Friction coefficient-force plot with two curves (a) generated by the plot command, and with three curves (b) when the third curve was added using the `hold on` command; some specifiers and property settings were used to plot both graphs.

from the $y(x)$ expression. In latter case, the vector of y-values should be calculated at the given vector of x-values (see Application examples, Section 4.4).

Note:

- To close a single Figure window, type and enter the close command in the Command Window. If more than one Figure windows are open, to close them, use the close all command.
- With each input of the plot command, the previous plot is deleted.

4.2.1.1 Using the Hold On/Off Command for Generating Multiple Curves

Aside from the plot command drawing two or more curves in the same plot, there is a method for adding new or multiple curves to an existing plot. To perform this, after creating the first plot, type the hold on command, and then enter a new plot command with the point coordinates of the new curve. To stop the hold on process, enter the hold off command that will cause the next curve appearing without any previously drawn curves. For example, the new series of friction coefficient values obtained for one more material of some mechanical part: μ = mu^3 = 1.3, 1.0, 0.65, 0.52, and 0.47, specified at the same normal force values as the previous two μ-series, can be added to the existing graph (Figure 4.2a) by entering the following commands:

```
>>mu3 = [1.3 1.0 0.65 0.52 0.47];      % creates new mu-vector
>>hold on        % holds previous graph and waits for new line
>>plot(FN,mu3,':gv','LineWidth',3,'MarkerSize',14,
'MarkerEdgeColor',...
'k', 'MarkerFaceColor','y')% adds dotted green line to the plot
>>hold off                              % cancels the hold on mode
```

Generated plot is shown in Figure 4.2b.

4.2.2 Plot Formatting Using the Commands and the Plot Tools Editor

In addition to formatting the curve, we can format the graph with a grid, axis labels, suitable axes ranges, text, title, and legend. The introduction of these elements is possible by either using special commands or interactively using the Plot editor (see Section 4.2.2.2) strictly in the Figure window.

4.2.2.1 Formatting with Commands

The formatting commands are effective when the plot is created with a computer program; these commands should be entered after the plot command or when the graph has already been built. Following is the list of the most common commands:

- grid on/off – displays/hides the grid lines on/from the current plot;
- axis([x_min,x_max,y_min,y_max]) – adjusts the x and y axes in accordance with the coordinate limits written in the square brackets as four-element vector;

- axis equal – sets the same scale for the x and y axes (the ratio x/y, width-to-height, is called the aspect ratio);
- axis square – sets the shape of the graph as square;
- axis tight – sets the plot axes limits to the range of the data;
- axis off – removes the axes and background from the plot.
- xlabel('text') – provides the desired text for the x axis;
- ylabel('text') – provides the desired text for the y axis;
- title('text') – provides the text to the plot title;
- text(x,y,'text') – introduces text/number/character starting from the point with coordinates x and y which the user must specify;
- legend('string1','string2',...,'Location', location_area) – reproduces, within the frame, an explanatory text written in the 'string1','string2',...; the 'Location' - location_area pair is a property (optional, default is the upper-top corner) that specifies the area, where the plot legend should be placed. For example: legend('line 1','line 2','Location', 'NorthWest') displays the legend frame with two line strings (line 1 and line 2) placed within the plot frame on the upper-left corner, or another example legend('line 1','line 2','Location','Best') displays the legend frame with two line strings (line 1, line 2) placed inside the plot at the best possible location (having least conflict with curve/s within the plot).

The text/lines/strings in the above commands can be formatted by applying some special word/s or character/s (called modifiers) that should be written inside the string immediately after the backslash, in the form \ModifierName{ModifierValue} (when entered, appeared in blue). Some useful modifiers are given in Table 4.5.

As an example of graph formatting, below is the live ExFormat script (Figure 4.3) that plots the friction coefficient-force graph as in Figure 4.2b using the above formatting commands; in addition, the script shows how the plot looks in a live program, which could not have been done in the previous chapter before studying graphics.

Text can be formatted in another way by including a pair – property name and value – in the command just after the text string. As in case of the `plot` command (see Subsection 4.1.), to format a text, several property pairs can be included, and each of these pairs must be written after comma and with separating commas in the form 'Name', Value. For example, we can write text(45,40,'1','fontsize',12) instead of text(45,40,'\fontsize{12}1').

To produce the two-line text in the above string-containing commands, the text of each line should be written between the quotes, the lines should be divided by the ";", and all of this must be placed in curly brackets, for example, >> xlabel({'viscosity';'nondim'})

4.2.2.2 Interactive Plot Formatting

In addition to the programmatic graph editing, it is possible to edit the graph interactively using the assortment of buttons of the Figure window (Figure 4.4).

To start formatting the previously created plot, the Plot Edit button  should be clicked firstly. After this the other format buttons can be used. Properties of

TABLE 4.5
Some Modifiers for the Text/String Formatting

Modifier	Purpose	Example
\fontsize{number}	sets size of the letters of the text following the modifier	`\fontsize{14}` – sets the letter/s size 14
\fontname{name}	sets name of the font to the text following the modifier	`\fontname{Courier New}` - sets the Courier New font
\name of the Greek letter	specifies a Greek letter	`\sigma` sets the lowcase σ `\Sigma` sets the capital Σ
\b	sets the bold font to the text following the modifier	`\b mechanics` Displays the word "mechanics" written after the modifier as **mechanics**
\it	sets the Italic (tilted) font to the text following the modifier	`\it mechanics` – sets the word "substance" (written after the modifier) to be tilted: *mechanics*
\rm	sets the normal (Roman) font to the text following the modifies	`\rm mechanics` – displays in the Roman font the word "mechanics" written after the modifier: mechanics
_ (underscore)	subscripts	A_ d – the "_" sign sets the d-letter to be subscripted: A_d To extend this sign to more than one letter, the curled brackets are used, e.g., `12_{dec}`is displayed as 12_{dec}
^ (caret)	superscripts	^oC – the ^ symbol sets 'o' to be superscripted: ^{o}C To apply this sign to more than one letters, the curled brackets are used, e.g., `e^{-(x-5)}` is displayed as $e^{-(x-5)}$

axes, curves, markers, lines, as well as the entire figure, can be changed using the pop-up menu, summoned by clicking the Edit option in the Figure menu. The title, axes labels, texts, and legend can be activated using the pop-up menu, which appears after clicking on the Insert option of the menu.

The "Property Inspector" window can be opened by clicking the "Open Property Inspector" button or by double-clicking a curve, plotted points, or axes. This provides the appropriate sections of the "Property Inspector" for modifying or adding various characteristics to the desired component. For more information about available interactive plot editing tools, input the doc plottools command in the Command Window.

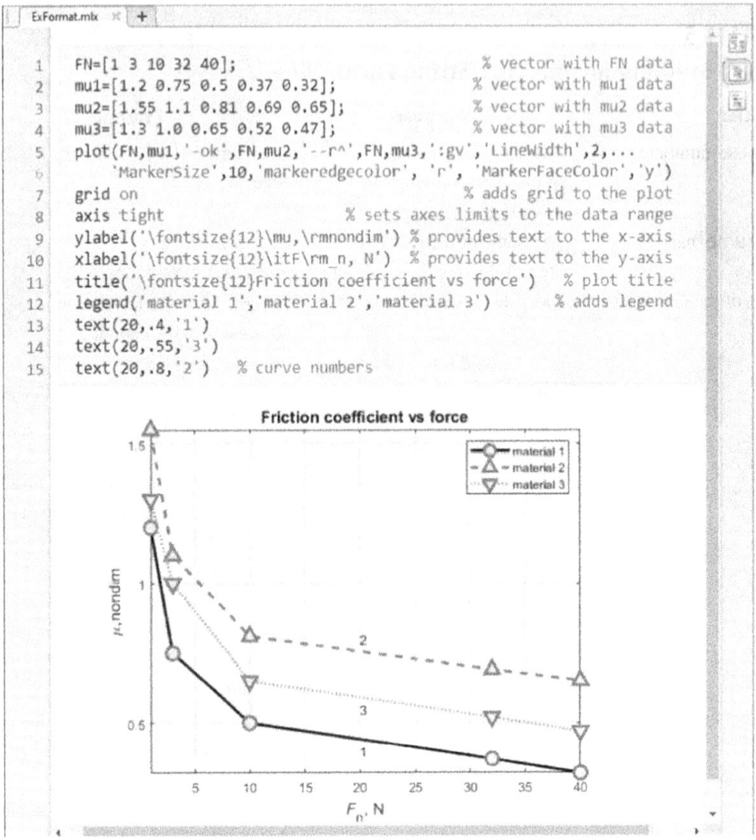

ExFormat.mlx
FN=[1 3 10 32 40]; % vector with FN data
mu1=[1.2 0.75 0.5 0.37 0.32]; % vector with mu1 data
mu2=[1.55 1.1 0.81 0.69 0.65]; % vector with mu2 data
mu3=[1.3 1.0 0.65 0.52 0.47]; % vector with mu3 data
plot(FN,mu1,'-ok',FN,mu2,'--r^',FN,mu3,':gv','LineWidth',2,...
'MarkerSize',10,'markeredgecolor', 'r', 'MarkerFaceColor','y')
grid on % adds grid to the plot
axis tight % sets axes limits to the data range
ylabel('\fontsize{12}\mu,\rmnondim') % provides text to the x-axis
xlabel('\fontsize{12}\itF\rm_n, N') % provides text to the y-axis
title('\fontsize{12}Friction coefficient vs force') % plot title
legend('material 1','material 2','material 3') % adds legend
text(20,.4,'1')
text(20,.55,'3')
text(20,.8,'2') % curve numbers
Friction coefficient vs force
material 1
material 2
material 3
μ,nondim
F_n, N
1
2
3

FIGURE 4.3 Live ExFormat script with commands that format the curve and the generated graph.

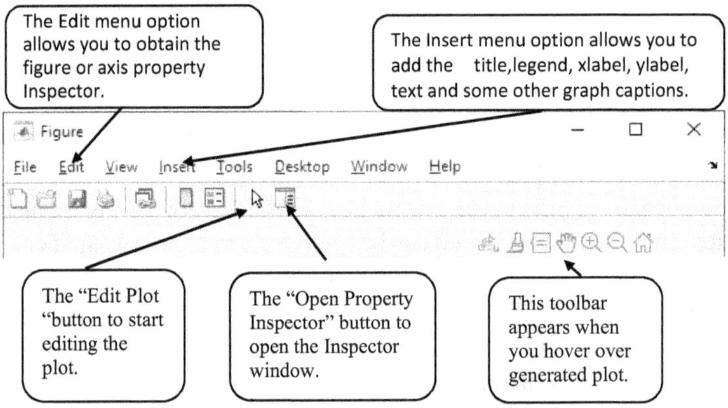

The Insert menu option allows you to add the title,legend, xlabel, ylabel, text and some other graph captions.

The "Edit Plot "button to start editing the plot.

This toolbar appears when you hover over generated plot.

FIGURE 4.4 The menu and bar buttons for plot editing from the Figure window.

4.2.3 Several Plots in the Same Figure Window

It is often necessary to compose several graphs on the same page or, in other words, in the same Figure window. To perform this, the **subplot** command is used; the command breaks the Figure window (and, accordingly, the printed page) into *m*-by-*n* rectangular panes. The command has two forms:

```
subplot(m,n,P)    and    subplot mnP
```

where m and n are the rows and columns of the panes and P is the current pane number where the plot will be located – see Figure 4.5 showing the panes with the corresponding **subplot** commands. The P within the first command form can be written as vector containing two or more numbers of the adjacent panes, which makes possible the arbitrary arrangement of the plots in the Figure window (e.g. as in Figure 4.5b).

Some examples of the subplot command:

subplot(3,2,4) or subplot 324 Breaks the page into six rectangular panes arranged in 3 rows and 2 columns and makes pane 4 current.

subplot(2,2,[2 4]) Breaks the page into three rectangular panes with first and third panes of regular sizes and enlarged third pane that combines the second and fourth panes (top and bottom right panes); the enlarged pane is current.

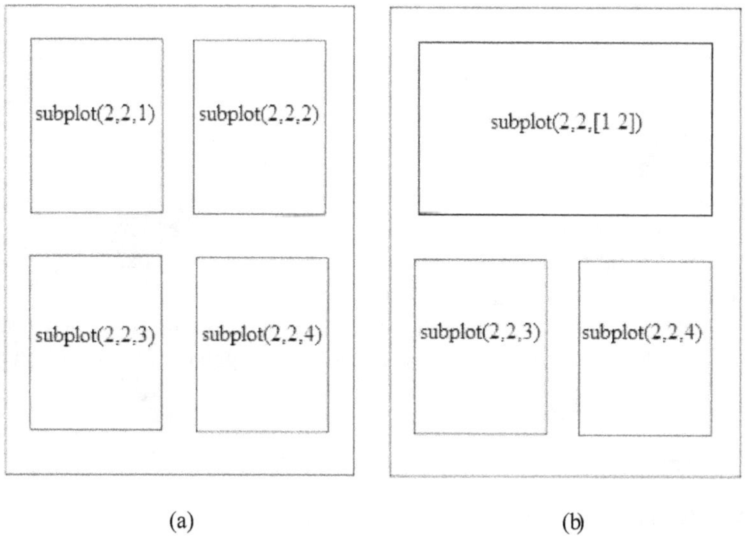

FIGURE 4.5 The **subplot** commands in the appropriate panes of the Figure window, which is broken into four (a) and three (b) rectangular panes.

subplot(2,3,3) or subplot 233 Breaks the page into six rectangular panes arranged in 2 rows and 3 columns and makes pane 3 current.
subplot(2,1,1) or subplot 211 Breaks the page into two rectangular panes arranged in a single column and makes the first pane current.
subplot(1,2,2) or subplot 122 Breaks the page into two rectangular panes arranged in a single row and makes the second pane current.

The required graph should be placed in the current pane with the plot command. As an example, generate three plots on one page and place three previous plots with three, one, and two $F_N(\mu)$ curves (as per data used for generating Figures 4.3 and 4.4) in three panes accordingly to Figure 4.5b.

The MATLAB® commands for interactively generating three plots are given in the same Figure window:

```
>>FN = [1 3 10 32 40];                                  % FN data
>>mu1 = [1.2 0.75 0.5 0.37 0.32];                       % mu1 data
>>mu2 = [1.55 1.1 0.81 0.69 0.65];                      % mu2 data
>>mu3 = [1.3 1.0 0.65 0.52 0.47];                       % mu3 data
>>subplot(2,2,[1 2])              % makes panes 1 and 2 current
>>plot(FN,mu1,'-ok',FN,mu2,'-.vr',FN,mu3,'--^g',
'LineWidth',2, 'MarkerSize',7,'MarkerEdgeColor',
'r','MarkerFaceColor','y'), grid on,xlabel('\mu,
nondim'),ylabel('F_N, N'),legend('material 1', 'material 3',
'material 2')
>>subplot(2,2,3)                         % makes pane 3 current
>>plot(FN,mu1,'-ok','LineWidth',2,'MarkerSize', 7,
'MarkerEdgeColor', 'r', 'MarkerFaceColor','y'), grid on,
xlabel('\mu, nondim'),ylabel('F_N, N')
>>subplot(2,2,4)                         % makes pane 4 current
>>plot(FN,mu1,'-ok',FN,mu2,'-.vr','LineWidth',2, 'MarkerSize
',7,'MarkerEdgeColor', 'r', 'MarkerFaceColor','y'), grid on,
xlabel('\mu, nondim'),ylabel('F_N, N'), legend('material 1',
'material 2')
```

The resulting plot is shown in Figure 4.6. Some marker sizes, colors, axis labels, and legend have changed from previously created separate graphs with this data.

The above commands written in the Live Editor window and saved with the name ExThreePlots, after launching, produce three plots in Output Inline mode as shown in Figure 4.7.

4.3 THREE-DIMENSIONAL PLOTS

MATLAB® provides three main groups of commands for representing lines, meshes, and surfaces in three-dimensional space. These commands as well as the various formatting commands are described below.

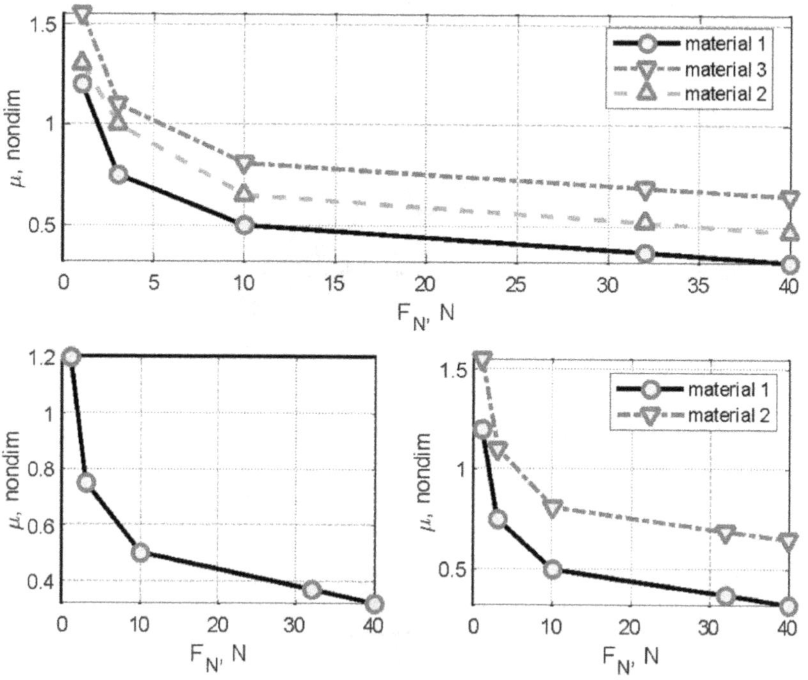

FIGURE 4.6 Figure window arranged in three panes containing three different $\mu(F_N)$ plots.

4.3.1 Generating 3D Lines

As in two-dimensional, in three-dimensional graphs, adjacent points are connected with a straight line forming a specific 3D line. Each 3D point has three coordinates in 3D space. To generate a 3D line the plot3 command is used:

```
plot3 (x,y,z)
```

Some extended form of this command is

```
plot3(x,y,z,'Line_Specifiers','Property_Name',Property_
Value,…)
```

where - x, y, and z are vectors of the same size, containing the coordinates of each point; the Line_Specifiers, Property_Name, and Property_Value have the same sense as in the corresponding plot command (Section 4.2, Tables 4.1, 4.2, 4.3, and 4.4).

The previously described formatting commands - grid on/off, text, title, xlabel and ylabel – and, in addition, the zlabel commands can be used to add grid, text, title, and axis labels to the 3D plot.

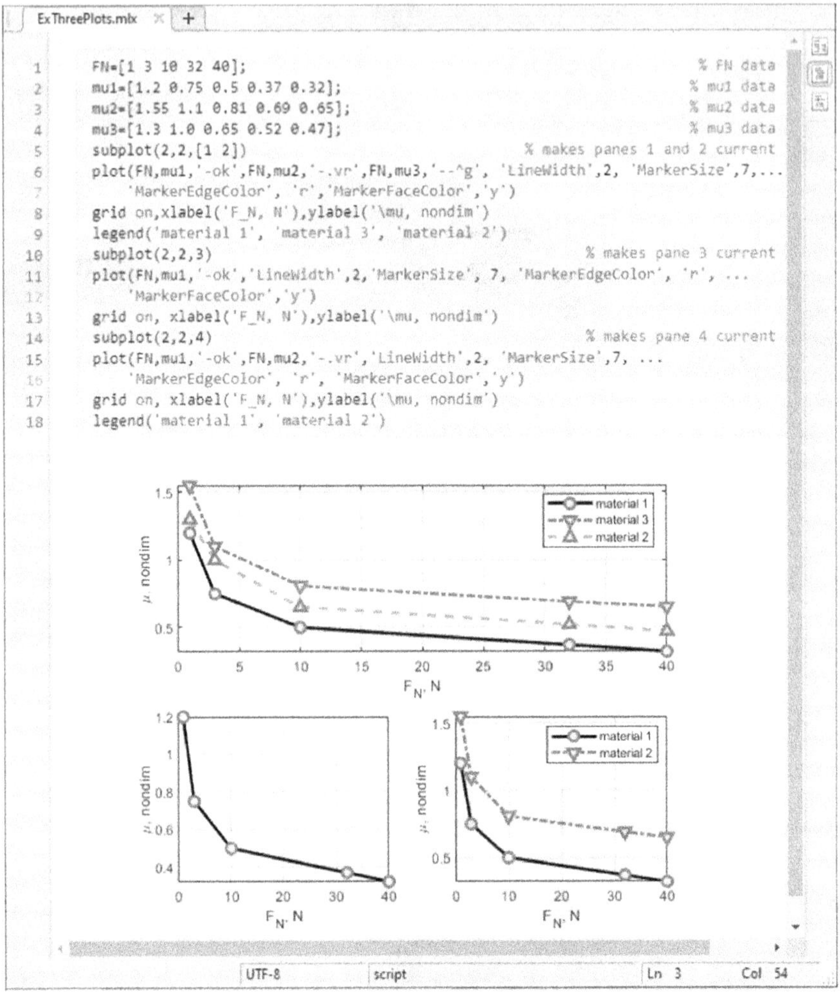

ExThreePlots.mlx
FN=[1 3 10 32 40]; % FN data
mu1=[1.2 0.75 0.5 0.37 0.32]; % mu1 data
mu2=[1.55 1.1 0.81 0.69 0.65]; % mu2 data
mu3=[1.3 1.0 0.65 0.52 0.47]; % mu3 data
subplot(2,2,[1 2]) % makes panes 1 and 2 current
plot(FN,mu1,'-ok',FN,mu2,'-.vr',FN,mu3,'--^g', 'LineWidth',2, 'MarkerSize',7,...
'MarkerEdgeColor', 'r','MarkerFaceColor','y')
grid on,xlabel('F_N, N'),ylabel('\mu, nondim')
legend('material 1', 'material 3', 'material 2')
subplot(2,2,3) % makes pane 3 current
plot(FN,mu1,'-ok','LineWidth',2,'MarkerSize', 7, 'MarkerEdgeColor', 'r', ...
'MarkerFaceColor','y')
grid on, xlabel('F_N, N'),ylabel('\mu, nondim')
subplot(2,2,4) % makes pane 4 current
plot(FN,mu1,'-ok',FN,mu2,'-.vr','LineWidth',2, 'MarkerSize',7, ...
'MarkerEdgeColor', 'r', 'MarkerFaceColor','y')
grid on, xlabel('F_N, N'),ylabel('\mu, nondim')
legend('material 1', 'material 2')
material 1
material 3
material 2
μ, nondim
F_N, N
UTF-8
script
Ln 3
Col 54

FIGURE 4.7 Live script that generates three plots on the same page using the **subplot** commands.

For an example, design a 3D plot showing a line and write this following program with name ExPlot3:

```
T = linspace(0,pi,501);
x = sin(t).*cos(15*t);
y = sin(t).*sin(15*t);
z = cos(t);
plot3(x,y,z,'k','LineWidth',3)
grid on
xlabel('x'),ylabel('y'),zlabel('z')
```

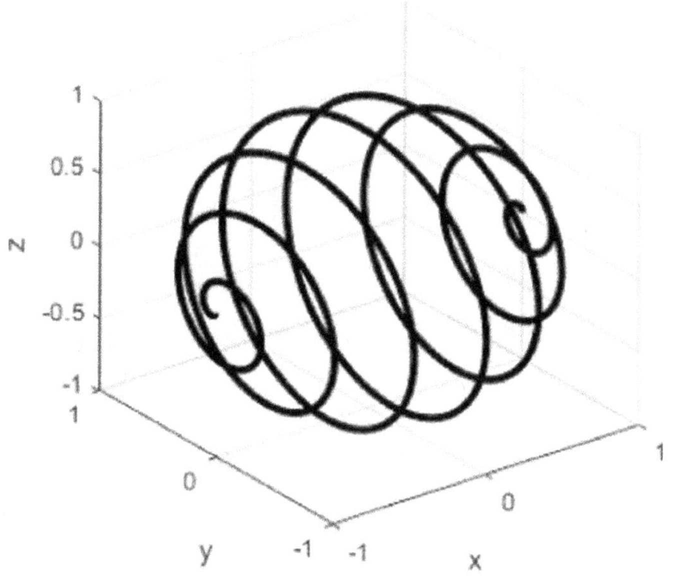

FIGURE 4.8 Sphere-shaped spiral generated with the plot3 command.

The first four commands compute the coordinates *x*, *y*, and *z* for 501 *t* values in the range 0 up to π. The extended form of the plot3 command is used here, with the line color specifier 'k' (black) and the line width property pair - 'LineWidth', 3 – that increases the line width to three points (1/24 inch or approximately 1.1 mm); the last commands are formatting commands that generate the grid and axes labels.

After running, the resulting 3D plot appears in the Figure window with the generated line in the form of a sphere-shaped spiral (Figure 4.8).

4.3.2 Generating Mesh in 3D Coordinates

Surfaces in MATLAB® can be drawn with the two basic commands: mesh and surf. To understand construction of the 3D surface, it is useful to clarify how the 3D mesh is built. In general, the 3D mesh is comprised of the points (mesh nodes) and the lines between them. Each point in 3D space has three coordinates (*x*, *y*, and *z*), which are used to reconstruct a surface. The *x*,*y* coordinates of the point are ordered in a rectangular grid of the *x*,*y* plane. The x and y coordinates of the plane grid can be represented in two size-equal matrices, one of which contains *x* and the other – *y* coordinates. For each grid *x*,*y*-node, the *z*-coordinate should be given or calculated by any *z*(*x*,*y*) expression and assembled then into a *z*-matrix.

An example of 3D points connected with the lines forming a mesh is shown in Figure 4.9.

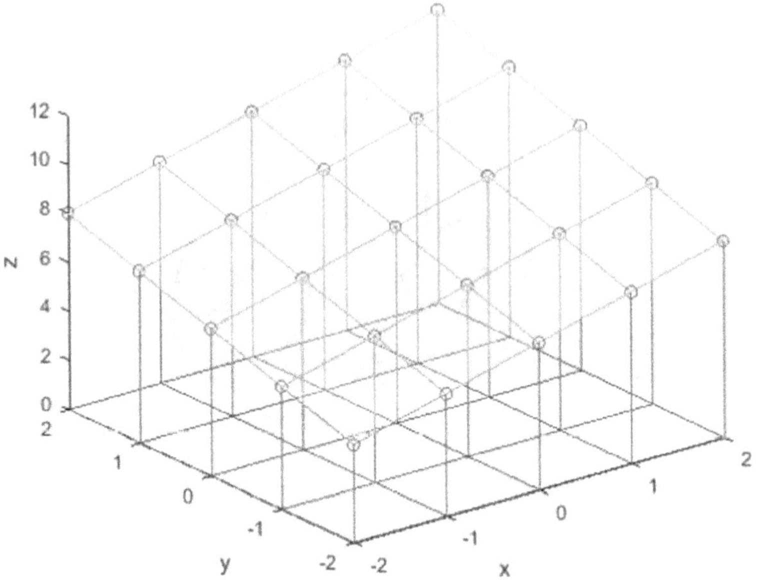

FIGURE 4.9 3D points (marked by "o") connected with the mesh lines, and rectangular grid in the *x,y*-plane (shown in blue) (The plot generated with the `stem3(x,y,z)` and `mesh(x,y,z)` commands; used expression z=8+x+y.)

The area of the *x* and *y* coordinates for which the *z*-coordinates must be obtained is called the *domain*. By writing all the *x*-values along each *iso-y* line, we obtain the *X*-matrix; as you can see from the above plot, this matrix should have the same coordinates in each column. A similar operation yields a *Y*-matrix with the same coordinates in each row. When we have expression $z = f(x,y)$, the matrix with *z*-coordinates can be obtained for every *x,y* point of the domain using the element-by-element calculations. After the *X*, *Y*, and *Z* matrices are defined, the whole surface can be plotted.

To produce the *X* and *Y* matrices, a special **meshgrid** command can be used. Two simplest forms of this command are:

```
[X,Y]= meshgrid(x,y)   or   [X,Y]= meshgrid(x)
```

where x and y are the vectors representing *x*- and *y* axis division respectively; X and Y are the matrices of the domain grid coordinates. The second form of this command can be used when the x- and y vectors are equal.

In case of Figure 4.9, the *X* and *Y* matrices can be produced with the following commands:

```
>>x = -2:2;
>>[X,Y] = meshgrid(x)
X =
```

```
-2 -1   0   1   2
-2 -1   0   1   2
-2 -1   0   1   2
-2 -1   0   1   2
-2 -1   0   1   2
Y =
-2 -2  -2  -2  -2
-1 -1  -1  -1  -1
 0  0   0   0   0
 1  1   1   1   1
 2  2   2   2   2
```

When the *X*, *Y*, and *Z* matrices are generated, a mesh in the 3D space can be drawn. The following command generates a mesh formed by colored lines:

```
mesh(X,Y,Z)
```

where the X and Y matrices are defined using the meshgrid command for the specified vectors *x* and *y*, while the Z matrix must be specified for each X, Y – node or calculated for these matrices using the given $z = f(x,y)$-expression.

To summarize: to build a 3D graph, you need to follow these steps. First, create a grid in the *x,y* domain using the meshgrid command. Then calculate or specify the *z* values for each *x,y* node of the grid. Finally, generate the mesh graph using the mesh command.

As an example, we will plot a 3D mesh graph for the Nusselt number *Nu* for turbulent flow over a plate (Beitz and Kuttner, 2011)

$$Nu = \frac{0.0.37Re^{0.8}Pr}{1+2.443Re^{0.1}\left(Pr^{2/3}-1\right)}$$

where *Re* and *Pr* are the Reynolds and Prandtl numbers, respectively.

Problem: Create the *Nu(Re,Pr)* plot for $Re = 5{\cdot}10^5 \ldots 10^7$ and $Pr = 20 \ldots 2000$, and assign 15 *Re* and *Pr* values each. Add title and axes labels to the plot.

The program named ExMesh that generates the 3D mesh plot reads:

```
Re = linspace(5e5,1e7,15);              % assigns Re vector
Pr = linspace(20,2e3,15);              % assigns Pr vector
[RE,PR] = meshgrid(Re,Pr);  % generates the Re and Pr matrices
Nu = .037*RE.^.8.*PR./(1+2.443*RE.^-.1.*(PR.^(2/3)-1));% Nu
matrix
mesh(RE,PR,Nu)                         % generates the mesh plot
title('Nusselt for turbulent flow')
xlabel('Reynolds number')
ylabel('Prandtl number')
zlabel('Nusselt number')
```

The resulting plot is shown in Figure 4.10.

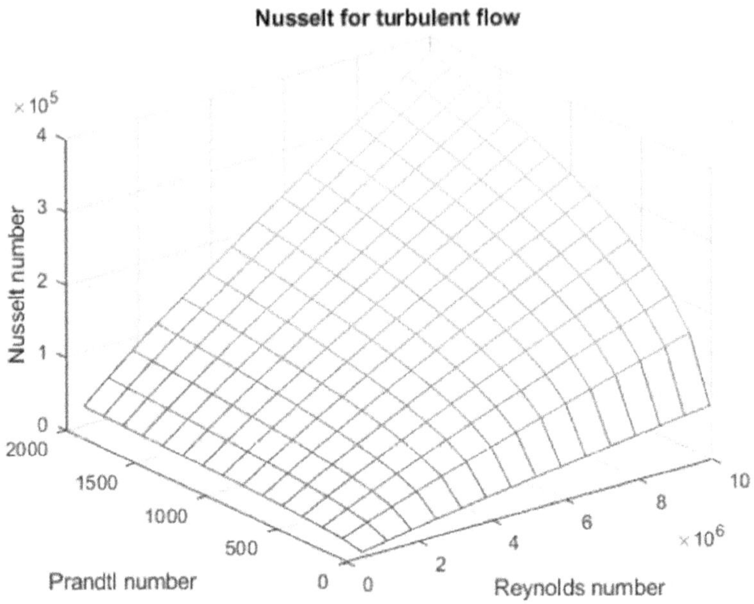

FIGURE 4.10 3D mesh plot depicting Nusselt numbers vs Reynolds and Prandtl numbers for turbulent flow over a plate.

4.3.3 Generating Surface in Plot

Three-dimensional mesh generated with the mesh command consists of colored lines and white (uncolored) surfaces between the lines. To create a colored surface, the surf command is used. The simplest form of this command is:

```
surf(X,Y,Z)
```

where X, Y, and Z are the same matrices as in the mesh command.

To demonstrate the use of the command, we apply the above ExMesh program with the surf(RE,PR,Nu) command instead of the mesh(RE,PR,Nu). With this change, the program generates the following graph (Figure 4.11):

Note:

- In the surf and mesh commands, vectors of x and y values can be used in place of the X and Y matrices, but of course the Z values must be specified as a matrix; so in the above example the mesh(RE,PR,Nu) command with three matrices can be written as mesh(Re,Pr,Nu) with two vectors Re and Pr and one matrix Nu;
- The surf and mesh commands can also be implemented in the surf(Z) or mesh(Z) form. However, it must be borne in mind that in this case the Z

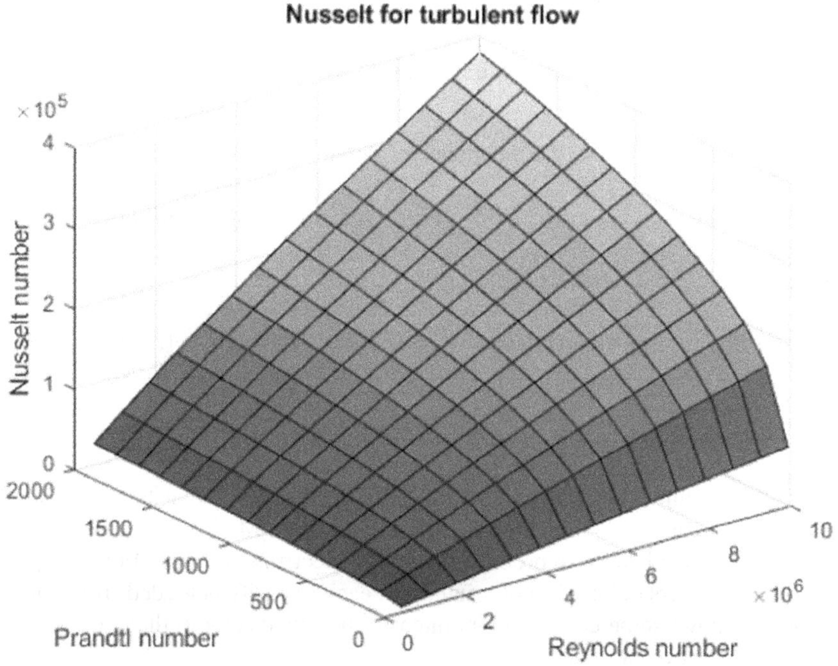

FIGURE 4.11 3D surface plot depicting Nusselt numbers vs Reynolds and Prandtl numbers for turbulent flow over a plate.

matrix values are plotted versus this matrix row numbers (x-coordinates) and column numbers (y-coordinates);

- When the surf and mesh commands are executed, the grid appears automatically in the plot and can be removed using the grid off command.

4.3.4 Formatting 3D Plots

As stated above, many of the 2D plot formatting commands, namely grid, title, xlabel, ylabel, text, and axis, can be used and have been partially used, to format 3D plots. However, there are various additional commands that frequently used to format 3D plots. Among the most commonly used commands of this group, there are commands for coloring the graph and creating various graph projections.

4.3.4.1 About Figure Colors

Color plays an important role in 2D and especially 3D graphs. When the above 3D graphics commands were entered, the surface color was generated by default according to the range of z-values. Nevertheless, to change/set some other color map the colormap command can be used:

TABLE 4.6
Some Typical Colors (Alphabetically) and Its Three-Number Representation

Color	Triplet (Vector c in colormap)	Color	Triplet (vector c in colormap)
aquamarine	[0.49 1 0.83]	gray	[0.5 0.5 0.5]
black	[0 0 0]	green	[0 1 0]
blue	[0 0 1]	magenta	[1 0 1]
copper	[1 0.62 0.4]	red	[1 0 0]
cyan	[0 1 1]	yellow	[1 1 0]
gold	[0.804 0.498 0.196]	white	[1 1 1]

```
colormap(c)
```

where c is a three-element row vector that specifies the intensity of the color using the so-called triplet - three numerical values. In MATLAB®, a red-green-blue, RGB triplet is used to produce the whole spectrum of colors. In this triplet, the three numbers define the color intensity; each intensity is graded from 0 to 1. Table 4.6 shows some colors and its intensity triplets (vector c) that can be used by the colormap command.

For example, the color of the mesh lines in Figure 4.9, assigned by default, can be changed to green with the colormap([0 1 0]) command, which should be entered after the plot creation.

Another possible form of the command is

```
colormap color_map_name
```

where the color_map_name may be jet, cool, summer, winter, spring, and some others. This form allows to implement some common color combinations (maps): For example, the colormap summer changes the current colors to shades of green and yellow. By default, the parule name is used as the color_map_name. More information about the available color maps can be defined by entering the >>doc colormap in the Command Window.

4.3.4.2 About Projections of the 3D Graph

All 3D plots are shown from a certain viewpoint by default. The spatial orientation of a plot can be installed by the view command that has the form

```
view(azimuth,elevation)
```

where azimuth and elevation are the orientation angles that determine the eye position (viewpoint) relative to the generated plot. azimuth is the horizontal (*x*,*y*-plane) rotation angle measured relative to the negative *y*-axes direction and is considered positive in counterclockwise direction. elevation is the vertical angle that defines the geometric height of the eye (viewpoint) position above

TABLE 4.7
View Projections and Corresponding Orientation Angle

View Projection	Azimuth, Degree	Elevation, Degree	Command
Top view – the *x,y*-projection	0	90	`>>view(0,90)` or `>>view(2)`
Front view – the *x,z*- projection	0	0	`>>view(0,0)`
Side view – the *y,z*- projection	90	0	`>>view(90,0)`
Default three-dimensional view	–37.5	30	`>>view(3)`
Mirror view relative to the default view	37.5	30	`>>view(37.5,30)`
Rotate the view around the x-axis by 180°	180	90	`>>view(180,90)`

the *x,y*-plane and is considered positive if it is located above the *x,y*-plane. Both azimuth and elevation values should be specified in degrees; their default values are azimuth = –37.5°, and elevation = 30°.

Orientation angles for various view projections, and examples of the view command are presented in Table 4.7.

Generate six plots of various views for the Nusselt number, *Nu(Re,Pr)* – see previous example.

Problem: Create a live script that builds the following Nusselt number projections on the same page: the default (as per Figure 4.9), mirrored to default, the top and the front projections, front and side *y,z* projections. Compose the six required plots in three rows and two columns.

The ExView live script program that solves the problem and generates the necessary plots are (Figure 4.12):

The ExView script commands perform the following operations:

1. Assign values to the Re and Pr vectors with the two linspace commands;
2. Create the domain grid matrices RE and PR using the meshgrid command;
3. Calculate the *Nu* number for each pair of the RE and PR values by the given *Nu(RE,PR)* expression;
4. Break the Figure window (page) into 3×2 panes and make the first pane current using the subplot command; generate the first plot using the surf command at default viewpoint;
5. Introduce the title and axes labels, and set the axes limits to the data range (axis tight);
6. The 4 and 5 actions, each for new current pane, are repeated five times for the following views: mirror (view(37.5,30)), top (view(2)), front (view(0,0)), side (view(90,0)), and rotate (view(180,90)).

Note: Desired view of any created plot can be set interactively by clicking the "rotate 3D" button . This button, among others, appears immediately above

ExView.mlx

```
Re=linspace(5e5,1e7,15);                              % assigns Re vector
Pr=linspace(20,2e3,15);                               % assigns Pr vector
[RE,PR]=meshgrid(Re,Pr);                  % generates the Re and Pr matrices
Nu=0.037*RE.^0.8.*PR./(1+2.443*RE.^-0.1.*(PR.^(2/3)-1));    % Nu marix
subplot(3,2,1),surf(RE,PR,Nu)                         % default 3D view
xlabel('Re'),ylabel('Pr'),zlabel('Nu')
title('Default view'),axis tight
subplot(3,2,2),surf(RE,PR,Nu),view(37.5,30)           % mirror 3D view
xlabel('Re'),ylabel('Pr'),zlabel('Nu')
title('Mirror view'),axis tight
subplot(3,2,3),surf(RE,PR,Nu),view(2)                 % top view
xlabel('Re'),ylabel('Pr'),zlabel('Nu')
title('Top view'),axis tight
subplot(3,2,4),surf(RE,PR,Nu),view(0,0)               % front fiew
xlabel('Re'),ylabel('Pr'),zlabel('Nu')
title('Front view'),axis tight
subplot(3,2,5),surf(RE,PR,Nu),view(90,0)              % side 3D view
xlabel('Re'),ylabel('Pr'),zlabel('Nu')
title('Side view'),axis tight
subplot(3,2,6),surf(RE,PR,Nu),view(180,90) % rotate around the x view
xlabel('Re'),ylabel('Pr'),zlabel('Nu')
title('Rotate (around x) view'),axis tight
```

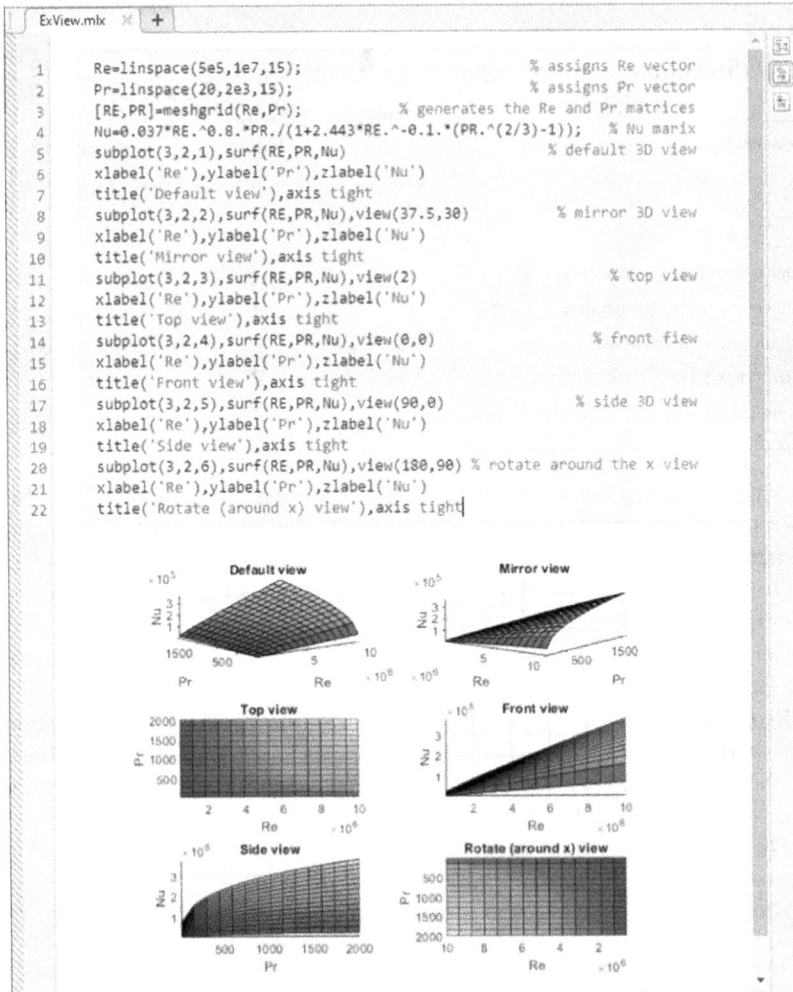

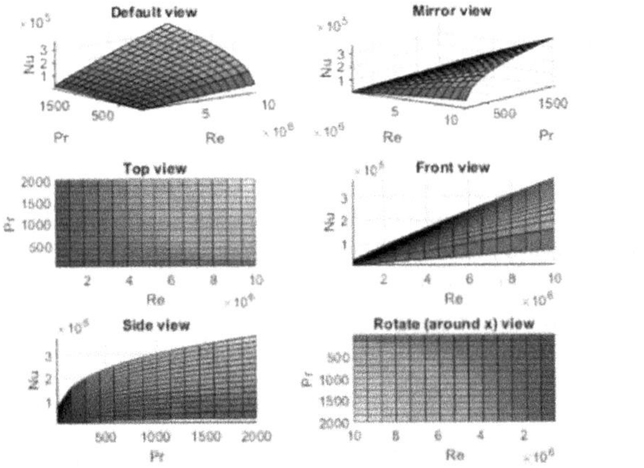

FIGURE 4.12 The ExView live script and various *Nu*(*Re*,*Pr*) surface projections.

the plot after placing the mouse cursor over the plot area. After clicking on the "Rotate 3D" button, the cursor should be placed on plot and you can start the rotation. The values of azimuth, Az, and the elevation, El, angles appear in the bottom-left corner of the window.

4.4 SUPPLEMENTARY 2D AND 3D PLOT COMMANDS

Additionally to the graphics commands described above, basic MATLAB® provides many other commands for 2D and 3D plotting. By entering the commands >>help graph2d, >>help graph3d or >>help specgraph in the Command Window, you can get a complete list of the available graphics commands.

Table 4.8 presents supplementary commands for 2D and 3D plotting that can be useful for graphic representation in engineering practice. The table shows one of the possible forms of commands, usually the simplest, with brief explanations, examples, and the resulting plots.

TABLE 4.8*
Additional Commands for Generating 2D and 3D Graphs

Command Form	Purpose and Parameters of the Command	Example with Generated plot/s

`bar(x,y)`	Displays the vertical bars of the y values at the locations specified by x	`>>x = 0.1:0.1:0.7;`
		`>>y = [1,2,5,4,2,2,1];`
		`>>bar(x,y)`
		`>>xlabel('Wear rate, mm^3/m')`
		`>>ylabel('Frequencies, units')`
		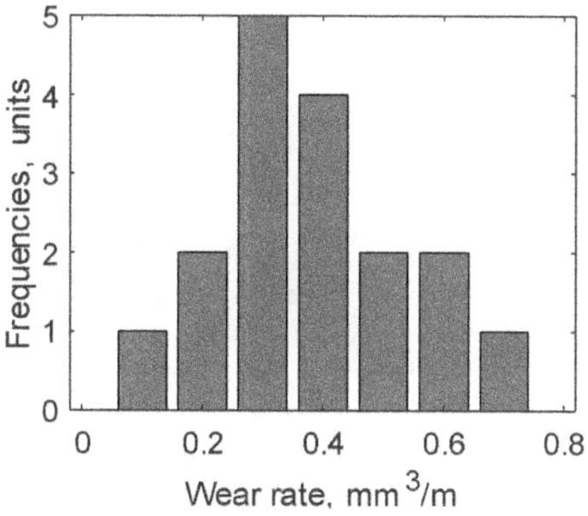

`barh(x,y)`	Displays the horizontal bars of the y values at the location specified by x	`>>x = 0.1:0.1:0.7;`
		`>>y = [1,2,5,4,2,2,1];`
		`>>barh(x,y)`
		`>>xlabel('Frequencies, units'`
		`>>ylabel('Wear rate, mm^3/m')`
		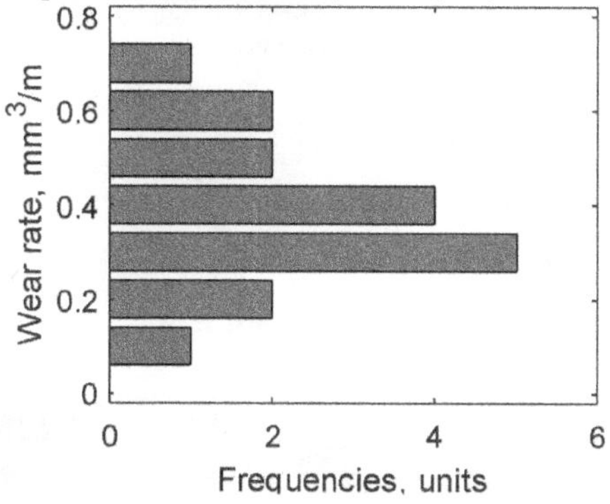

(*Continued*)

TABLE 4.8 (*Continued*)
Additional Commands for Generating 2D and 3D Graphs

Command Form	Purpose and Parameters of the Command	Example with Generated plot/s

`bar3(Y)`	Generates 3D-bar plot by the data grouped in columns in the **Y** matrix	`>>Y = [8 4 3`
		`3 7 10`
		`4 4 6];`
		`>>bar3(Y)`
		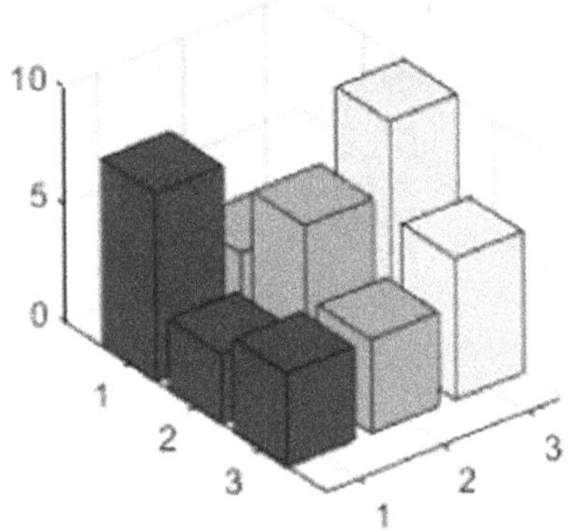

`contour (X,Y,Z,n)`
or
`c=contour (X,Y,Z,n)`
with addition
`clabel(c)`
	Displays the *x*,*y*-plane of iso-lines of the Z(X,Y)-surface; Z is the height with respect to the *x*,*y*-plane; n is the number of contour lines.
	The second command form displays the level values c of the iso-lines.
		`>>x=linspace(-2,2,20);`
		`>>[X,Y]=meshgrid(x);`
		`>>Z=1e3*X.^2.*Y.*exp(-X.^2-Y.^2);`
		`>>contour(X,Y,Z,7)`
		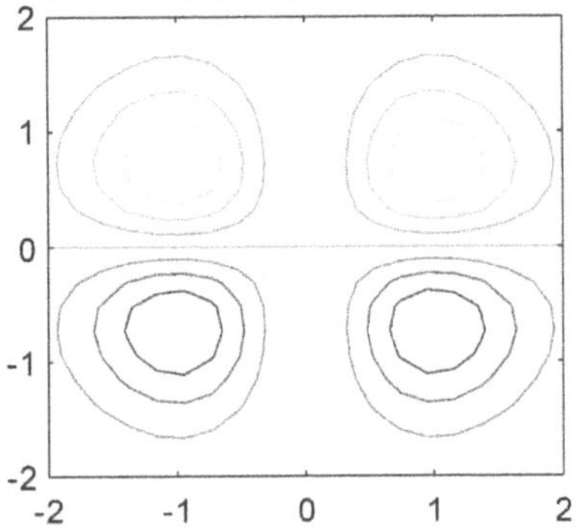
		`>>% or with height values`
		`>>c=contour(X,Y,Z,7)`
		`>>clabel(c)`

(*Continued*)

TABLE 4.8 (*Continued*)
Additional Commands for Generating 2D and 3D Graphs

Command Form	Purpose and Parameters of the Command	Example with Generated plot/s
		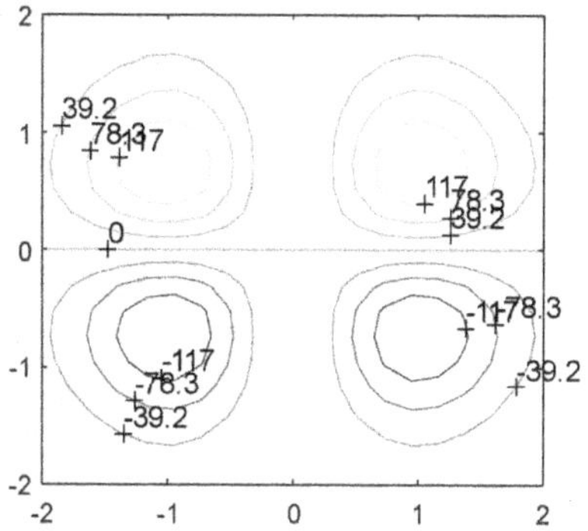

`contour3 (X,Y,Z,n)`	displays x,y-planes of iso-lines of the Z(X,Y) surface in 3D; `n` is the number of contour lines	`>>x=linspace(-2,2,20);`
		`>>[X,Y]=meshgrid(x);`
		`>>Z=1e3*X.^2.*Y.*exp(-X.^2-Y.^2);`
		`>>contour3(X,Y,Z,7)`
		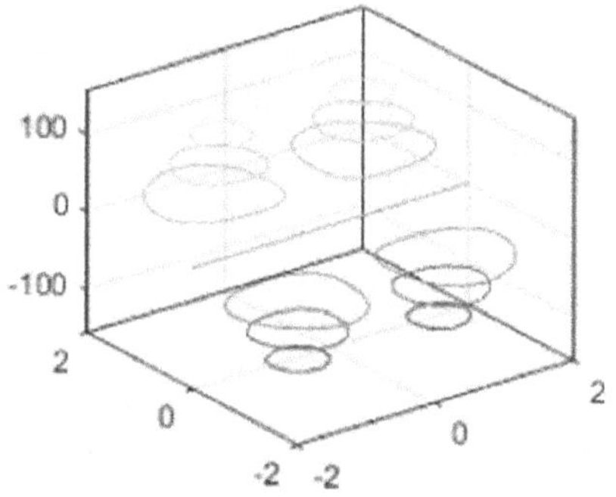

cylinder or `cylinder(r)`	Draws an ordinary or a profiled cylinder; the profile is given by the r expression	`>>t=0:pi/10:2*pi;`
		`>>r=atan(0.1*t);`
		`>>cylinder(r)`
		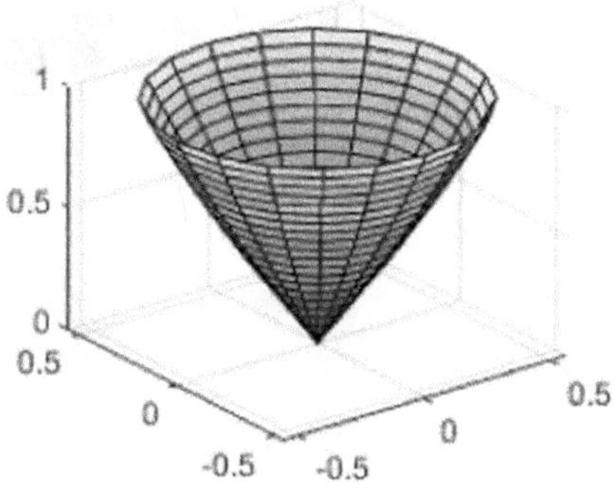

(*Continued*)

TABLE 4.8 (*Continued*)
Additional Commands for Generating 2D and 3D Graphs

Command Form	Purpose and Parameters of the Command	Example with Generated plot/s
`errorbar(x,y,e)` or `errorbar(x,y, e_up,e_low)`	plots *x,y* curve with two side-equal error bars e; or with unequal error bars,e_up and e_low; x,y,e and x,y,e_up,e_low are size-equal vectors	`>>w=[15 24 35 38 64];` `>>s=[0 24.5 41 83 115];` `>>e=0.05*w;` `>>errorbar(s,w,e)` 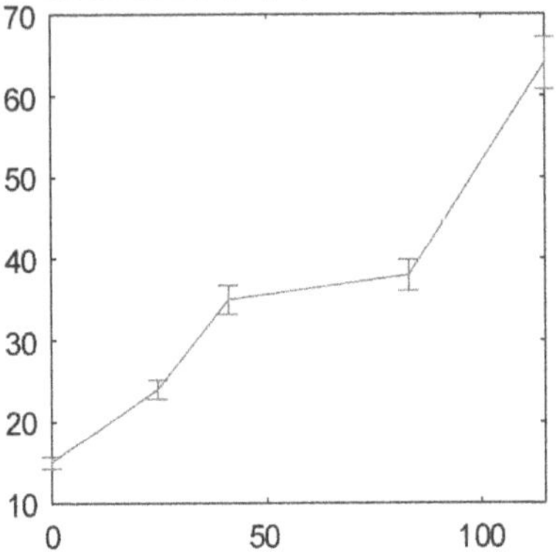
figure or `figure(n)`	Generates a new figure window named as Figure 1 or Figure n, where n is the number of figure	`>> figure(3)` 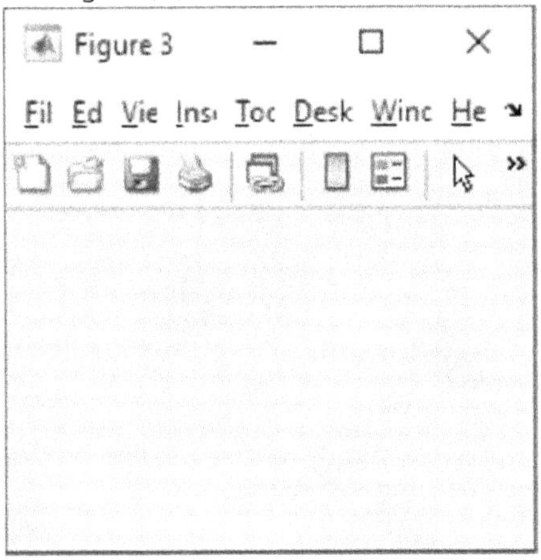

(*Continued*)

TABLE 4.8 (*Continued*)
Additional Commands for Generating 2D and 3D Graphs

Command Form	Purpose and Parameters of the Command	Example with Generated plot/s
`fplot(@(x) function, limits)`	Plots a function y=f(x) with specified limits for *x*-axis (the limits of *y*-axis may be added); function is the function written with the @(x)sign	`>>fplot(@(x)70*exp(-.2*x),[0,20])` 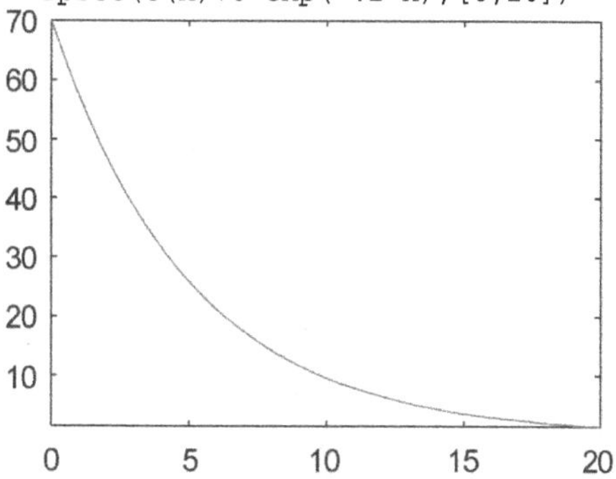

`loglog(x,y)`	Generates a y(x) plot with a log scaled (base 10) on both the *x*- and the *y*-axes	`>>x=linspace(1,200);`
		`>>y=sqrt(x.^3);`
		`>>loglog(x,y)`
		`>>grid on`
		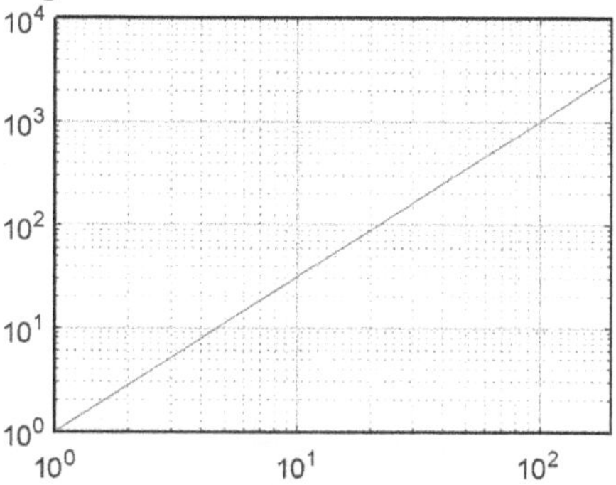

`pie(x)`	Generates a pie-like chart by the x data. Each element in x is represented as a pie slice	`>> x = [88 17 92 97 68 55];`
		`>> pie(x)`
		`>> title('Grades')`
		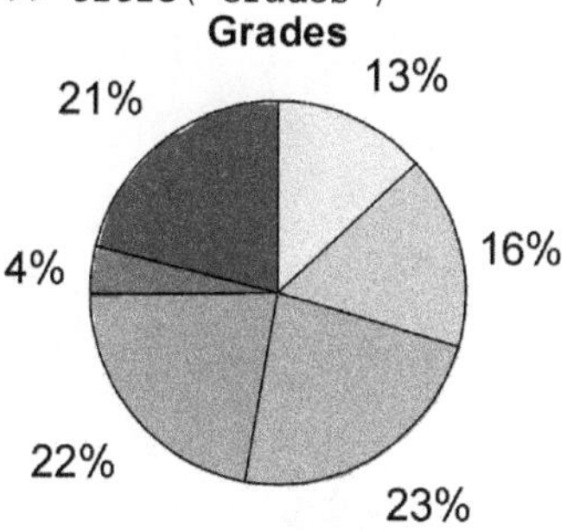

(*Continued*)

TABLE 4.8 (*Continued*)
Additional Commands for Generating 2D and 3D Graphs

Command Form	Purpose and Parameters of the Command	Example with Generated plot/s

`pie3(x,explode)`	Generates a 3D pie-like chart; explode is a vector that specifies an offset of a slice from the center of the chart; 1 denotes an offseted slice and 0 a plain slice	`>>x=[88 17 92 97 68 55];`
		`>>explode=[0 0 1 1 0 0];`
		`>>pie3(x,explode)`
		`>>title('Grades')`
		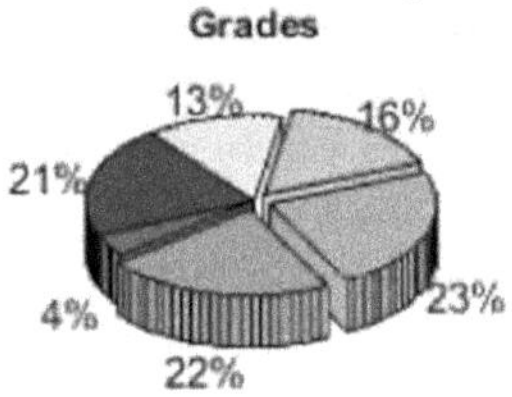

`polar (theta,rho)`	Generates a plot in polar coordinates; theta and rho are the angle and radius respectively	`>>theta=linspace(0,2*pi);`
		`>>rho=atan(0.3*theta);`
		`>>polar(theta,rho)`
		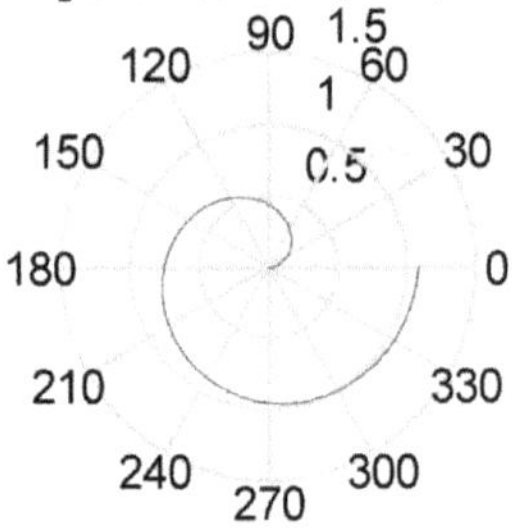

`semilogy(x,y)` or `semilogx(x,y)`	Generates with a log-scaled *x*- or *y*-axis; x and y are the point coordinates	`>>mu=[1.517.993.701.524.41];`
		`>>T=0:30:120;`
		`>>semilogy(T,mu,'-o')`
		`>>xlabel('Temperature,^oC')`
		`>>ylabel('Bromobenzene Viscosity,mPa.s')`
		`>>grid on`
		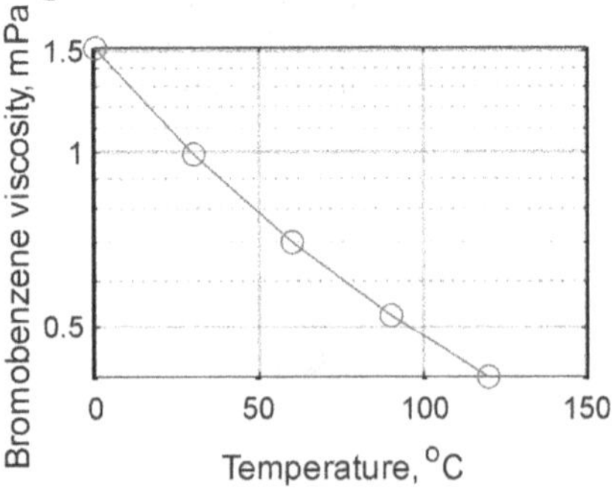

(*Continued*)

TABLE 4.8 (*Continued*)
Additional Commands for Generating 2D and 3D Graphs

Command Form	Purpose and Parameters of the Command	Example with Generated plot/s

`sphere`
or
`sphere(n)`
	Generates a sphere with 20×20 (default) or nxn mesh cells respectively	`>> sphere(15), axis equal`
		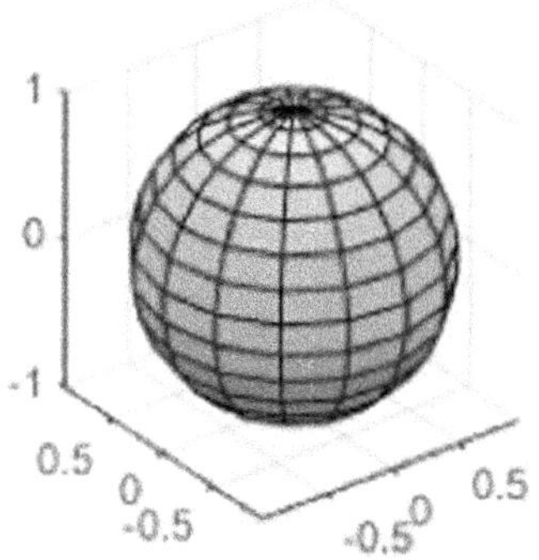

`stairs(x,y)`	Generates a stairs-like plot of the discrete y data given at the specified x points	`>> x=1995:5:2020;`
		`>>y=[7.4 7.8 8.1 8.4 8.7 9.1];`
		`>> stairs(x,y)`
		`>> xlabel('Year')`
		`>> ylabel('People, mln')`
		`>>axis([1995 2020 7 9.5])`
		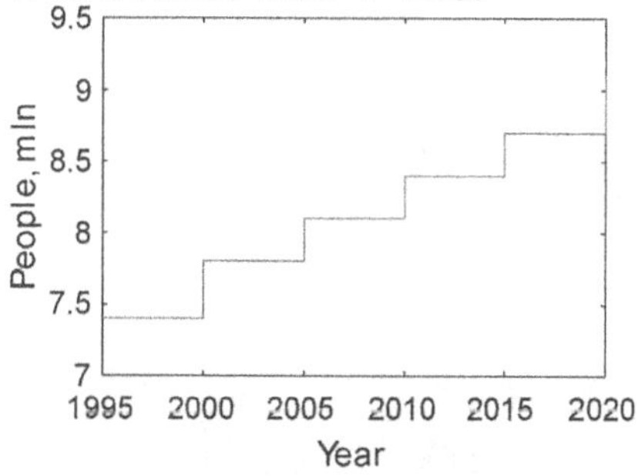

(*Continued*)

TABLE 4.8 (*Continued*)
Additional Commands for Generating 2D and 3D Graphs

Command Form	Purpose and Parameters of the Command	Example with Generated plot/s

`stem(x,y)`	Displays data as stems extending from a baseline along the *x*-axis. A circle (default) terminates each stem	`>> x=10:10:100;`
		`>> y=[4,5,2,5,4,1,2,3,5,5];`
		`>> stem(x,y)`
		`>> xlabel('Grades')`
		`>> ylabel('Number of Grades')`
		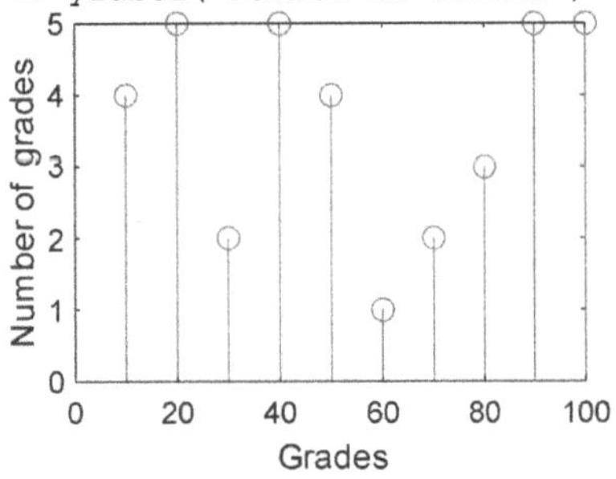

`stem3(x,y,z)`	Generates a 3D plot with stems from the x,y - plane to the z- values; a circle (default) terminates each stem	`>>x = 0.25:0.25:1;`
		`>>[X,Y]=meshgrid(x);`
		`>>z= (X.*Y.^2./(X.^2+Y.^2)).^1.2;`
		`>>stem3(X,Y,z)`
		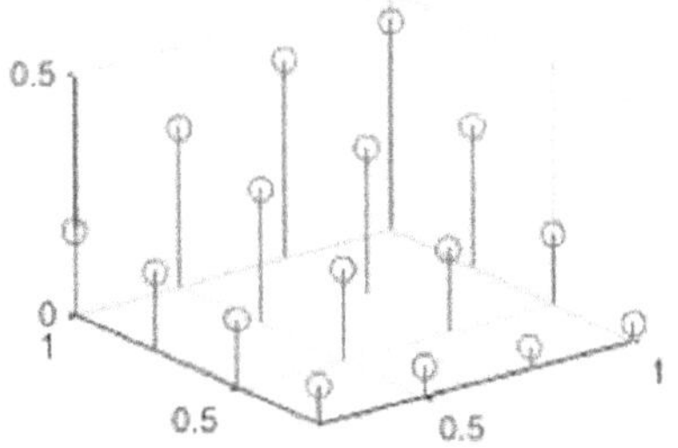

`surfc(X,Y,Z)`	Generates surface and contour plots together; X, Y, Z – matrices with the *x*, *y*, *z* coordinates	`>>x=-2:.2:2;`
		`>>[X,Y] = meshgrid(x);`
		`>>Z=1e3*X.*Y.^2.*exp(-X.^2-Y.^2);`
		`>> surfc(Z);`
		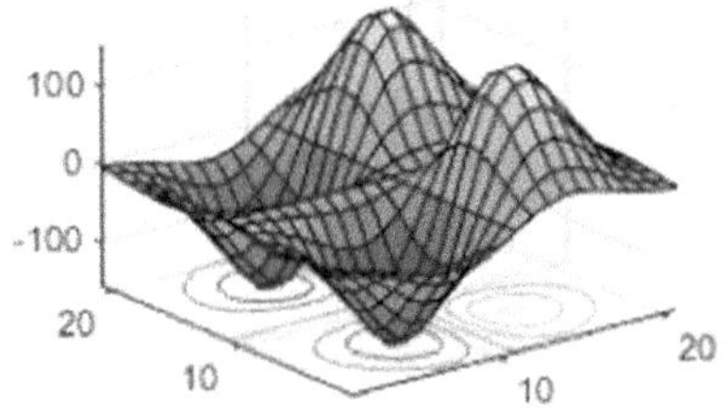

(*Continued*)

TABLE 4.8 (*Continued*)
Additional Commands for Generating 2D and 3D Graphs

Command Form	Purpose and Parameters of the Command	Example with Generated plot/s

`waterfall (X,Y,Z)`	Generates the same Z(X,Y) plot as the mesh command (Section 4.2.2) but without the column lines	`>>x=linspace(-5,5,10);`
		`>> [X,Y]=meshgrid(x);`
		`>>e=Y.*sin(X);`
		`>>waterfall(X,Y,e)`
		`>>xlabel('x'),ylabel('y'),zlabel('z')`
		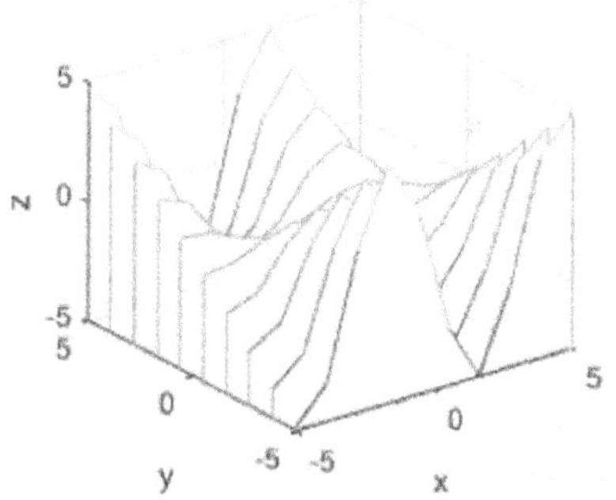

`yyaxis left`
or
`yyaxis right`
	Activate the left-side and right-side *y*-axis to create graph with two different *y*-axis	>>y_left = [1684 831 260 73 30];x1=[0 10 30:30:90];
		>>y_right = [1159 1185 1211 1239]*1e-6;x2 = 0:30:90;
		>>yyaxis left
		>>plot(x1,y_left,'o-')
		>>yyaxis right
		>>plot(x2,y_right,'s-')
		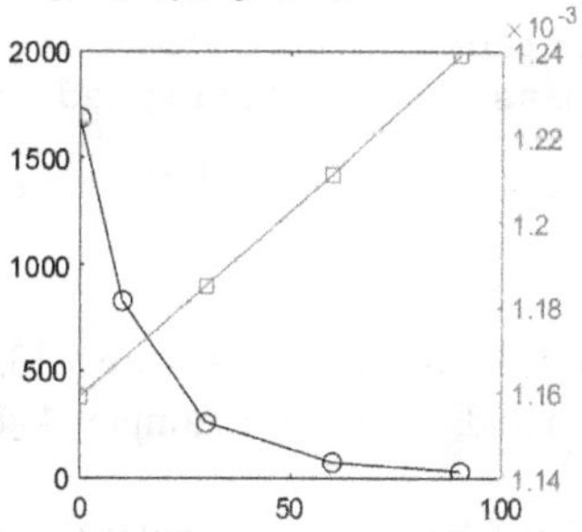

[a] Based on Table 3.3 in Burstein, 2015.

4.5 APPLICATION EXAMPLES

4.5.1 Surface Tension of Water

For ordinal water, surface tensions γ changed with temperature. The values: γ = 74.23 72.74 71.19 69.59 67.93 66.24 64.47 62.68 60.82 and 58.92 mN/m were measured with the uncertainties ±0.5% of γ, mN/m. The temperatures at which each of these values were observed are respectively t = 10, 20, 30, 40, 50, 60, 70, 80, 90, and 100 °C. The data have been fitted by the expression

$$\gamma = 235.8\left(1-\frac{T}{T_c}\right)^{1.256}\left[1-0.625\left(1-\frac{T}{T_c}\right)\right]$$

where T_c equal to 647.096 K, $T = 273.15 + t$, K.

Problem: Write script that generates 2D plot with the measured data with their error bars showing the error limits at each point. Add to plot the curve calculated by the above expression. Provide plot with the grid, captions, and legend.

The program ApExample_4_1 solving the problem are:

```
Gam_e = [74.23 72.74 71.19 69.59 67.93 66.24 64.47 62.68
60.82 58.92];
t = 273.15+(10:10:100);          % vector of temperatures in K
Tc = 647.096;
error = 0.005;                   % error in parts of percent
Gam_t = 235.8*(1-t./Tc).^1.256.*(1-0.625*(1-t./Tc));
e = error *Gam_e;               % absolute surface tension errors
errorbar(t,Gam_e,e,'o')           % plots experimental points
hold on
plot(t,Gam_t)                        % plots calculated curve
xlabel('Temperature, K'),ylabel('Surface tension, mN/m')
title('Surface tension for water')
legend('measurements','fitted curve')
grid on
axis tight                       % sets axes to the data limits
hold off
```

The resulting plot is shown in Figure 4.13:

The commands of the ApExample_4_1 program operate as follows:

- The first four commands enter the data, and the constants are used further; at that, the temperatures in degrees Celsius are converted to the Kelvin units, and the error is presented in parts of percent;
- The fifth command calculates surface tension Gam_t with the above expression;
- The sixth command calculates the error for each of the measured points;

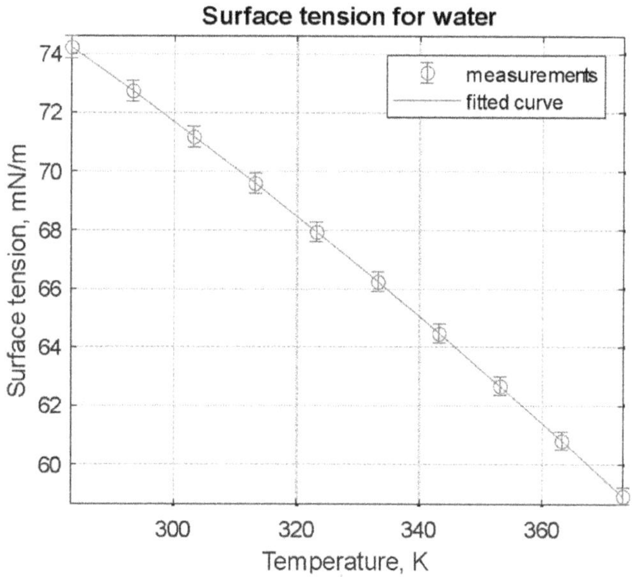

FIGURE 4.13 Surface tension plot generated by the ApExample_4_1 command.

- The errorbar command generates a graph of data points with an error bar at each point;
- The next two commands set the graph as current and add the fitting curve to it;
- Final commands format the graph and make it out of the "current" mode.

4.5.2 A Bandpass Filter

The ratio of the output and input voltages *RV* in a bandpass filter that passes the signal is determined by the expression

$$RV = \frac{\omega RC}{\sqrt{\left(1-\omega^2 LC\right)^2 + (\omega RC)^2}}$$

where ω is the input signal frequency, *R* is the resistance of the resistor, *C* is the capacity of the capacitor, and *L* is the inductance of the inductor.

Problem: Create a live script that calculates *RV* versus ω that specified in the range 20 …1000 at *C* = 5 μF, *R* = 200 ∧, and set it up for three values *L* = 8, 12, and 28 mH. Generate a graph containing three *RV*(ω) curves, each for one of the specified *L*. Provide the generated graph with the grid, captions, and legend.

The ApExample_4_2 live script, which solves this problem and generates a plot, is presented in Figure 4.14.

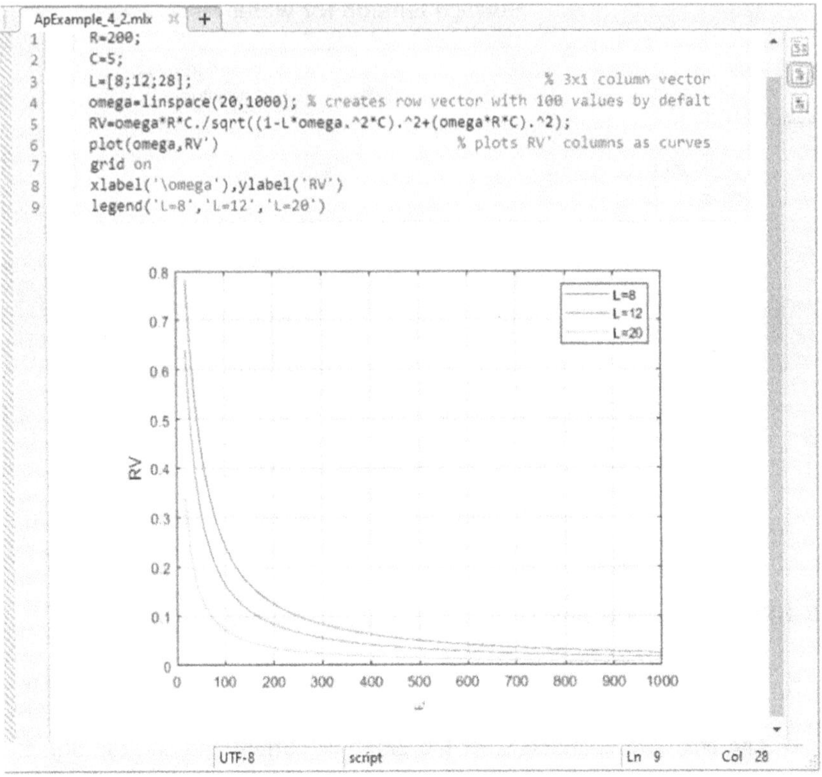

FIGURE 4.14 Live script ApExample_4_2 in "Output inline" mode with generated three $RV(\omega)$ curves.

The commands of this live script act as follows:

- The first four commands enter data; the inductance is entered as [3×1] column vector and is a [1×100] row vector.
- The fifth command calculates the filter ratio RV with the above expression; here the exponentiation and division operations are performed element by element. Nevertheless, the product $L\omega^2$ is performed with the matrix multiplication rule – the inner dimensions of the L and ω^2 vectors must be equal, in our case they are [3×1] and [1×100], and the resulting RV matrix has the [3×100] dimension.
- The sixth command generates a graph with three curves, for this the [3×100] RV matrix is converted to the [100×3] matrix (using the transpose operator "′") since the plot command plots the curves by the columns of the matrix;
- The final commands format the graph.

4.5.3 Vibrational Modes of a Rectangular Membrane

A two possible mode shapes of a rectangular membrane clamped on its outer boundaries can be calculated by the following expression

$$W_{m,n} = \sin\sin\left(\frac{m\pi x}{L_x}\right)\sin\sin\left(\frac{n\pi y}{L_y}\right)$$

where m and n are the membrane modes, L_x and L_y are the membrane dimensions, x and y are the membrane points coordinates.

Problem: Write a script that calculates w for modes $m = 2$, $n = 3$ and $m = 3$, $n = 2$ with $L_x = 1$ and $L_y = 2L_x$ and generates (in the same Figure window) the $w_{2,3}$ and $w_{3,2}$ surface plots, each of which accompanied by contour lines; use 30 x and y values each. Add axis labels and a title to plots.

The following program, ApExample_4_3, after running, performs calculations and generates the $w_{23}(x,y)$ and $w_{32}(x,y)$ plots at the same Figure window – Figure 4.15.

```
Lx = 1;
Lz = 2*Lx;
x = linspace(0,Lx,30);                                    % vector x
y = linspace(0,Lz,30);                                    % vector y
[X,Y] = meshgrid(x,y);                              % X and Y matrix
m = 2;
n = 3;                                                 % mode (2,3)
w23 = sin(m*pi*X/Lx).*sin(n*pi*Y/Lz);
subplot(1,2,1)
surfc(X,Y,w23)                % surface plot with contour lines
xlabel('x'),ylabel('y'),zlabel('Mode shape, w_{23}')
title('Rectangular membrane mode (2,3)')
axis equal                    % to keep the scale for both axes
m = 3;
n = 2;                                                 % mode (2,3)
w32 = sin(m*pi*X/Lx).*sin(n*pi*Y/Lz);
subplot(1,2,2)
surfc(X,Y,w32)                % surface plot with contour lines
xlabel('x'),ylabel('y'),zlabel('Mode shape, w_{32}')
title('Rectangular membrane mode (3,2)')
axis equal                    % to keep the scale for both axes
```

The commands of this program act as follows:

- The first five commands generate X,Y grid in the domain L_xx L_y;
- The next three command calculate w for the mode (2,3);
- The `subplot` and `surfc` commands make the first part of figure window current and generate a 3D plot with contour lines;

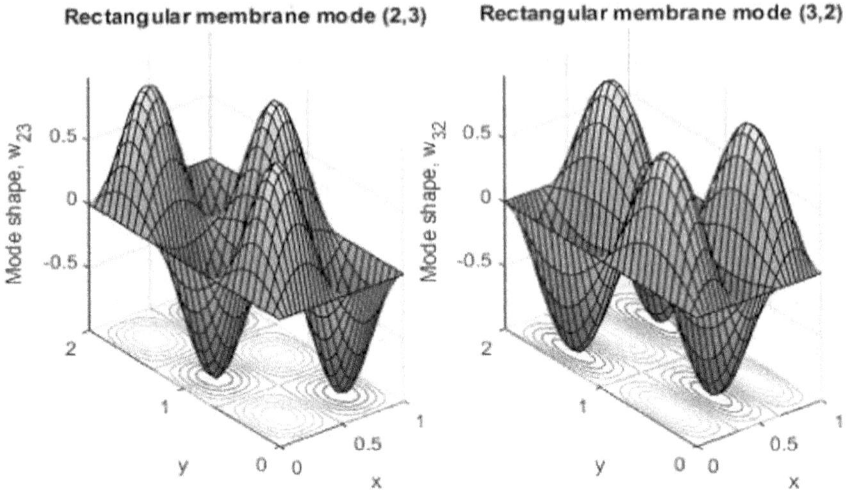

FIGURE 4.15 Two mode shapes of a rectangular membrane with contour lines in the *x*,*y* plane.

- The next three lines contain the commands that add caption, title, and scale axes; note that the zlabel command uses curled parentheses to represent two digits as a subscript using a single "_" character;
- The following commands then calculate *w* for mode (3,2), and then perform the same operations as for mode (2,3) to plot the result.

4.5.4 Flow in a Rectangular Channel

Flow rate Q (in m) in an open rectangular channel can be calculated by Manning's equation

$$Q = \frac{kdw}{n}\left(\frac{wd}{w+2d}\right)^{2/3}\sqrt{S}$$

where d (in m) is the depth of water, w (in m), is the width of the channel, S, m/m, is the slope of the channel, is the channel roughness coefficient, and $k = 1$ for our units.

Problem: Compose a live script that calculates Q and generates two $Q(w,d)$ views at the same page, one is the default view and the second is the mirror view. Assume $n = 0.05$, $S = 0.001$ m/m, $w = 0.1, \ldots, 9$ m, $d = 0.1, \ldots, 5$ m; use 10 w and d values each. Add axis labels and a title to each plot.

The ApExample_4_4 live script, which solves this problem and generates two required views of the $Q(w,d)$ surface, is presented in Figure 4.16.

```
 ApExample_4_4.mlx ✕ +

1 n=0.05;
2 S=0.001;
3 k=1;
4 w=linspace(0.1,9,10);
5 d=linspace(0.1,5,10);
6 [W,D]=meshgrid(w,d);
7 Q=k*D.*W./n.*(W.*D./(W+2*D)).^(2/3)*sqrt(S);
8 subplot(2,1,1)
9 surf(W,D,Q) % default view
10 xlabel('Chanel width, m'),ylabel('Water depth, m')
11 zlabel('Flow, m^3/s')
12 title('Q(w,d). Default view')
13 subplot(2,1,2)
14 surf(W,D,Q)
15 view(37.5,30) % mirror view
16 xlabel('Chanel width, m'),ylabel('Water depth, m')
17 zlabel('Flow, m^3/s')
18 title('Q(w,d). Mirror view')
```

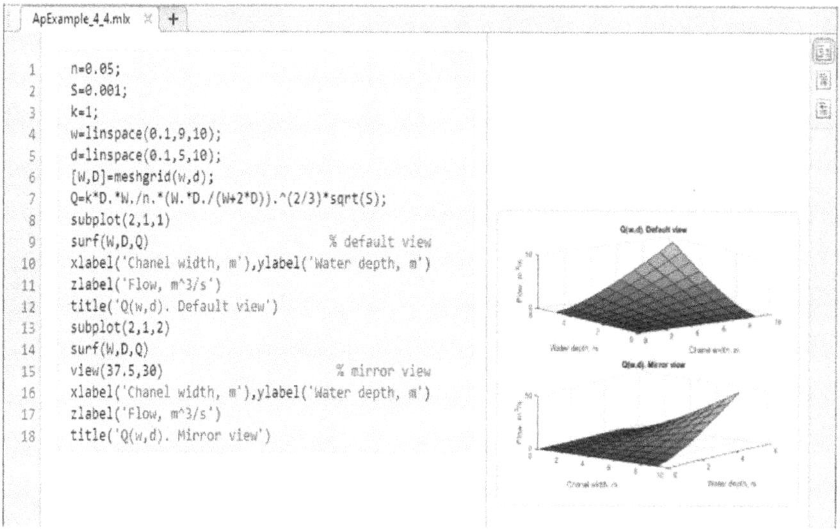

FIGURE 4.16 Live script ApExample _ 4 _ 4 in the "Output on right" mode with two views of the $Q(w,d)$ surface.

The commands of this live script are as follows:

- The first five commands enter data; channel width and water depth are entered as ten-element vector each, using the linspace commands;
- The sixth command generates matrices W and D for the x,y plane;
- The seventh command calculates the flow rate Q; here math operations with matrices W and D are performed element by element;
- All commands starting from the eighth generate two different views (with the view commands) in the same Figure window (with the subplot commands); the mirrored view is created assuming the azimuth angle of 37.5° and elevation angle of 30°; part of this group of commands formats graphics.

5 ODE Solvers for Initial and Boundary Value Problems

5.1 INTRODUCTION

In the natural sciences, various processes and phenomena are described by differential equations (DEs). Therefore, they are studied in colleges and universities; are used by students, graduates, and engineers; and occupy in fact a central place in science, technology, and engineering. Accordingly, DEs solving refer to the basic MATLAB®. Due to the fact that actual DEs are often impossible to solve analytically, the sole possibility is a numerical solution. Unfortunately, there is no universal numerical method, and the basic MATLAB® provides special commands, called *solvers*, for solution two classes of differential equations, specifically ordinary, ODE, and partial, PDE. A brief description of the first class DE solvers is provided in this chapter, while the second class DEs are described in the subsequent chapters of the book. Solvers for ODEs are divided into two groups depending on the problems they solve – initial value problem (IVP) solvers and boundary value problem (BVP) solvers. When the initial value problem is solved, the initial value of the search function is specified, at some point in the solution range (termed the domain). Whereas when solving a boundary value problem, more than one, usually two, intra-domain values of the search function are specified. All materials are provided with application examples, which include:

- IVP ODE solution for undamped forced oscillations of the spring;
- BVP ODE solution for undamped oscillations of the spring-mass system;
- IVP ODE solution representing behavior of the current in the series *RLC* circuit;
- IVP solution of the ODE simulating the relative strengths of the military;
- BVP solution of the heat ODE for temperature distribution in the wire heater;
- BVP solution of Euler ODE for a uniformly loaded beam hinged at both ends and with an external moment applied to one end of the beam.

The following assumes sufficient familiarity with both ODE groups.

DOI: 10.1201/9781003200352-5

5.2 ODE SOLVER FOR IVP

The ODE solvers are intended for the following set of first order DEs:

$$\frac{dy_1}{dt} = f_1\left(t, y_1, y_2, \ldots, y_n\right)$$

$$\frac{dy_2}{dt} = f_2\left(t, y_1, y_2, \ldots, y_n\right)$$

$$\ldots$$

$$\frac{dy_n}{dt} = f_n\left(t, y_1, y_2, \ldots, y_n\right)$$

where 1, 2, …, n are the numbers of the first order ODEs; y_1, y_2, …, y_n are the dependent variables while t is the independent varying between the starting t_s and final t_f values. Note: The variable x or some other can be used instead of t.

To solve a high-order ODE with the MATLAB® solver, such an equation must be reduced to first order creating a set of first-order equations. It can be done, for example, for a second-order ODE as follows: the first derivative dy_1/dt is denoted as y_2 after which y_2 is substituted into the original equation. Such derivative replacements can be used to reduce a higher-order ODE to a first-order ODE set.

To illustrate what was stated, let us reduce the order, for example, the DE of the second, third, and fourth orders:

- The second-order equation

$$\frac{d^2y}{dt^2} + a\frac{y}{t}\left(\frac{dy}{dt}\right)^{1/3} = b\sqrt{t}$$

Denote, first, y as y_1 and then $\frac{dy_1}{dt}$ as y_2; substitute now y_2 into the equation above, thus we obtain two first-order differential equations:

$$\frac{dy_1}{dt} = y_2 \qquad \frac{dy_2}{dt} = -a\frac{y_1}{t}y_2^{1/3} + b\sqrt{t}$$

The solution of these equations is the $y_1(t)$ and $y_2(t)$ functions that should be defined. The $y_1(t)$ dependence is the solution to the original second-order equation, and $y_2(t)$ is the derivative of this solution.

- The third order equation

$$\frac{d^3y}{dt^3} + a\left(t^2 - 2.71\frac{y}{t}\right)\frac{d^2y}{dt^2} - b\frac{t}{y} = c\,(\sin 2t + \cos 2t)$$

Denote y as y_1, $\frac{dy_1}{dt}$ as y_2, $\frac{d^2y_2}{dt^2}$ as y_3 and substitute y_1, y_2, and y_3 in the equation above, this yields

$$\frac{dy_1}{dt}=y_2,\ \frac{dy_2}{dt}=y_3,\ \frac{dy_3}{dt}=-a\left(t^2-2.71\frac{y}{t}\right)y_3+b\frac{t}{y_1}+c\left(\sin2t+\cos2t\right)$$

The solution of this system is the $y_1(t)$, $y_2(t)$, and $y_3(t)$ functions; between them, the $y_1(t)$ dependence is a solution to the original third order equation, and $y_2(t)$ and $y_3(t)$ are the first and second derivatives of this solution, respectively.

- The fourth-order equation

$$a\left(2t^2-\pi\right)\frac{d^4y}{dt^4}+b\frac{d^2y}{dt^2}+c\left[\left(\frac{dy}{dt}\right)^2-\frac{y}{t}\right]=d\cdot y^3$$

Analogously, the previous examples can be transformed into the set of the fourth first-order equations:

$$\frac{dy_1}{dt}=y_2,\ \frac{dy_2}{dt}=y_3,\ \frac{dy_3}{dt}=$$

$$y_4,\ \frac{dy_4}{dt}=-\frac{b}{a\left(2t^2-\pi\right)}y_3-\frac{c}{a\left(2t^2-\pi\right)}\left[y_2^2-\frac{y_1}{t}\right]+\frac{d}{a\left(2t^2-\pi\right)}y_1^3$$

The solutions of the above system are the $y_1(t)$, $y_2(t)$, $y_3(t)$, and $y_4(t)$ functions. The $y_1(t)$ dependence is a solution of the original fourth-order equation, and $y_2(t)$, $y_3(t)$ and $y_4(t)$ are the first, second, and third derivatives of the solution, respectively.

There is no unified numerical method allowing to solve any ODE. Therefore, the MATLAB® IVP and BVP solvers include a number of commands that actualize different numerical methods for solving actual ODEs.

5.2.1 About Numerical Methods for Solving DEs

The basic idea of numerical methods for solving ODEs is to convert the DE into an algebraic form that is achieved by representing the derivative $\frac{dy}{dt}$ as a ratio of small but finite differences, i.e.:

$$\frac{dy}{dt}\simeq\lim_{\Delta t\to0}\frac{\Delta y}{\Delta t}=\frac{y_{i+1}-y_i}{t_{i+1}-t_i}$$

where Δt and Δy are the differences of the argument t and function y respectively, $i = 0, 1, 2,\ldots$, n is the number of the points t taken in the range $[t_0, t_n]$, n is the number of the last point t. The approximation of the derivative as the ratio of differences seems to be correct for very small, $\Delta t \rightarrow 0$, but nonzero (finite) distances between two adjacent points of the argument: $i + 1$ and i. In accordance with the IVP, the numerical solution of a first-order differential equation consists in specifying, at first, the initial value y_0 at the first point; calculate then the value of the derivative for t_0 from the original expression of the ODE

$$\left.\frac{dy}{dt}\right|_0 = f(t_0, y_0)$$

and after that determine the value of the function y_1 for the next point t_1 as

$$y_1 = y_0 + \left.\frac{dy}{dt}\right|_0 t$$

Assuming the constant argument difference Δt and using the defined y_1, we can calculate the derivative at $t_1 = t_0 + \Delta t$ as

$$\left.\frac{dy}{dt}\right|_1 = f(t_1, y_1)$$

and then determine the second point ($t_2 = t_1 + \Delta t$) function value y_2

$$y_2 = y_1 + \left.\frac{dy}{dt}\right|_1 t$$

Such calculations can be repeated to obtain all values y_i in the range $t_0 \ldots t_n$.

To demonstrate the effectiveness of this simple scheme, we solve the first-order Newton's law ODE for cooling of a preheated substance:

$$\frac{dT}{dt} = -\alpha(T - T_s)$$

where T and t are temperature, C, and time, minute, respectively; α is the substance-dependent coefficient, 1/min, T_s is the surrounding temperature, C.

Finite difference equation for this ODE is:

$$T_{i+1} = T_i - \alpha(T_i - T_s)\Delta t$$

where Δt is the time step assumed to be constant over the calculating range.

Problem: Develop a live script named EulerMethodExample that calculates temperatures by finite differences and plots $T(t)$ for 100 times in the range 0 ... 30 min. Add to the graph the 20 T-values calculated using the theoretical solution $T = T_s + (T_o - T_s)e^{-\alpha t}$. Consider $T_s = 25$ °C, $T_o = 80$ °C, and $\alpha = 0.21$/min.

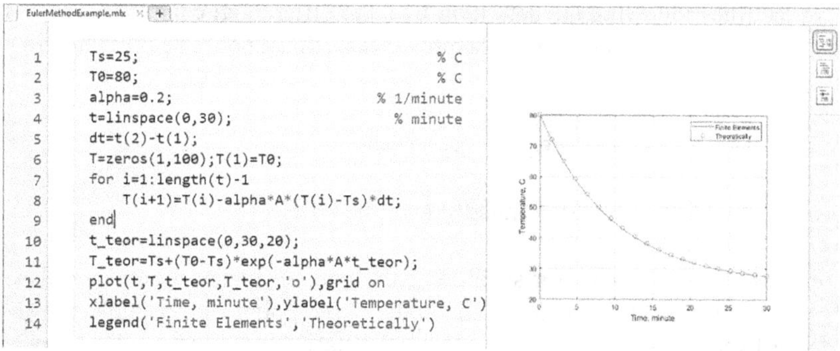

EulerMethodExample.mlx
Ts=25; % C
T0=80; % C
alpha=0.2; % 1/minute
t=linspace(0,30); % minute
dt=t(2)-t(1);
T=zeros(1,100);T(1)=T0;
for i=1:length(t)-1
T(i+1)=T(i)-alpha*A*(T(i)-Ts)*dt;
end
t_teor=linspace(0,30,20);
T_teor=Ts+(T0-Ts)*exp(-alpha*A*t_teor);
plot(t,T,t_teor,T_teor,'o'),grid on
xlabel('Time, minute'),ylabel('Temperature, C')
legend('Finite Elements','Theoretically')

FIGURE 5.1 The live script EulerMethodExample demonstrates finite differences and analytical solutions of cooling equation (Newton's low).

The live script EulerMethodExample that solves this problem with the resulting graph is shown in Figure 5.1.

As can be seen, the results of finite differences are in good agreement with the theoretical ones that indicate the effectiveness of the finite-difference method. This approach, pioneered by Euler, is used with various improvements and complications in advanced numerical methods such as the Runge-Kutta, variable order differences, Rosenbrock and Adams, etc.

The following describes the MATLAB® commands that implement the finite difference approach and are available in the ODE solver.

5.2.2 ODE Solver Commands

The commands for the ODE solution are summarized as follows:

```
[t,y] = odeNNNss(@ODE_fun,t_span,y0)
```

where:

- NNNss is a two or three-digit number with one or two letters (optionally); this part of the ODE solver command denotes the numerical method realized in this command for solving ODE, for example, the ode45 command uses the Runge-Kutta method, and the ode15s command uses the variable order step size method.
- @ODE_fun is the name of the user-defined function where the solving first-order DEs are written. The definition line of the user-defined function should look like

```
function dy = fun_ode (t,y)
```

where t, y, and dy are the names of the argument, the search function, and the search function differential, respectively.

In the lines following the definition line, the differential equation/s, rewritten as first-order PDE/s (as shown in the previous section) should be presented in the form:

```
dy = [1st_ODE_rhs; 2nd_ODE_rhs; …];
```

where rhs denotes the right-hand side of the ODE.

- t_span – a row vector specifying the interval of the argument t that changes from the starting t_s and up to final t_f points, for example, the [1 26] vector specifies a t-interval with the starting and final t-values equal to 1 and 26, respectively. The t_span vector can contain more than two values intended to display the solution at these values, for example, [1:4:17 26] means that results of solution lie in the t-range of 1…26 and should be displayed at t-values equal to 1, 4, 7, 10, 12, and 14. The values specified in t_span do not affect the steps used in the solution method: the solver command automatically chooses/changes the step to ensure the solution accuracy. The default tolerance is 0.000001 absolute units.
- y0 is a vector of initial conditions, for example, for the set of two first-order differential equations with initial function values $y_1 = 0$ and $y_2 = 4$, this vector is written as y0 = [0 4]; y0 can be given also as a column vector, e.g. y0 = [0;4].
- y and t output arguments are the column vectors containing the obtained y-values at the corresponding t-values; in the case of the second or higher order of ODEs, y is the n-column matrix, each column contains the y values defined for each first-order ODE. In case the odeNNS commands are used without the output arguments, the solver shows an automatically generated graph of the obtained solution.

Both regular and so-termed stiff problems can be solved by the commands of the ODE solver. A DE is stiff if it contains terms that form any singularities, for example, function jumps, holes, discontinuities, etc. This leads to a discrepancy in numerical solution even at very small size of solution step. Unlike stiff equations, regular DE is characterized by stable convergence of the solution. Most commonly used commands for solving these problems are ode45 and ode15s: the ode45 command is used for solving non-stiff ODEs and ode15s for stiff ODEs.

Unfortunately, it is impossible to know *a priori* if the problem is stiff or not, and therefore, we cannot determine in advance which ODE command should be used. When we solve an ODE that describes a real technology or phenomenon, it will help us select a certain command to this ODE solution. For example, rapidly changing or momentary processes or technologies using fast chemical reactions or even explosions can be apparently modeled using stiff equations. In these cases, the ode15s command should be used. Sometimes the ratio of the maximal to minimal values of the ODE coefficients is used as a criterion of the stiffness. If this ratio is larger than 1000, then the problem can be classified as stiff. Unfortunately, this criterion is empirical and therefore not always correct. In general, if the ODE type is *a priori* unknown, then it is rational to try first the `ode45` command and only then the `ode15s` command.

5.2.2.1 Steps to Solve ODE

Demonstrate the steps to be taken to use the solver with an example solution of the ODE for the undamped forced oscillations of a spring

$$m\frac{d^2u}{dt^2}+ku=f(t)$$

where u is the displacement from the equilibrium point, t – time, m is the attached mass, $k = mg/L$ is a spring constant, L – stretch length, g – gravitational acceleration, and $f(t)$ is a forcing function. Depending on inputted u and $f(t)$ function, this equation can be adopted for many different phenomena, including the behavior of materials within some viscoelasticity simulations, RC circuits of the various test apparatuses, and other technological and physical processes.

Problem: Compose a user-defined function ExODE without parameters that solves the above ODE. Assume m = 3 kg, L = 0.4 m, g = 9.8 m/s^2, and $f(t)=10\cos(\omega_0 t)$ (oscillations grows with t, resonance study case), $\omega_0=\sqrt{\frac{k}{m}}$, t changed in range 0 … 5. Initial displacement and velocity are 0.1 m and −0.05 m/s, respectively. Plot resulting graph of $u(t)$.

Solution steps are:

1. The order of the original differential equation should be lowered to the first-order ODE (if it is higher than first order) and presented as set of the equations. Each first-order equations should be presented in the form $\frac{dy}{dt}=f(t,y)$ in the range $t_s \le t \le t_f$ (where t_s and t_f are the starting and final argument points) and must be accompanied by the initial condition $y = y0$ at $t_s = t_0$.

   To lower the order of the equation to be solved, rewrite it as:

$$\frac{d^2y}{dt^2}=\frac{10}{m}\cos(\omega_0 t)-\frac{k}{m}y$$

   where u was denoted as y.

   Now designating y as y_1 and $\frac{dy_1}{dt}$ as y_2, we obtain the following two ODEs of the first order:

$$\{\frac{dy_1}{dt}=y_2$$

$$\frac{dy_2}{dt}=\frac{10}{m}\cos(\omega_0 t)-\frac{k}{m}y_1$$

2. In the second step, the user-defined function with solving equation should be created. The definition line of this function must include input arguments t and y and output argument naming the left side of the first order equations above, for example, dy. In the next lines, the right-hand side/s of the ODE/s should be written as vector separated with the semicolon (;) or in the separate lines. In our example:

```
function dy = myODE(t,y)
m = 3; L = 0.4; g = 9.8;
k = m*g/L;omega0 = sqrt(k/m);
dy = [y(2);10/m*sin(sqrt(k/m)*t)- k/m*y(1)];
```

   Note: the tilde (~) character can be written instead *t*, e.g. function dy = myODE(~,y), if *t* is absent in the right part of the differential equation (not in our case),

3. At the final step, one of the ODE solver commands should be chosen for implementation. For our example, we can select the ode45 solver because there are no specific recommendations about the ODE stiffness. Thus

```
[t,y] = ode45(@myODE,[0:5],[0.2 0.1]) %y0_1 = 0.2,
y0_2 = 0.1
```

   Writing the commands above together with the new one for plotting, we generate the ExODE user-defined function as follows:

```
function ExODE
%                          Solution of the second order ODE
%                                              to run: >> ExODE
close all
[t,y] = ode45(@myODE,[0 5],[0.1-0.05])
plot(t,y(:,1),t,y(:,2),'--')
grid on
xlabel('Time, s'),ylabel('u and du/dt')
legend('u','u''','location','best')
axis tight
function dy = myODE(t,y)
m = 3; L = 0.4; g = 9.8;
k = m*g/L;omega = sqrt(k/m);
dy = [y(2);10/m*sin(omega*t)- k/m*y(1)];
```

   After running, the t-vector and two-column matrix y appear in the Command Window and graph - in the Figure window (Figure 5.2):

```
>> ExODE
t =

          0
     0.0010
```

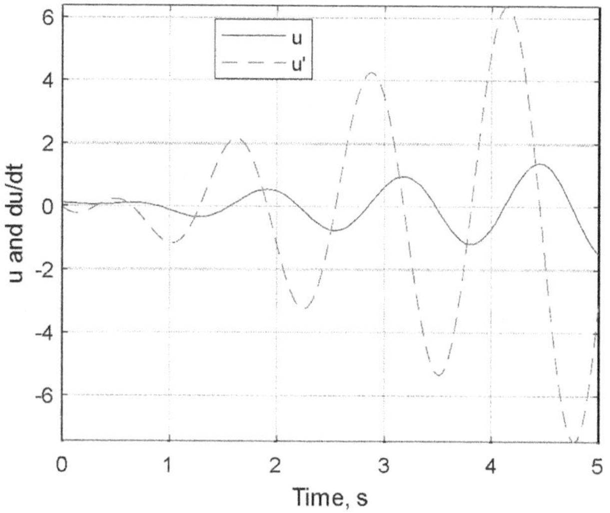

FIGURE 5.2 Solution for ODE of the undamped forced oscillations.

```
      0.0021
      0.0031
...
   y =
   0.1000-0.0500
   0.0999-0.0525
   0.0999-0.0550
...
```

In the ExODE function, the calculations start with initial values of $y_1 = 0.1$ and $y_2 = -0.05$ at $t = 0$; the starting and final times of the process are 0 and 24, respectively. The given commands solve the equation and display the one-column vector of the *t* values (that automatically chosen in the specified t_span range) and two-column matrix with the resulting *y* values. To plot the solution, the plot command and plot-formatting commands are used. Graph shows the displacement *u* and velocity d*u*/d*t* as function of time.

5.2.2.2 ODE Solver Commands with Additional Parameters or with Anonymous Function

When ODEs include some parameters that can be varied depending on the actual problem (e.g., in the discussed example, the *m*, *g*, *L*, *k*, and *ω* variables are constants used in the ODE), the ODE solver commands can be written in an extended form allowing you to pass these parameters to the ODE_fun function:

```
[t,y] = odeNNNss(@ODE_fun,t_span,y0,[ ], prm1,prm2,…)
```

where:

- empty brackets [ ] denote an empty vector intended for so-termed options that can be used to control the integration process.[1] Since this vector is not assigned, the default option values are used, which provide satisfactory solutions for the vast majority of practical cases. Therefore, we do not specify these options.
- prm1,prm2,... – the names of the parameters, the value of which we are going to pass into the ODE_fun function.

If the parameters prm1,prm2,... are named in the odeNNNss solver, they must also be named in the ODE_fun definition line. Let us show, for example, how the user-defined function ExODE, rewritten for more generality as function with input arguments *m*, *L*, t_span, and y0, can be modified for introducing the m, *k*, and ω coefficients to the ODE45 and ODE_fun functions. The program is as follows:

```
function ExODE(m,L,t_span,y0)
%                               Solution of the second order ODE
%                          run: >> ExODE(3,.4,[0 5],[0.1; -0.05])
g = 9.8;
k = m*g/L;
omega = sqrt(k/m);
[t,y] = ode45(@myODE,t_span,y0,[],m,k,omega)
plot(t,y(:,1),t,y(:,2),'--')
grid on
xlabel('Time, s'),ylabel('u and du/dt')
legend('u','u''','location','best')
axis tight
function dy = myODE(t,y,m,k,omega)
dy = [y(2);10/m*sin(omega*t)- k/m*y(1)];
```

To run this function, the following command should be entered in the Command Window:

```
>> ExODE(3,.4,[0 5],[0.1; -0.05])
```

The results are the same as in Section 5.2.2.1.

The advantage of the used forms of the user-defined function ExODE and the extended ODE-solver command is their greater versatility, for example, these functions can be used for any *m, L, t_span, y0, k*, and ω without introducing their values into the main function and into the myODE sub-function.

Others, perhaps more appropriate and simpler ODE solver command forms use an anonymous function:

---

[1] Enter the `help odeset` command to review the possible options.

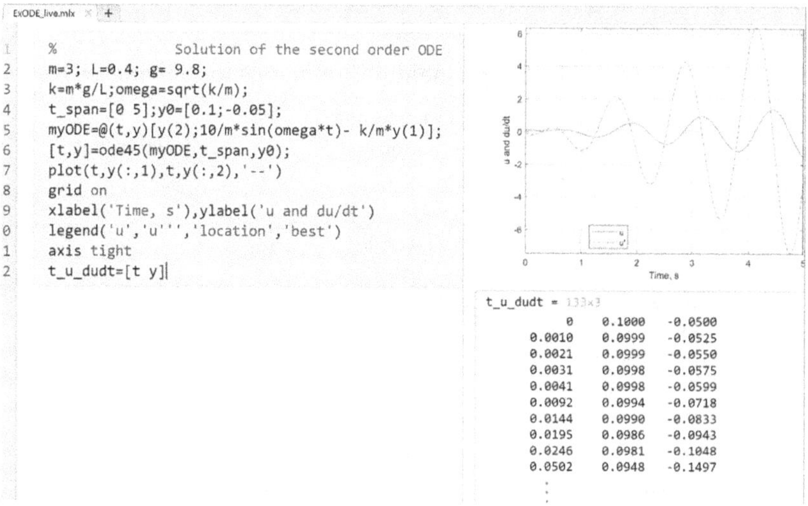

ExODE_live.mlx
% Solution of the second order ODE
m=3; L=0.4; g= 9.8;
k=m*g/L;omega=sqrt(k/m);
t_span=[0 5];y0=[0.1;-0.05];
myODE=@(t,y)[y(2);10/m*sin(omega*t)- k/m*y(1)];
[t,y]=ode45(myODE,t_span,y0);
plot(t,y(:,1),t,y(:,2),'--')
grid on
xlabel('Time, s'),ylabel('u and du/dt')
legend('u','u''','location','best')
axis tight
t_u_dudt=[t y]
u and du/dt
Time, s
u
u'
t_u_dudt = 133×3
0 0.1000 -0.0500
0.0010 0.0999 -0.0525
0.0021 0.0999 -0.0550
0.0031 0.0998 -0.0575
0.0041 0.0998 -0.0599
0.0092 0.0994 -0.0718
0.0144 0.0990 -0.0833
0.0195 0.0986 -0.0943
0.0246 0.0981 -0.1048
0.0502 0.0948 -0.1497

FIGURE 5.3 Live script ExODE with an anonymous function as ODE_fun argument of the ode45 command.

```
[t,y]=odeNNNss(@(t,y)ODE_rhs,tspan,y0)
```

or command pair

```
ODE_fun=@(t,y)ODE_rhs
[t,y]=odeNNNss(ODE_fun,tspan,y0)
```

where ODE_rhs is the column vector containing the right-hand side/s of the ODE/s to be solved. The other arguments of these commands are the same as in the previously discussed commands.

When some of these forms are used, it is not necessary to create a separate sub-function containing the ODE/s and there is no need to pass parameters to it. The live function script ExODE_live illustrates the usage of the second form with anonymous function (Figure 5.3).

The last presented forms of ODE solver commands are preferable when we solve short differential equations (which can be put on one line) having coefficients that can be varied by user.

5.2.3 Supplemental Commands of the ODE Solver

In addition to the ode45 and ode15s commands described above, the solver has other commands for solving ODEs; some of them are listed in Table 5.1. For each solver command, the table represents its name, the implemented numerical method with recommended solver use case, an ODE example with the solution command, and a graphical representation of the solution. The input variables of

TABLE 5.1[a]
Initial Value Problem: The Supplemental Commands for Solving ODEs

Command Name	Method and Recommendation	ODE Example, and Solver Command	Solution Plot
ode23	**Method**: Runge-Kutta **Recommended**: for non-stiff and moderately stiff ODEs. Quicker, but less precise than ode45	ODE: $y' = 2t^2\cos(t)/y$, with y0 = 1, ts = 0, tf = 2. `>>ode23 (@(t,y)2*t.^2.*cos(t)./y,[0 2],1);`	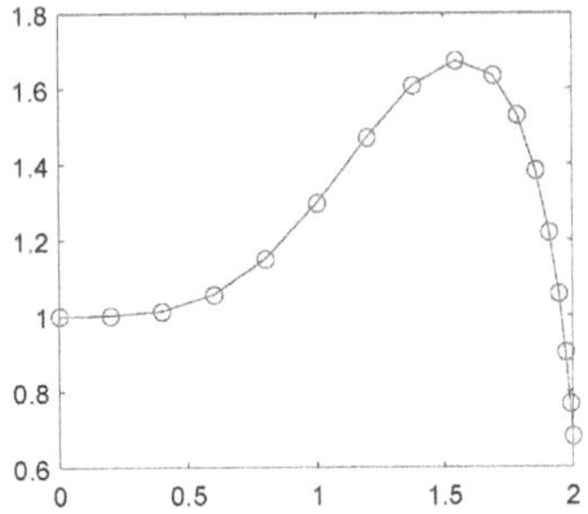
ode113	**Method**: Adams **Recommended**: for non-stiff ODEs and for problems with stringent error tolerances, or for solving computationally intensive problems	ODE: the same as for ode23. `>>ode113 (@(t,y) 2*t.^2.*sin(t)./y, [0 2],1);`	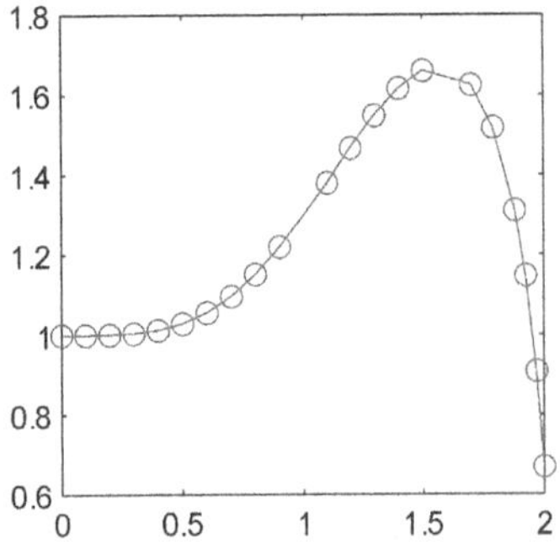
ode23s	**Method**: Rosenbrock **Recommended**: for stiff ODEs and when ode15s is slow	ODE: $y'=-10y$ with y0 = 0.7, ts = 0, tf = 1. `>>ode23s (@(t,y) 10*y, [0 1],0.7);`	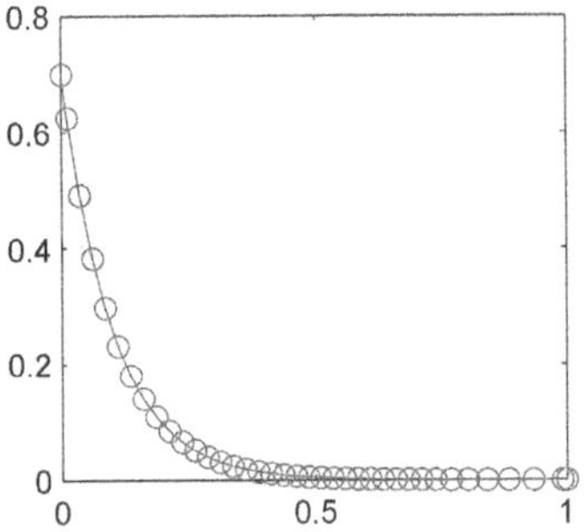
ode23t	**Method**: Trapezoidal rule **Recommended**: for moderately stiff ODEs and for differential algebraic equations (DAEs)	ODE: the same as for ode23s. `>>ode23t (@(t,y)10*t.^2, [0 1],0.7);`	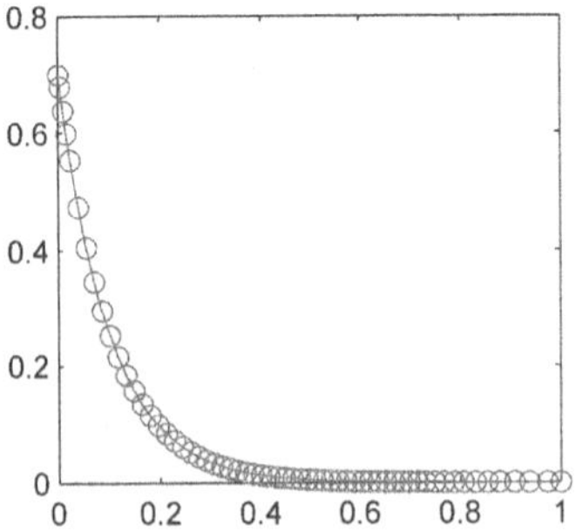

(Continued)

TABLE 5.1[a] (*Continued*)

Initial Value Problem: The Supplemental Commands for Solving ODEs

Command Name	Method and Recommendation	ODE Example, and Solver Command	Solution Plot
ode23tb	**Method**: Trapezoidal rule/ second order Backward differentiation formula,TR/BDF2 **Recommended**: for stiff ODEs; sometimes more effective than ode15s	ODE: the same example as for the ode23s. `>>ode23tb (@(t,y)10*t.^2, [0 1],0.7);`	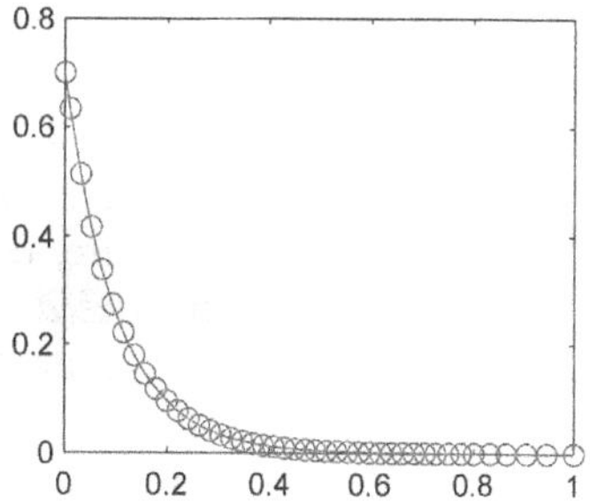

[a] Based on Table 6.1, Burstein (2020).

the presented commands are identical to those described previously and therefore are not explained in the table. Solver commands are used without output parameters, so the plot is generated automatically, which does not require graphical commands.

5.3 SOLVER FOR BOUNDARY VALUE PROBLEM OF ODEs

MATLAB provides facilities for solving two-point BVP when, unlike the IVP, at least two values of the function must be specified at two endpoints of the solution range. Problems of this kind arise in the field of strength or materials mechanics, fluid dynamics, heat and mass transfer mechanics, physics, electricity, and many other practical applications. The problem is accurately formulated in such a way that the ODE can be solved as an explicit system of first-order ODEs

$$y_1'(x) = f(x, y_1, y_2, \ldots, y_n)\; y_2'(x) = f(x, y_1, y_2, \ldots, y_n)\; \ldots\; y_n'(x) = f(x, y_1, y_2, \ldots, y_n)$$

where x and y are independent and dependent variable respectively, y' is the dy/dx derivative, and the indices 1, 2, …, n denote the first order ODE number.

The ODE to be solved must be accompanied by boundary conditions, g, specified at two points a and b of the x range $[a,b]$

$$g\big(y(a), y(b)\big) = 0$$

Below is described the bvp4c command designed to solve a boundary value problem using the fourth-order polynomial collocation method.

5.3.1 The bvp4c Command

The most common, simplest form of this command is

```
xy_sol = bvp4c(@ODE_fun,@BC,y_init)
```

where input and output arguments are:

- @ODE_fun is the name of the user-defined function containing the single or set of ODEs to solve;
- @BC and y_init are, respectively, the name of function containing *y* values at boundaries *a* and *b*, and the name of a structure with initial (guess) *y* values specified along the [*a*,*b*] interval;
- xy_sol is the name of the so-called structure containing fields with the *x* and *y* values defined in the solution. The structure variable is a more advanced formation than those used previously, and further, we will give only the minimum information required to use structure in the bvp4c and other structure-containing commands that are used in the book. To obtain the *x* or *y* values of the xy_sol structure, the following form should be used: X = xy_sol.x assigns the *x* values contained in the xy_sol structure to X, or Y = xy_sol.y assigns the y values of the xy_sol structure to Y.

The following description explains how to represent the input and output arguments in the bvp4c command.

The definition line of the user-defined function ODE_fun should be identical to the ODE_fun definition line of the odeNNNss commands (Section 5.2.2):

```
function dy=ODE_fun(x,y)
```

with dy as a column vector containing the right-hand side/s of the ODE/s.

The extended bvp4c command form with anonymous function @(x,y)ODE_rhs is also possible (Section 5.5.2.2).

The definition line for user-defined BC function should be as follows:

```
function res = BC(ya,yb)
```

where

- ya and yb are the column vectors containing *y*(*a*) and *y*(*b*) values for each ODE to be solved;
- res is an outputted column vector that should be formed so:

```
res= [ya(1)-y_at_point_a;yb(1)-y_at_point_b]
```

here y_at_point_a and y_at point_b are the values of *y* at boundary points *a* and *b* and ya(1) and yb(1) are the names of *y* at points *a* and *b*. For example,

- if the boundary values are $y = 0.21$ at point a and y = 3.4 at point b, they should be rewritten as y-0.21 = 0 and y-3.4 = 0 and the res vector can be look as follows: res = [ya(1) −0.21;yb(1) −3.4];
- if the boundary values are $y = 0.21$ at point a and y = 0 at point b, they should be rewritten as y-0.21 = 0 and y = 0 and the res vector can be look as follows: res = [ya(1) −0.21;yb(1)];
- if in addition to the above y values at the boundaries, we have the derivatives y' = 0.05 at point y = 0 thus the res vector is: res = [ya(1) −0.1;yb(1); ya(2) −0.05].

The y_init argument of the bvp4c function can be specified using the bvpinit command specially developed for this purpose as follows:

```
y_init = bvpinit(x,y)
```

where x and y is the column vectors with the initial values x and y along points from the interval $x\epsilon[a,b]$ that are used by the bvp4c command as the guess values. Note: The number of the initial points can be few and is not influenced by the number of points adapted by the solver inside the solution process; initial values should satisfy the boundary conditions.

To evaluate the defined y values contained in the sol structure at the desired points x, the deval function is used.

```
y = deval(xy_sol,x)
```

where x must be specified in the solution range, and y is a matrix with rows each of which contains y obtained from the xy_sol at specified x.

Note that when the ODE has a singularity point, the "'SingularTerm', S-matrix" pair property must be added in the bvpset command. Detailing information about this command and possible property pairs can be obtained by entering the doc bvpset command; see also the "Sole BVP with Singular Term" section of the PDE toolbox documentation.

5.3.2 Solution Steps with bvp4c

To demonstrate the use of the bvp4c function, solve the equation for free undamped oscillations of the spring-mass system (compare with the problem in Section 5.2.2.1):

$$\frac{d^2y}{dt}+\frac{k}{m}y=0$$

where y is the displacement of the spring-mass system due to the stretch force; m is the mass value; k is the spring stiffness constant; t is time.

Problem: (based on Burstein, 2021): Solve this equation in the range $t = t_s \ldots t_f$ s with $y = 0.27$ m at $t_s = 0$ and $y = 0$ at $t_f = 4.85$ s. Coefficients in the ODE are $k = 36$ N/m and $m = 7.3$ kg. Compose a live ExBVP script that implements the solution and generate the $y(t)$ graph.

Solution steps:

1. Represent the original second-order ODE as a set of two first-order ODEs. Re-designate t as x, y as y_1 and denote y_1' as y_2, thus we have:

$$\frac{dy_1}{dx} = y_2 \quad \frac{dy_2}{dx} = -\frac{k}{M} y_1$$

2. Create now a function containing ODEs to be solved using the bvp4c command

```
function dy=ODE_fun(x,y)
k = 36;M = 7.3;
dy = [y(2); -k/M*y(1)];
```

3. Create BC function with boundary conditions

```
function res=BC(ya,yb)
res = [ya(1) -0.1;yb(1)];
```

4. Now specify the y_init argument with initial y values, for example, at five points within the x range using the bvpinit command:

```
y_init = bvpinit(linspace(0,4.85,5),[0.27;0])
```

   where 0.27 and 0 assume as the initial values for y_1 and y_2 that are constant for each initial x point.

5. In this step, we get the obtained y values from the xy_sol structure at 100 x points and plot the results using the following commands:

```
x = linspace(0, 5);
y = deval(sol,x);
plot(x,y(1,:),'^')
```

   The complete ExBVP live script containing the commands and the generated graph is shown in Figure 5.4.

Note that singular and multipoint BVP can be solved with the bvp4c command, the more information can be obtained by entering the doc bvp4c command in the Command Window.

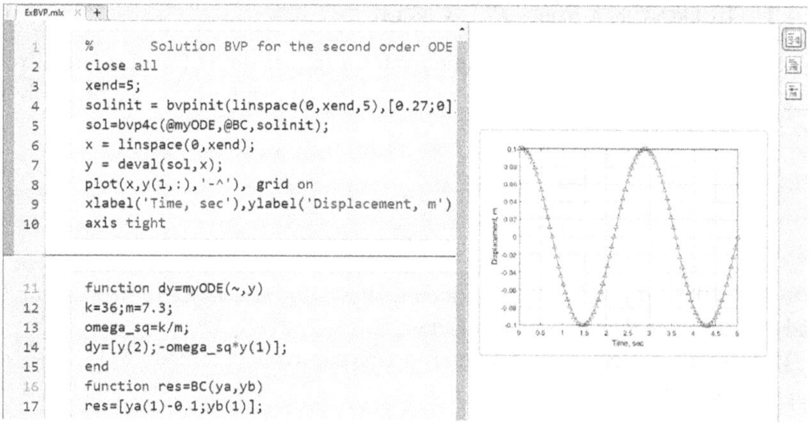

ExBVP.mlx
% Solution BVP for the second order ODE
close all
xend=5;
solinit = bvpinit(linspace(0,xend,5),[0.27;0]
sol=bvp4c(@myODE,@BC,solinit);
x = linspace(0,xend);
y = deval(sol,x);
plot(x,y(1,:),'-^'), grid on
xlabel('Time, sec'),ylabel('Displacement, m')
axis tight
function dy=myODE(~,y)
k=36;m=7.3;
omega_sq=k/m;
dy=[y(2);-omega_sq*y(1)];
end
function res=BC(ya,yb)
res=[ya(1)-0.1;yb(1)];
Displacement, m
Time, sec

FIGURE 5.4 ExBVP live script with the bvp4c function that solves BVP for ODE describing free undamped oscillations.

5.3.3 About the BVP5c Command

Form of this command is identical to the bvp4c command

```
xy_sol = bvp5c(@ODE_fun,@BC,y_init)
```

All comments concerning the bvp4c command arguments are valid for bvp5c. So if in the ExBVP live script you change only the digit 4 to 5 in the name bvp4c, the script works and shows the same results.

The bvp5c command solves BVPs of the same class as bvp4c using the fifth-order polynomial collocation method and provides an accuracy that may be more appropriate for the requirements of accuracy and smoothness of the solution.

5.4 APPLICATION EXAMPLES FOR IVP OF THE ODEs

The programs in the following examples are composed as live scripts or as user-defined functions with arguments, in the first case they have one help line with a brief definition of the program, and in the second case they have a definition line and a line representing the command for running program in the Command Window. Usually, the codes of each program have some differences, even if the program uses the identical IVP or BVP solutions; this is done to demonstrate the various possibilities in implementing the ode or bvp solvers and in presenting the results of calculations. There are little or no explanatory comments within the programs / functions because it is assumed that the user reading this chapter has experience with MATLAB® or has gained sufficient experience after studying the previous chapters. Therefore, when necessary, explanations are given directly in the text.

5.4.1 Current in a Series *RLC* Circuit

When the switch is closed at the time $t = 0$ (in sec), in an *RLC* series circuit with an unchanging voltage source, the current I (in A) can be calculated by solving the following ODE equation

$$L\frac{d^2I}{dt^2}+R\frac{dI}{dt}+\frac{1}{C}I=0$$

where L is the inductance of the inductor, H; R is the resistance of the resistor, Ω; and C is the capacitance of the capacitor, F.

Problem: Develop a user-defined function named ApExample_5_1 that solves this two-order ODE at times from the range 0 …2 s. Assume $L = 2.99$ H, R = 9.8 and 180 Ω, $C = 81{\cdot}10^{-6}$ and $1190{\cdot}10^{-6}$ F for underdamped response and $L = 2.9$ H, $R = 9.8$ and 180 Ω, $C = 81{\cdot}10^{-6}$ and $1190{\cdot}10^{-6}$ F for overdamped (based on A. Gilad, 2015). The initial values are $I = 0$ A and $\frac{dI}{dt} = 8$ A/s at $t = 0$ s. The user-defined function should contain only input parameters, namely the L, R, C, time range t_span, and initial values I_0. Use an anonymous function for the given first-order ODE set to pass it to the ode-solver command. The function should generate the resulting $I(t)$ plot.

To solve the second-order equation of this problem, denote I as y and transform it into the two first-order equations.

$$\{\frac{dy_1}{dt}=y_2 \quad \frac{dy_2}{dt}=-\frac{R}{L}y_2-\frac{1}{CL}y_1$$

To provide the solution, the appropriate ODE command should be chosen. Unless otherwise indicated, it should be the ode45 command.

The user-defined function ApExample_5_1 for the solution is

```
function ApExample_5_1(L,R,C,t_span,i0)
%                     calculates current in the RLC circuit
%Run >> ApExample_5_1(2.99,[9.8 180],[81 1190]*1e-6,[0
2],[0;8])
for i=1:length(R)
ODE_fun=@(t,y) [y(2); -R(i)/L*y(2)-1/(C(i)*L)*y(1)];
[t,y]=ode45(ODE_fun,t_span,i0);
plot(t,y(:,1))
hold on
end
grid on
xlabel('Time, sec'),ylabel('Current, A')
legend('Underdamped','Overdamped')
hold off
```

The ApExample_5_1 user-defined function is developed as follows:

- the first function line defines the function name and the input variables L, R, C, t_span, and i0;
- the help part of the function is presented in two lines after the function definition command and gives a brief purpose of the function and command for running the function;
- then the for … end loop is organized, providing two passes of the loop with the required pair of R and C values in each;
- the next two lines contain the anonymous function ODE_fun with the ODEs (see Section 5.2.2.2) and the ode45 command with a call to this function and other required arguments.
- commands for the loop ending, plotting, and formatting the resulting graph are contained in the final function lines;

The ApExample_5_1 function should be saved in a file with the same name. After that, in the Command Window, you must type the following command:

```
>>ApExample_5_1(2.99,[9.8 180],[81 1190]*1e-6,[0 2],[0;8])
```

After entering, the resulting graph will appear – Figure 5.5.

5.4.2 Relative Strengths of Military Forces

The strength of the military can be modeled with Lanchester's ODEs for aimed fire:

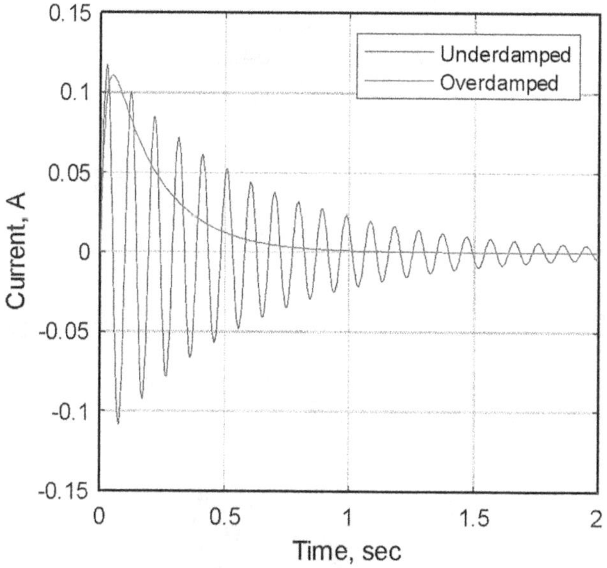

FIGURE 5.5 Solution of the second-order ODE for the current in the RLC circuit.

$$\frac{dA}{dt} = -\beta B \quad \frac{dB}{dt} = -\alpha A$$

where A and B are the number of soldiers, for example, in the Red and Blue military forces, respectively; t is the combat time; α and β are the coefficients characterizing the offensive firepower of the soldiers of the Red and Blue troops, respectively.

Problem: Develop a live script named ApExample_5_2 solving the set of the above ODEs for $A = 1000$, $B = 800$, $\alpha = 0.7$, and $\beta = 0.9$. Consider the time in the range 0 … 2 and display results on the graph for the time range in which the number of soldiers in at least one of the forces is greater than zero.

We have a set of two equations of the first order; therefore, the order of the ODE does not need to be lowered. Denoting A as y_1 and B as y_2, we obtain

$$\frac{dy_1}{dt} = -\beta y_2$$

$$\frac{dy_2}{dt} = -\alpha y_1$$

To perform the solution, the appropriate ode-solver command should be selected. Unless otherwise stated, this should be the ode45 command.

The live script function ApExample_5_2 with the resulting graph is presented in Figure 5.6

The live script ApExample_5_2 is designed as follows:

- The first three lines of the script assign values to the variables A0, B0, alpha, beta, and t_span;
- The next two lines contain the anonymous function ODE_fun with the ODEs (see Section 5.2.2.2) and the ode45 command that calls this function and receives other required arguments as well;

ApExample_5_2.mlx

```
%       Calculates survivors in battle
A0=1000;B0=800;
alpha=0.7;beta=0.9;
t_span=[0 2];
ODE_fun=@(t,y)[-beta*y(2);-alpha*y(1)];
[t,y]=ode45(ODE_fun,t_span,[A0;B0]);
Ai=find(y(:,1)>=0);
Bi=find(y(:,2)>=0);
if Ai(end)<Bi(end)
    i_plot=Ai(end);
else
    i_plot=Bi(end);
end
plot(t(1:i_plot),y(1:i_plot,:))
grid on
xlabel('Time'),ylabel('Survivors')
legend('"Red" military forces', ...
    '"Blue" military forces')
axis tight
```

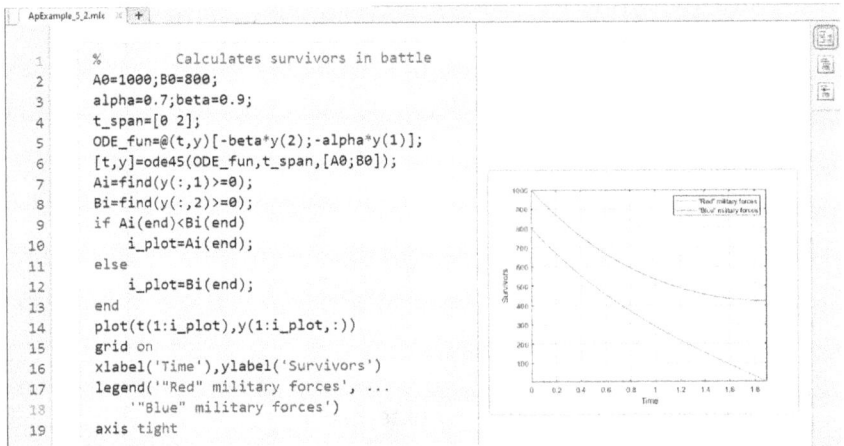

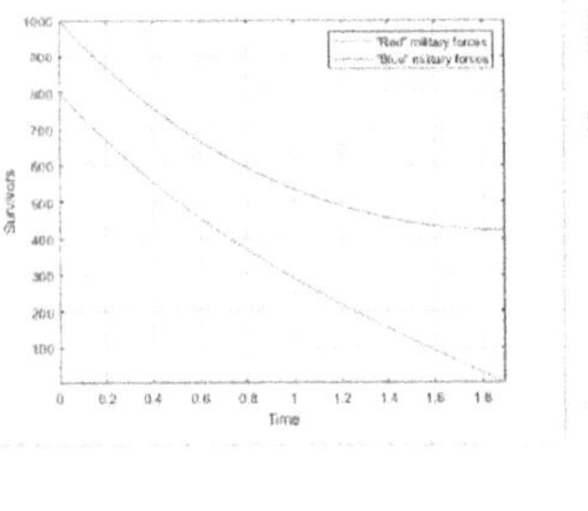

FIGURE 5.6 The ApExample_5_2 live script solving Lanchester's ODEs for aimed fire.

- The following two lines contain the find commands that determine the indices of the elements of the vector y(:,1) and y(:,2) containing values greater than zero;
- The if … else … end command finds a smaller index and assigns it to i_plot for further use in the plot command to show only positive survivors found;
- Further function lines content commands for plotting and formatting the resulting graph.

The calculation results show that the Blue forces are destroyed at time 1.8 despite the fact that the offensive firepower of the Blue soldiers is greater than that of the Red ones ($\beta>\alpha$).

5.5 APPLICATION EXAMPLES FOR BVP

5.5.1 Temperature Distribution in a Wire Heater

The steady-state ODE that governs the temperature T (in K) distribution in a long homogenous wire that is heated by the passage of electric current is:

$$\frac{d^2T}{dx^2}+\frac{E}{k}=0$$

where x is the wire transverse coordinate that changes along wire diameter from 0 to $2r_w$, m; r_w – is the wire radius, m; E – rate of heating, W/m^3; k – thermal conductivity coefficient, W/(m·K).

Problem: Develop a live script named ApExample_6_3 that uses the bvp4c command to calculate the temperature distribution in a wire heater with radius $r_w = 0.005$ m, $k = 15$ W/(m·K) and $E = 50 \cdot 10^6$ W/m^3. Consider the boundary conditions as follows: $T = 381$ K at $x = 0$ and $x = 2r_w$. Use the anonymous function with ODEs to be solved. The script should generate a graph $T(x)$ and display the T values at the points $x = 0, 2, 4, \ldots, 2r_w$ mm.

To solve the second-order equation of this problem, denote T as y and convert it into the two first-order equations

$$\frac{dy_1}{dx}=y_2 \quad \frac{dy_2}{dt}=-\frac{E}{k}$$

The live script function ApExample_5_3 with the resulting graph is shown in Figure 5.7

The live script ApExample_5_3 is designed as follows:

- The first three lines of the script assign values to the variables k and E;
- The next three lines containing the bvpinit command with the initial values y_1 and y_2 given at 5 points, the anonymous function ODE_fun with the two first-order ODEs, and the ode45 command that calls this and the boundary condition functions and gets the initial values;

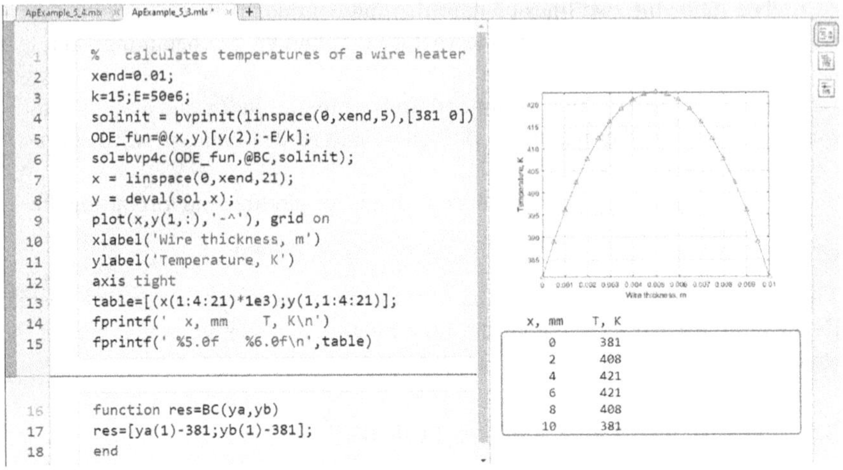

ApExample_5_4.mlx
ApExample_5_3.mlx *
% calculates temperatures of a wire heater
xend=0.01;
k=15;E=50e6;
solinit = bvpinit(linspace(0,xend,5),[381 0])
ODE_fun=@(x,y)[y(2);-E/k];
sol=bvp4c(ODE_fun,@BC,solinit);
x = linspace(0,xend,21);
y = deval(sol,x);
plot(x,y(1,:),'-^'), grid on
xlabel('Wire thickness, m')
ylabel('Temperature, K')
axis tight
table=[(x(1:4:21)*1e3);y(1,1:4:21)];
fprintf(' x, mm T, K\n')
fprintf(' %5.0f %6.0f\n',table)
function res=BC(ya,yb)
res=[ya(1)-381;yb(1)-381];
end
Temperature, K
Wire thickness, m
x, mm T, K
0 381
2 408
4 421
6 421
8 408
10 381

FIGURE 5.7 The ApExample_5_3 live script solving the heat equation for a wire heater.

- The following two lines contain commands that extract the y values at 20 x-points;
- Further four lines contain commands for generating and formatting the resulting graph;
- The table and fprintf commands provide a two-column titled table where x is converted from m to mm and with integer values of x and T;
- The user-defined BC function in the last part of the script represents the temperatures at the boundary a and b points that are given in the form required by the bvp4c solver.

5.5.2 Uniformly Loaded Beam Hinged at Both Ends

The dimensionless equation of a uniformly loaded Euler beam has form (based on Magrab et al., 2015)

$$\frac{d^4 w}{\partial x^4} = q(x)$$

where w is the transverse displacement, $0 \le x \le 1$ – coordinate, q is the load that does not change along the x (uniformly loaded beam) or change with x. All variables in this equation are considered dimensionless.

We consider a beam hinged at both ends with an external moment M_r applied to its right end. The boundary conditions for this case are as follows: the transverse displacement w and moment $\frac{d^2 w}{dx^2}$ are equal to zero for the left end of the beam, and w is equal to zero and the moment $\frac{d^2 w}{dx^2}$ is equal to M_r at its right end.

Problem: Compose a user-defined function named ApExample_5_4 that has no in/output and uses the bvp4c command to calculate the transverse displacement of a uniformly loaded, $q(x) = 1$, simple supported (hinged) beam for $y(0) = \frac{dw^2(0)}{dx^2} = 0$ (left end), and $y(1) = 0$ and $M_r = \frac{d^2w(1)}{\partial x^2} = 0.7$ (right end). Use the bvp4c command with the ODEs contained in the anonymous function; assume that initial values for displacement w, slope $\frac{dw}{dx}$, moment $\frac{d^2w}{dx^2}$, and shear force $\frac{d^3w}{\partial x^3}$ are 0.05 each. The script should display the w-values at the points $x = 0, 0.1, 0.2, \ldots, 1$ and generate the $w(x)$ graph with the displacement line four points wide.

To solve the above fourth-order ODE, designate w as y and transform it to the four first-order ODEs.

$$\frac{dy_1}{dx} = y_2$$

$$\frac{dy_2}{dx} = y_3$$

$$\frac{dy_3}{dx} = y_4$$

$$\frac{dy_4}{dx} = q$$

The required boundary conditions are: $y_1(0) = y_3(0) = 0$, $y_1(1) = 0$, and $y_3(1) - M_r = 0$.

The script ApExample_5_4 solving the problem are:

```
function ApExample_5_4
% Calculates displacement of the uniformly loaded simple
supported beam
% To run: >>ApExample_5_4
q0=1;
solinit=bvpinit(linspace(0,1,5),0.5*ones(1,4));
ODE_fun=@(x,y) [y(2),y(3),y(4),q0];
xy_sol=bvp4c(ODE_fun,@BC,solinit);
x = linspace(0,1,21);
y = deval(xy_sol,x);
plot(x,y(1,:),'LineWidth',4), grid on
xlabel('Coordinate, nondim')
ylabel('Displacement, nondim')
title({'Uniformly loaded beam';'hinged at both ends'})
axis tight
table=[(x(1:2:21));y(1,1:2:21)];
fprintf('x,nondim w,nondim\n')
fprintf(' %5.1f %7.4f\n',table)
function res=BC(ya,yb)
Mr=0.7;
res=[ya(1);ya(3);yb(1);yb(3)-Mr];
```

The ApExample_5_4 user-defined function is designed as follows:

- The first line defines the user-defined function ApExample_5_4;
- Two help lines following the first line are a short description of the function and the running command;
- The q value is assigned to the q0 variable in the fourth line of the script;
- In the next line, the initial values of the variables y_1, y_2, y_3, and y_4 are assigned at five equally-spaced x points each using the bvpinit command;
- The following two commands are, respectively, an anonymous function as the first input argument to the bvp4c command containing the ODEs, and the bvp4c command itself;
- Then 21 values of the x and y are obtained using the linspace and deval commands, respectively; these values are intended to be used in the plot command and in the resulting table;
- The next five lines include commands to graphical representation of the obtained displacement values; in accordance with the problem requirements the 'LineWidth' property is used in the plot command;
- Then, three commands are written to display the resulting table; the two fprintf commands create a table title and –column table with the obtained x and y values;
- Finally, the boundary condition function BC is written (used by the bvp4c command). The function contains the value of the applied moment

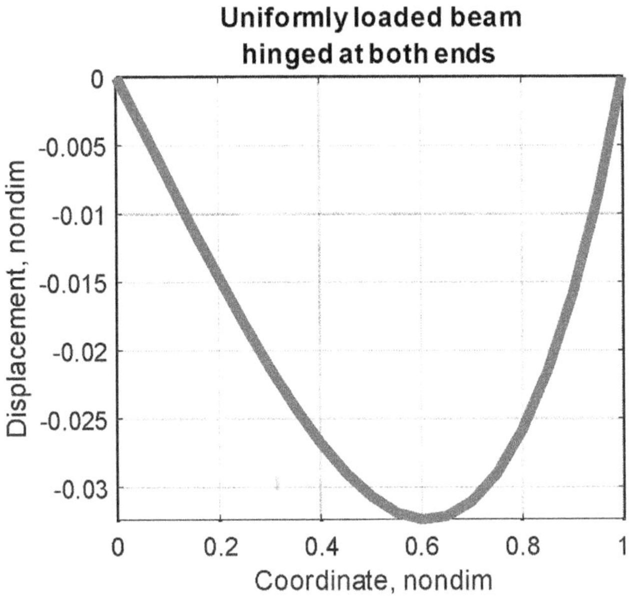

FIGURE 5.8 Displacement of a uniformly loaded simple supported beam with an external moment applied at the right end.

M_r and the vector res with four specified boundary conditions at the boundaries a (left) and b (right); the y_3 value at the boundary b is $y_b = M_r$, thus it is written as $y_b - M_r = 0$.

After entering the command >> ApExample_5_4, the following table and the resulting graph (Figure 5.8) appear:

```
>> ApExample_5_4
x,nondim w,nondim
0.0 0.0000
0.1-0.0075
0.2-0.0147
0.3-0.0213
0.4-0.0268
0.5-0.0307
0.6-0.0324
0.7-0.0311
0.8-0.0259
0.9-0.0159
1.0 0.0000
```

6 Partial Differential Equations and Programmatic Tool of the PDE Toolbox

6.1 INTRODUCTION

To solve two-dimensional (2D) and some three-dimensional (3D) partial differential equations and PDEs describing a real process and/or technical object, a special Partial Differential Equation Toolbox™ was designed and included in the MATLAB® software. The toolbox contains the programmatic and PDE Modeler tools. The numeric method used to solve PDEs, called the *finite elements method* (FEM), was developed in the 1950s to solve some problems in aeronautical engineering and has been extended to other areas of technology. This chapter presents the programmatic tool of the PDE Toolbox. A numerical solution in finite elements by the collocation method for a one-dimensional second-order PDE is demonstrated below. The minimum required set of PDE Toolbox commands for step-by-step solution of elliptic, parabolic, hyperbolic, and eigenvalue PDEs with Dirichlet and/or Neumann boundary conditions is presented. The use of these commands is explained by solving Poisson's equation (electrostatics) for a square plate with a rectangular slot and the eigenvalue PDE for a square membrane. The PDE programmatic tool is applied to solve such problems as transient heat transfer in a small aluminum plate, wave equation for a drumhead, and eigenvalue PDE for an elliptical membrane. The programs presented in this chapter are written primarily in the Live Editor for better clarity.

6.2 ABOUT NUMERICAL METHODS USED FOR SOLUTION OF THE PDEs

As stated above, the method applied by the PDE Toolbox for PDE solutions is called the *finite element method* (FEM). This method is more complicated than the finite difference method (Chapter 5) but more suitable for the real object geometry and for accounting for heterogeneous properties of materials. There are several schemes used to solve PDEs with finite elements, including collocation, least squares, Galerkin, and variational. The simplified PDE solution collocation scheme used by FEM is as follows: The body shape in the *x*,*y* plane, referred to as the domain, is divided into small triangles. For each of the triangles, a trial

DOI: 10.1201/9781003200352-6

second-order polynomial solution is assumed; the polynomial coefficients are determined from triangle nodes; then the polynomials with defined coefficients are used to calculate the solution values at the triangle nodes inside each of the triangles. This is a rough description of a 2D differential equation. To give the reader an idea of numerical finite element calculations, we present here a simple example with a point collocation scheme that solves a 1D second-order ordinal differential equation using firstly a one finite element and then two finite elements:

$$\frac{d^2y}{dx^2} = x + y$$

Assume the boundary conditions $y = 0$ at $x = 0$ and $x = 1$.

Consider a solution at four arbitrary x points, sometimes called collocation points. Since we have four points, we use a cubic polynomial to approximate the solution:

$$y = a_1 + a_2 x + a_3 x^2 + a_4 x^3$$

Here, for coefficients a, we use indexing starting at 1 (not 0) as is practice for matrices in MATLAB®. Thus,

$$\frac{dy}{dx} = a_2 + 2a_3 x + 3a_4 x^2$$

$$\frac{d^2y}{dx^2} = 2a_3 + 6a_4 x$$

and the solving differential equation takes the following algebraic form:

$$-a_1 - a_2 x + a_3\left(2 - x^2\right) + a_4(6x - x^3) = x$$

From the first and second boundary conditions, we have $a_1 = 0$ and $a_1 + a_2 + a_3 + a_4 = 0$.

Therefore, the equation reads:

$$-a_1 x + a_2\left(2 - x^2\right) + a_3(6x - x^3) = x$$

Substituting the two other collocation points (x = 0.3 and 0.6) in this equation and considering both boundary conditions, we have the following set of four equations:

$$a_1 = 0$$

$$-0.3a_2 + 1.91a_3 + 1.773a_4 = 0.3$$

$$-0.6a_2 + 1.64a_3 + 3.384a_4 = 0.6$$

$$a_1 + a_2 + a_3 + a_4 = 0$$

This yields the flowing matrix representation:

$$[10000-0.31.911.7730-0.61.643.3841111]*[a_1a_2a_3a_4]=[00.30.60\ ]$$

Therefore, to obtain the a_1,$a_{2,}$, a_3, and a_4 coefficients, the commands are as follows:

```
>>A = [-0.3 1.91 1.773;-0.6 1.64 3.384; 1 1 1];B =
[0.3;0.6;0]; a = A\B
a =
       0
-0.1455
-0.0117
 0.1572
```

Thereby, the polynomial solution of the problem is

$$y=0.1455x-0.0117x^2+0.1572x^3$$

This solution together with solution obtained using the bvp4c function (this function will be described in Chapter 8) is presented in Figure 6.1.

This solution used the entire x range as one element. Obviously, we can increase the tolerance by increasing the number of the elements inside each of which the third-order polynomial approximates the solution. Let's see how it is in our example. Assume two finite elements in the x range: the first elements in x range 0 … 0.5 while the second is in 0.5 … 1. Take the following arbitrary points inside the elements $x_{element\_1}$ = 0, 0.2, 0.3 and 0.5 and $x_{element\_2}$ = 0.5, 0.6, 0.8 and 1.

The approximate solution for each of two elements is:

$$y=\{a_1+a_2x+a_3x^2+a_4x^3,\quad 0\le x\le 0.5\ a_5+a_6x+a_7x^2+a_8x^3,\quad 0.5\le x\le 1$$

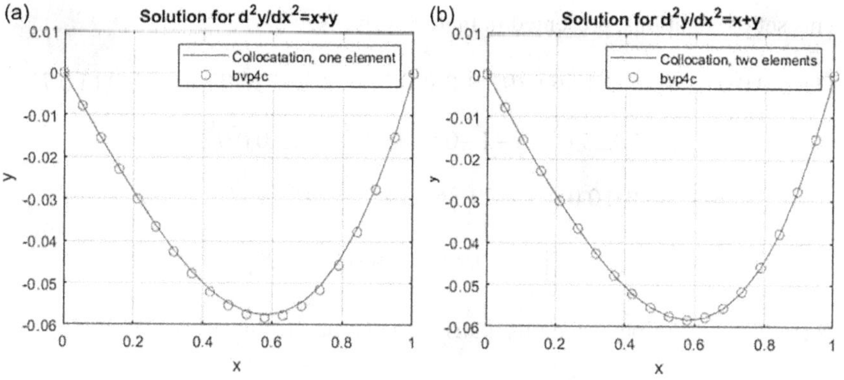

FIGURE 6.1 Solution of a differential equation by the collocation scheme of the FEM using one (a) and two (b) finite elements.

and

$$\frac{dy}{dx} = \{a_2 + 2a_3x + 3a_4x^2, \quad 0 \le x \le 0.5 \; a_6 + 2a_7x + 3a_8x^2, \quad 0.5 \le x \le 1$$

$$\frac{d^2y}{dx^2} = \{2a_3 + 6a_4x, \quad 0 \le x \le 0.5 \;\; 2a_7 + 6a_8x \;\; 0.5 \le x \le 1$$

The differential equation solution takes now the following forms:

$$\{-a_1 - a_2x + a_3\left(2 - x^2\right) + a_4(6x - x^3)$$

$$= x, \quad 0 \le x \le 0.5 \; -a_5 - a_6x + a_7\left(2 - x^2\right) + a_8(6x - x^3) = x, \quad 0.5 \le x \le 1$$

At point $x = 0.5$, in accordance to the continuity requirements: $\left(\frac{dy}{dx}\right)_{element\_1} = \left(\frac{dy}{dx}\right)_{element\_2}$ and $y_{\text{element\_1}} = y_{\text{element\_2}}$. Taken together, including the boundary conditions, the following set of equations results in:

$$a_1 = 0$$

$$-0.2a_2 + 1.96a_3 + 1.192a_4 = 0.2$$

$$-0.3a_2 + 1.91a_3 + 1.773a_4 = 0.3$$

$$a_2 + a_3 + 0.754a_4 - a_6 - a_7 - 0.75a_8 = 0$$

$$a_1 + 0.5a_2 + 0.25a_3 + 0.125a_4 - a_5 - 0.5a_6 - 0.25a_7 - 0125a_8 = 0$$

$$-0.6a_6 + 1.64a_7 + 3.384a_8 = 0.6$$

$$-0.8a_6 + 1.36a_7 + 4.288a_8 = 0.8$$

$$a_5 + a_6 + a_7 + a_8 = 0$$

This set of equations, presented in matrix form, is

$$[1 0 0 0 0 0 0 0 0 -0.2\ 1.96 1.192\ 0\ 0\ 0\ 0\ 0 -0.3\ 1.91\ 1.773\ 0 0 0 0 0 1 1 0.75 0$$

$$-1 -1 -0.75 1 0.5\ 0.25\ 0.125\ -1\ -0.5\ -0.25\ -0.125\ 0 0 0 0 0$$

$$-0.6\ 1.64\ 3.384\ 0 0 0 0 0\ -0.8\ 1.36\ 4.288\ 0 0 0 0 1 1 1 1]$$

$$*\left[a_1\ a_2\ a_3\ a_4\ a_5\ a_6\ a_7\ a_8\ \right] = [0\ 0.2\ 0.3\ 0\ 0\ 0.6\ 0.8\ 0\ ]$$

For a matrix solution, write this matrix in Command Window as:

```
>> A = [1 0 0 0 0 0 0 0
   0 -0.2 1.96 1.192 0 0 0 0
   0 -0.3 1.91 1.773 0 0 0 0
```

```
    0 1 1 0.75 0 -1 -1 -0.75
    1 0.5 0.25 0.125 -1 -0.5 -0.25 -0.125
    0 0 0 0 0 -0.6 1.64 3.384
    0 0 0 0 0 -0.8 1.36 4.288
    0 0 0 0 1 1 1 1];
>>B=[0 0.2 0.3 0 0 0.6 0.8 0]';a=A\B
aa =
          0
    -0.1486
    -0.0021
     0.1463
    -0.0036
    -0.1264
    -0.0483
     0.1783
```

Therefore, the solution of the differential equation is

$$y=\{-0.1486x-0.0021x^2+0.1463x^3,$$

$$0\leq x<0.5\ -0.0036-0.1264x-0.0483x^2+0.1783x^3,\quad 0.5\leq x\leq 1$$

The results of the calculation by these equations are presented in Figure 6.1b. As can be seen, the curve obtained by the two-element collocation scheme practically coincides with the points obtained using the bvp4c command (fourth-order polynomial collocation method).

The described method was used for a one-dimensional differential equation with the elements represented as linear pieces of the x line. In the case of two dimensions, the finite elements discretizing the 2D shape can be represented as triangles of different sizes; such discretization gives the best fit to the complex figures of actual engineering.

More detailed explanations of the FEM used for 1D, 2D, and 3D transient PDEs can be obtained in courses on numerical analysis and are beyond the scope of this book.

6.3 PDE TOOL AND TYPES OF DIFFERENTIAL EQUATIONS THAT CAN BE SOLVED

The Partial Differential Equation Toolbox™ included in the MATLAB® software provides a tool to solve various 2D and simple 3D problems in structural mechanics, heat transfer, linear static analysis, diffusion, and electro- and magneto-statics as well as general PDEs for many other engineering applications. The problems are solved by applying finite element analysis. There are two ways of solving second order PDEs: programmatically by special commands and with a PDE Modeler interface, which automatically generates programs that solve elliptic, parabolic, hyperbolic, and eigenvalue PDEs with Dirichlet or Neumann boundary conditions.

The general standardized form of solving PDEs that is adopted in the toolbox is

$$m\frac{\partial^2 u}{\partial t_2}+d\frac{\partial u}{\partial t}-\nabla(c\nabla u)+au=f$$

and for the eigenvalue PDEs

$$-\nabla(c\nabla u)+au=\lambda du \quad \text{or} \quad -\nabla(c\nabla u)+au=\lambda^2 mu$$

These equations can be solved with the Dirichlet or Neuman boundary conditions whose generalized forms are:

Dirichlet, are sometimes called essential boundary conditions:

$$hu=r$$

Neumann, sometimes called natural boundary conditions:

$$\vec{n}\,(c\nabla u)+qu=g$$

In the presented equations:

u is the generalized name of dependent variables that can be, for example, the temperature *T* or any other variable used in a specific PDE;

t is the independent time variable;

$\nabla=\frac{\partial}{\partial x}+\frac{\partial}{\partial y}$ – denotes an operator termed *nabla*; note $\nabla(\nabla)=\nabla^2=\frac{\partial^2}{\partial x^2}+\frac{\partial^2}{\partial y^2}$ = is called Laplacian; in vector form, the Laplacian is u = *div*(*grad* u);

x and *y* are independent variables represented by the coordinates;

m, *d*, *c*, *f*, *a*, *λ*, *h*, *r*, *q*, and *g* are the coefficients that can be constant or vary depending on *x*, *y*, *t*, and *u*;

$\vec{n}$ is the outward unit normal vector used to indicate the normal direction of the flux *du*/*d*x and *du*/*dy* in respect to the boundary surface;

div – divergence, differential operator that can be defined as *div*(**F**) = ∇·*F* where **F** is a vector field with Cartesian components $\mathbf{F}_x$, $\mathbf{F}_y$ and $\mathbf{F}_z$;

grad – gradient, operator that is some equivalent to nabla that can be written as *grad*(f) = ∇f.

The above general PDE forms are written for the second-order PDEs categorized as elliptic, parabolic, hyperbolic, and eigenvalue.

Elliptic equations have the following forms:

$$-\nabla(c\nabla u)+au=f$$

parabolic:

$$d\frac{\partial u}{\partial t} - \nabla(c\nabla u) + au = f$$

hyperbolic:

$$m\frac{\partial^2 u}{\partial t_2} - \nabla(c\nabla u) + au = f$$

and the eigenvalue equations presented above in two possible form.

The above PDE types and boundary conditions (BC) are presented in Tables 6.1 and 6.2. The tables show the specific equation type and coefficients that bring each PDE or BC type to the corresponding standard form. There are also examples showing how to convert actual PDE to the PDE Toolbox standard form.

Note that the system of equations can also be solved with the PDE Toolbox; for this purpose, each of the PDEs and accompanying BCs should be presented in the standard form.

These notations will be further clarified when the actual PDEs are solved using the programmatic tool or PDE Modeler (Chapter 7).

6.4 PROGRAMMATIC TOOL OF THE PDE TOOLBOX

The PDE programmatic tool as well as Modeler uses the new data type termed *object*. This kind of data is similar in use to a structure (see Section 5.3.1) and has the fields referred to in that case as properties, but not all tasks that can be executed with a structure can be done with an object, which is defined by *method/s*. These concepts are beyond the scope of this book. We only explain how to use the object/s when it is necessary.

6.4.1 Solution Steps and Commands for Their Implementation

To solve a PDE using the programmatic tool, the following steps should be executed:

1. Creation of a PDE model object;
2. Creation of 2D geometry from the command line;
3. Specification of boundary conditions and initial conditions, the latter only for time-dependent PDEs.
4. Presenting the solving PDE in a standardized form and specifying its coefficients;
5. Creation of a mesh of triangles;
6. Solution of the PDE;
7. Presentation of the solution in graphical and/or numerical form.

Here we describe the most commonly used commands for each of these steps followed by a step-by-step description of the solutions for elliptic and eigenvalue PDEs.

6.4.1.1 Command for Step 1

To create a model as a MATLAB object, the createpde command should be used in the form

```
model=createpde(N)
```

where N is the number of PDEs for which the model object is created; in case of a single equation, N can be omitted: model=createpde.

The command creates a PDE model with the specified properties that define the body geometry, PDE to be solved, boundary conditions, mesh od triangles, and solution options, for example:

```
>> model=createpde
model =
 PDEModel with properties:
           PDESystemSize: 1
         IsTimeDependent: 0
                Geometry: []
    EquationCoefficients: []
      BoundaryConditions: []
       InitialConditions: []
                    Mesh: []
           SolverOptions: [1×1 pde.PDESolverOptions]
```

Unfilled, empty properties (designated as []) should be filled using the commands entered in the subsequent steps.

6.4.1.2 Commands for Step 2

The PDE Toolbox uses Constructive Solid Geometry (CSG) modeling to describe the geometry of real objects In the CSG system, the geometry must be represented in three matrices: (a) basic shapes, generally denoted as the variable gd; (b) the name/s of basic shape/s denoted as the variable ns; and (c) the formula describing the interactions of the basic shapes denoted as variable sf.

a. Four basic shapes are used: circle, polygon, rectangle, and ellipse. The corresponding matrix contains size-equal columns for each shape that describe the body geometry; the following values should be entered into the column in the order they are written for basic form:

   For a **circle**: digit 1 (denotes a circle), the x-coordinate of the center, the y-coordinate of the center, the radius;

   For a **polygon**: digit 2 (denotes a polygon), the number n of segments, n-1 x-coordinates of the polygon edges, n-1 y-coordinates of the polygon edges;

   For a **rectangle**: digit 3 (denotes a rectangle), digit 4 (number of segments), four x-coordinates of the rectangle edges, four y-coordinates of the rectangle edges;

For an **ellipse**: digit 4 (denotes an ellipse), the x-coordinate of the ellipse center, y-coordinate of the ellipse center, the first semiaxis length (positive), the second semiaxis length (positive), angle (in radian) from x to the first semiaxis.

For example, a centered square 2 × 2 (dimensionless units) can be presented as R=[3 4 −1 −1 1 1 −1 1 1 −1]', while a circle with radius 0.4 (dimensionless) and center at (0,0) point can be represented as C=[1 0 0 0.4]'.

Because of the different length of columns with the basic shapes, the short columns should have zeros added to the longest column. So, for the above example, we have C=[C;zeros(length(R)-length(C),1)].

All shapes should be combined in the one matrix, for example, gd=[R,C].

b. Each basic shape must be named, with these names presented in size-equal rows that can be performed with the char command. The result must be transposed. For example, the commands ns=char('R','C') and ns=ns' produce the ns matrix of names to be used in the formula describing the resulting body geometry.
c. A formula set should be presented as a string containing the basic shape names and their interactions denoted with + for union, * for intersection, − for difference, and parentheses for grouping. So, continuing the example, the formula set sf='R −C' specifies a rectangle with a circled hole in the center.

Once the gd, ns, and sf matrices have been formed, the geometry can be finally created with the decsg command, the simplest form of which is:

```
dl = decsg(gd,ns,sf)
```

where all input arguments are clarified above and dl is the resulting geometry matrix that simplifies (decomposes) the geometry matrix gd considering formula set sf with the names ns. This constitutes the minimal geometry region and allows the commands on other solution steps to work with the created geometry. To include the geometry in the model and plot it with boundary and body face names, the geometryFromEdges and pdegplot commands can be used:

```
geometryFromEdges(model,g)
pdegplot(model, 'Name1',Value1, 'Name2', Value2,…)
```

where model and g are the above model object name and matrix of the decomposed geometry, respectively, and Name1, Value1, Name2, Value2,… are property-value pairs, which to label the edge and/or face must read 'EdgeLabels','on' and/or 'FaceLabels','on', respectively. Thus, the commands for our example are geometryFromEdges(model,g) and pdegplot(model,'EdgeLabels','on','Face Labels','on').

6.4.1.3 Command for Step 3

To describe the boundary conditions, the applyBoundaryCondition command is used. The simplest form of the command is:

```
applyBoundaryCondition(model,'BC_type','Edge',Edge_Number,
'Name1',Value1,…)
```

where model is as above, 'BC_type' is the 'dirichlet', 'neumann' or 'mixed' boundary conditions, 'Name1',Value1… are the variables of the specified boundary type (see Table 6.2). For example, these pairs for Neumann boundary conditions along the edge 3 are 'g',0 and 'q',0 with the command reading applyBoundaryCondition(model,'neumann','Edge',3,'g',0,'q',0).

Note that for the Dirichlet boundaries of the form u=0 (zero condition), the pair 'u',0 can be used instead of 'r',0 and 'h',1.

In case of time-dependent boundary conditions, the anonymous function can be used in the ValueN place.

In case of time-dependent PDEs, we need to specify the initial conditions for each body face. For these purposes, the setInitialConditions command is used:

```
setInitialConditions(model,u0,ut0,'Face',Face_number)
```

where the u0 and ut0 are the initial u and time-derivative du/dt values for the domain denoted by the 'Face',Face_number pair; For example, the setInitialConditions(model, 0, 0, 'Face',1) command specifies u=0 and du/dt=0 along the face 1 of the body.

Note:

- For a PDE containing only a first-order time derivative (as in a parabolic PDE), the ut0 parameter can be omitted;
- If one of the coefficients of the boundary and initial conditions is a function of coordinates, the coefficient should be represented as an anonymous or user-defined function with this function having location and state arguments. In the coefficient expression, we must use location.x and/or location.y instead of x and/or y, and state.u for current value of the solution u (see the next step). For example, if $ut0 = x/2$, then ut0=@(location,state)location.x/2.

6.4.1.4 Command for Step 4

At this step, the coefficients m, d, c, a, and f of the PDE converted to the standard form (Table 6.1) must be entered into the model. To do this, the specifyCoefficients command is used:

```
specifyCoefficients(model,'Coefficient1_name',Coefficient1_
value, …'face1_name',face1_value,…)
```

where model is as above,

TABLE 6.1
Various PDE Types and Their Correspondence to the Standard Form

PDE Type	Coefficients to Match the Standard Form	Example: Specific PDE to Be Solved	Example: Variables and Coefficients to Match the PDE Standard Form
Elliptic equation $-\nabla(c\nabla u)+au=f$	m=0, d=0	$-\nabla(k\nabla T)=s$	$u=$T, m=0, $a=0$, f=s, $c=k$, $d=0$
Parabolic equation $d\frac{\partial u}{\partial t}-\nabla(c\nabla u)+au=f$	m=0	$\frac{\partial T}{\partial t}=\nabla(k\nabla T)$+s	$u=$T, m=0, $a=0$, f=s, $c=k$, $d=1$
Hyperbolic equation $m\frac{\partial^2 u}{\partial t_2}-\nabla(c\nabla u)+au=f$	d=0	$d\frac{\partial^2 P}{\partial t_2}=C^2\nabla^2 P$	$u=P$, $c=C^2$, $a=0$, f=0, $d=0$
Eigenvalue equation $-\nabla(c\nabla u)+au=\lambda du$ or $-\nabla(c\nabla u)+au=\lambda^2 mu$	Matches the standard form of the eigenvalue equation	-$\nabla(\nabla u)=\lambda u$	$m=0$ (since $\lambda^2 mu$ –is absent), $c=1$, $a=0$, $d=1$, λ –in the specified range

'Coefficient1_name',Coefficient1_value, … are the property pairs with the name of the coefficient and its value, e.g., 'm',0 or 'd',0;

'face1_name',face1_value, … are the region name and region number pairs, e.g., 'face1',1.

Thus, a command can look like:

```
specifyCoefficients(model,'m',0,'d',0,'c',0.5,'a',0,'f',1,'
Face',1).
```

Note, as in the case with the coefficients of the boundary and initial conditions, if one of the PDE coefficients is a function of coordinates, then a function must have location and state argument. Thus, in the coefficient expression, we must use location.x and/or location.y instead of x and/or y. For example, if $f = u^2$, f=@(location,state)state.u.^2.

6.4.1.5 Commands for Step 5

Now the triangular mesh should be generated and entered in the model object for which the generateMesh command should be used. The two simplest forms of this command are:

```
generateMesh(model)    and    generateMesh(model,
'Name',Value,…)
```

where 'Name',Value, … – are the argument pairs, of which the most frequently used is the maximum mesh length 'Hmax' whose Value is usually 0.25 but is adjustable. Additional argument pairs and other forms of this command can be obtained by entering >> doc generateMesh in the Command Window.

To view the body with the generated mesh, the following command may be used:

```
pdeplot(model,'NodeLabels','on')
```

where the 'NodeLabels','on' property pair is used only when we want to see the numbers of the triangular mesh nodes.

6.4.1.6 Commands for Step 6

After completing the previous steps, the commands that solve the PDEs should be used. For a stationary PDE, the following command can be used:

```
results = solvepde(model)
```

For a time-dependent (but not eigenvalue) PDE:

```
results = solvepde(model,tlist)
```

For an eigenvalue PDE:

```
results = solvepdeeig(model,lambda_range)
```

In these commands, model and results are the model and result object, respectively; tlist is timed at which the defined values of the solution should be returned in the result object, for example, 0:15; lambda_range is the vector entering the eigenvalue range (λ in the eigenvalue PDE), for example, [-inf;100].

The numerical results for stationary and time-dependent PDEs are placed in the NodalSolution vector of the result structure. For an eigenvalue PDE, the defined eigenvalues and eigenvector values are located in the Eigenvalues and Eigenvector vectors of the results structure.

6.4.1.7 Commands for Step 7

To represent results in a 3D graphical form, an extended version of the pdeplot command (see step 5) is used:

```
pdeplot(model,'XYData',results.
NodalSolutions,'ZData',results.NodalSolution,
'ColorMap','hot')
```

where the pairs 'XYData', results. NodalSolutions and 'ZData', results. NodalSolution is defined z results of the solution at the x,y coordinates of the mesh nodes; the 'ColorMap', 'hot' pair is used when we want to change the default map of colors (the 'jet' and other color values can be used).

In addition, the 'Contour', 'on' pair can be used to draw contour lines on a 2D or 3D plot. Additional possible pairs can be obtained by typing >>doc pdeplot in the Command Window.

To display the results of the stationary problem solution at the desired points (not nodes), the interpolateSolution command is used.

```
uintrp = interpolateSolution(results,x,y)
```

where results are clarified at Step 5 while x and y are the coordinates at which we want to obtain the interpolated numbers; uintrp is the column vector containing the interpolated values; the reshape command can be used to convert the uintrp vector into a 2D matrix.

In the case of a time-dependent solution, this command has the form:

```
uintrp = interpolateSolution(results,x,y,1:length(tlist))
```

where tlist is the time vector and has the same sense as in the solvepde command.

In the case of an eigenvalue PDE, the command takes the form:

```
uintrp = interpolateSolution(results,x,y,Lambda_value)
```

where Lambda_value is the eigenvalue λ at which we want to obtain the interpolated uintrp values.

6.4.2 Elliptic PDE, Programmatic Solution, Example

The Poisson equation in the form typical for electrostatics has the form $-\nabla(\varepsilon\nabla V) = \rho$ where V is the potential and ε and ρ denote the dielectric constant and the charge density, respectively.

Problem: Write a live script with the name ExProgrammingPDE that solves the above PDE (elliptic) for a 2×2 square with a 0.4×0.8 centered rectangular slot at all boundaries of the square $V = 0$ and at all boundaries of the slot $V = 1$. Assume $\varepsilon = 0.5$ and $\rho = 1$. Represent the result in a 3D plot and in a table for a rectangular grid with x and y coordinates equal to –2:.0.25:2. For simplicity, all of the above values are assumed to be dimensionless.

To solve the problem, follow the above steps.

1. The first command of the PDE solution is model=createpde, which creates the model object.
2. Now we need to create and plot the geometry of the object as described in step 2 above. This step can be performed with the following commands:

```
R1 = [3,4,-1,-1,1,1,-1,1,1,-1]';                    % square
R2 = [3,4,-.2,-.2,.2,.2,-.8,.8,.8,-.8]'; % rectangular slot
gd = [R1,R2];                              % geometry matrix
sf = 'R1-R2';                                  % formula set
ns = char('R1','R2');                % geometric shape names
ns = ns';
g=decsg(gd,sf,ns);                      % decomposed geometry
geometryFromEdges(model,g);               % geometry to model
pdegplot(model,'EdgeLabels','on','FaceLabels','on')  %
plotting
axis equal
```

The resulting geometry is presented in Figure 6.2.
3. According to the conditions of the problem, the boundaries are the Dirichlet type on the outer and inner edges of the body. V notation is u for standardized Dirichlet boundaries (see Table 6.2). As can be seen from Figure 6.2, the outer boundaries are numbered 1, 2, 6, and 7 while the inner boundaries are numbered 3, 4, 5, and 8. |Accordingly, the commands that specify the boundary conditions in our model are

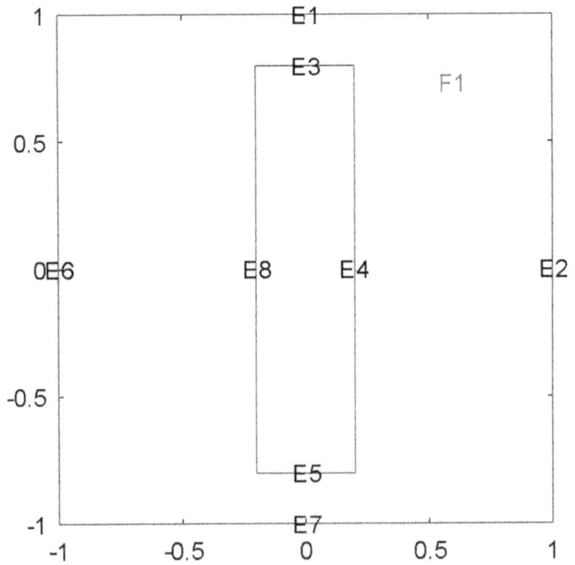

FIGURE 6.2 Square with a rectangular slot at its center; the origin is located in the center of the square; the boundaries are designated with the E1 … E8 labels while the face is labeled as F1.

TABLE 6.2
Various BC Types and Their Correspondence to the Standard Form

BC Type	Coefficients: To Match the Standard Form	Example: Specific BC	Example: Variables and Coefficients to Match the BC Standard Form
Dirichlet boundary $hu = r$	The *h*, *u*, and *r* values should be specified	T=20	u=T, h=1, r=20
Neumann boundary $\vec{n}\,(c\nabla u) + qu = g$	The *du/dx*, *du/dy* derivatives and *c*, *q*, *g*, and *u* value (at a boundary) should be specified	$\frac{dT}{dx} = 0$	$u = T$, $c = 1$ (this is the same c as in the standardized PDEs) $q = 0$, g=0

```
applyBoundaryCondition(model,'dirichlet','Edge',[1,2,6,
7],'h',1, 'r',0);
applyBoundaryCondition(model,'dirichlet','Edge',[3,4,5,
8],'h',1, 'r',1);
```

4. To specify the coefficients, we need to compare our PDE with the standardized PDE to determine the coefficient values. In standard notations $u = V$ and the coefficients are $m = 0$, $d = 0$, $c = 0.5$, $a = 0$, and $f = 1$. Therefore, the command that sets the PDE coefficient looks like

```
specifyCoefficients(model,'m',0,'d',0,'c',0.5,'a',0,'f',
1,'Face',1);
```

5. Following the fifth step above, generate a mesh with triangular elements and draw the mesh graph. This can be done with the following commands:

```
generateMesh(model);
pdeplot(model)
axis([-1 1 -1 1])
```

   The axis command has been added here to constrain the axes to the body shape. The generated mesh plot is shown in Figure 6.3.
6. Now we can solve the PDE with the command

```
res = solvepde(model)
```

7. The defined u values and mesh nodes are contained in the res.NodalSolution matrix.
8. As described in step seven above, the resulted graph and required table can be obtained using the following commands:

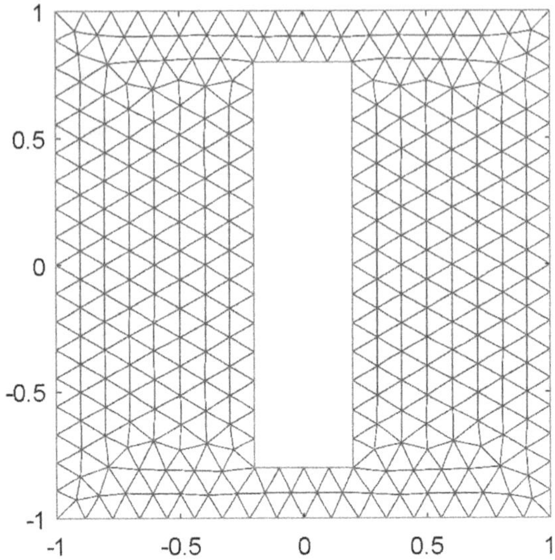

FIGURE 6.3 Generated triangular mesh.

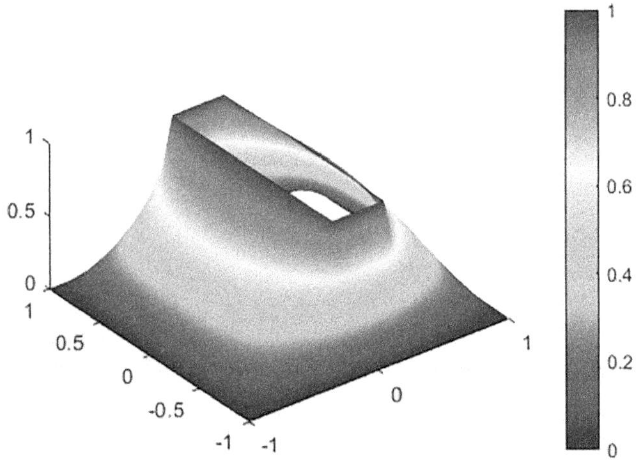

FIGURE 6.4 Solution of Poisson's equation (electric potential) for a square with a rectangular slot.

```
pdeplot(model,'XYData',res.NodalSolution,'ZData',res.
NodalSolution,'ColorMap','jet')
axis equal
[x,y]=meshgrid(-1:0.25:1);
uintrp=interpolateSolution(res,x,y);
uintrp = reshape(uintrp,size(x))
```

The resulting plot is represented in Figure 6.4.

ExProgrammingPDE.mlx | uintrp = 9×9

	1	2	3	4	5	6	7	8	9
1	0.0000	0.0000	-0.0000	-0.0000	-0.0000	0.0000	-0.0000	-0.0000	0.0000
2	-0.0000	0.2294	0.4438	0.8448	NaN	0.8448	0.4437	0.2294	0.0000
3	0.0000	0.3539	0.6529	0.9456	NaN	0.9456	0.6529	0.3539	-0.0000
4	0.0000	0.4075	0.7254	0.9642	NaN	0.9642	0.7254	0.4075	-0.0000
5	-0.0000	0.4216	0.7428	0.9681	NaN	0.9681	0.7428	0.4216	-0.0000
6	0.0000	0.4075	0.7254	0.9642	NaN	0.9642	0.7254	0.4075	0.0000
7	0.0000	0.3539	0.6529	0.9456	NaN	0.9456	0.6529	0.3539	-0.0000
8	0.0000	0.2294	0.4438	0.8448	NaN	0.8448	0.4438	0.2294	-0.0000
9	0	0	0.0000	0.0000	0.0000	0	0.0000	-0.0000	0

FIGURE 6.5 *V* values interpolated to a rectangular grid; NaN symbols refer to the rectangular slot.

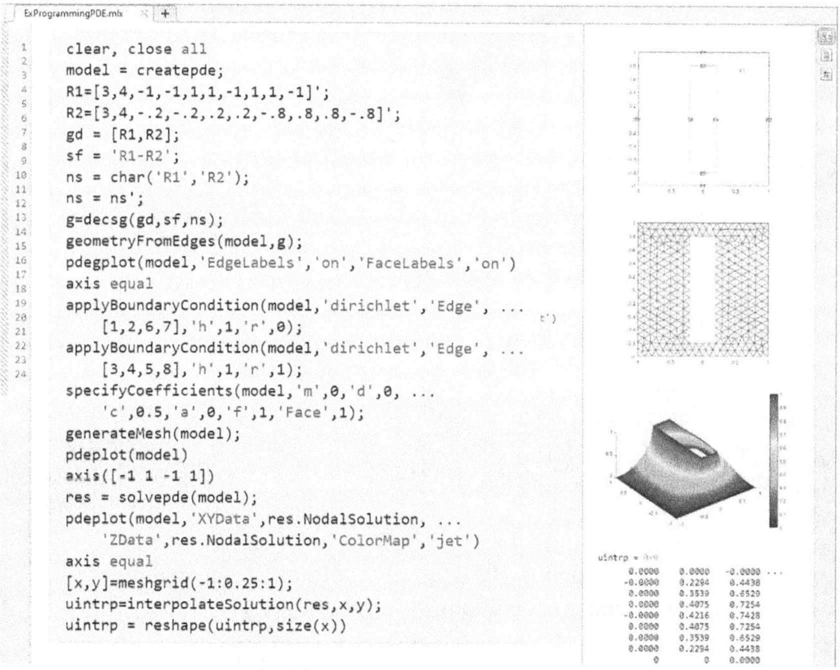
ExProgrammingPDE.mlx

```
clear, close all
model = createpde;
R1=[3,4,-1,-1,1,1,-1,1,1,-1]';
R2=[3,4,-.2,-.2,.2,.2,-.8,.8,.8,-.8]';
gd = [R1,R2];
sf = 'R1-R2';
ns = char('R1','R2');
ns = ns';
g=decsg(gd,sf,ns);
geometryFromEdges(model,g);
pdegplot(model,'EdgeLabels','on','FaceLabels','on')
axis equal
applyBoundaryCondition(model,'dirichlet','Edge', ...
    [1,2,6,7],'h',1,'r',0);
applyBoundaryCondition(model,'dirichlet','Edge', ...
    [3,4,5,8],'h',1,'r',1);
specifyCoefficients(model,'m',0,'d',0, ...
    'c',0.5,'a',0,'f',1,'Face',1);
generateMesh(model);
pdeplot(model)
axis([-1 1 -1 1])
res = solvepde(model);
pdeplot(model,'XYData',res.NodalSolution, ...
    'ZData',res.NodalSolution,'ColorMap','jet')
axis equal
[x,y]=meshgrid(-1:0.25:1);
uintrp=interpolateSolution(res,x,y);
uintrp = reshape(uintrp,size(x))
```

FIGURE 6.6 Live script with the commands and results of solving Poisson's equation for electric potential.

The resulting table is shown in Figure 6.5.

The complete live script with the resulting graphs and table is presented in Figure 6.6.

6.4.3 Eigenvalue PDE with a Programmatic Solution Example

The eigenvalue equation in a form typical for oscillations is $-\Delta u = \lambda u$ where u is the displacement and λ is the eigenvalue.

Problem: Write a live script named ExProgrammingEigenvalue that solves the above PDE for a 2×2 square membrane. The membrane is clamped at the top, bottom, and right edges, u = 0, while the derivative $du/dx = 0$ is given on the left edge. Consider a solution for the eigenvalues in the range $-\infty\ldots15$. Present results as 3D plots for the last two eigenvalues. All of the above values are assumed to be dimensionless.

To solve this problem, we repeated the steps described in previous example with elliptic PDE.

1. Enter the model=createpde command, which creates the model as a PDE object.
2. Now we create the geometry of our object and draw it with the following commands:

```
model = createpde;
R1=[3,4,-1,-1,1,1,-1,1,1,-1]';                        % square
gd = [R1];                                   % geometry matrix
sf = 'R1';                                       % formula set
ns = char('R1');                                 % square name
ns = ns';
g=decsg(gd,sf,ns);                         % decomposed geometry
geometryFromEdges(model,g);                  % geometry to model
pdegplot(model,'EdgeLabels','on','FaceLabels','on') %
plotting
axis equal
```

The resulting geometry is shown in Figure 6.7.

3. According to the problem conditions, we have the Dirichlet boundaries, u=0, on the top, bottom, and right boundaries – numbers 2, 3, and 4 (Figure 6.7) and the Neumann boundary, du/dx=0 on the left boundary – number 1 (Figure 6.7). For standardized Dirichlet boundaries (Table 6.2), we have h = 1 and r = 0 while for standardized Neumann boundary – g = 0 and q = 0. Bearing this in mind, the commands that specify the boundary conditions in our model are as follows:

```
applyBoundaryCondition(model,'dirichlet','Edge',2:4,
'h',1, 'r',0);
applyBoundaryCondition(model,'neumann','Edge',[1],'g',
0,'q',0);
```

4. To specify the coefficients, we need to compare our PDE with the standard form of the eigenvalue PDE (see Table 6.1) and determine the

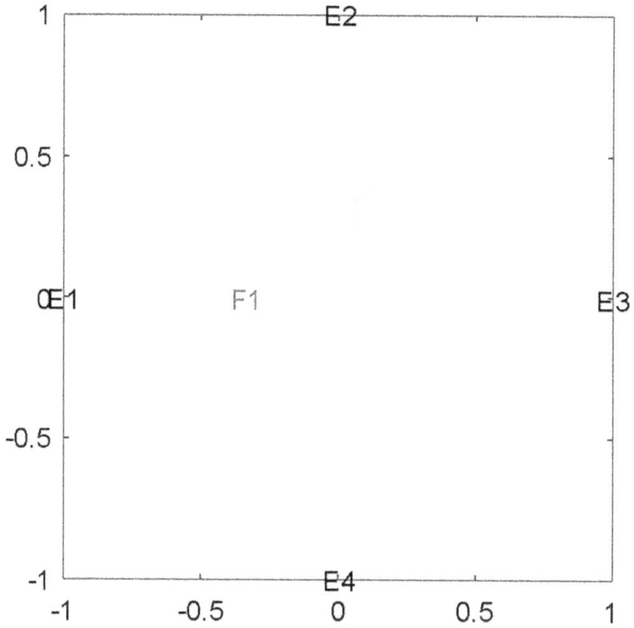

FIGURE 6.7 Square membrane shape, origin in the center of the square; the boundaries are designated by the E1 … E4 labels while the face is labeled by F1.

values of the coefficients. In standard notations, the coefficients are m = 0, d = 1, c = 1, a = 0, and f = 0. Therefore, the command that specifies the PDE coefficients is

```
specifyCoefficients(model,'m',0,'d',1,'c',1,'a',0,'f',0);
```

5. Now generate a mesh with triangular elements and build the mesh plot. This can be done with the following commands:

```
generateMesh(model,'Hmax',0.25);
pdeplot(model)
axis([-1 1 -1 1])
```

Here we used the 'Hmax',0.25 pair to reduce the number of triangles. The axis command has been added to constrain the axes to the membrane square. The generated mesh plot is shown in Figure 6.8.

6. We can solve now the eigenvalue PDE using the solvepdeeig command:

```
res = solvepdeeig(model,[0,15])
l = res.Eigenvalues
u = res.Eigenvectors;
```

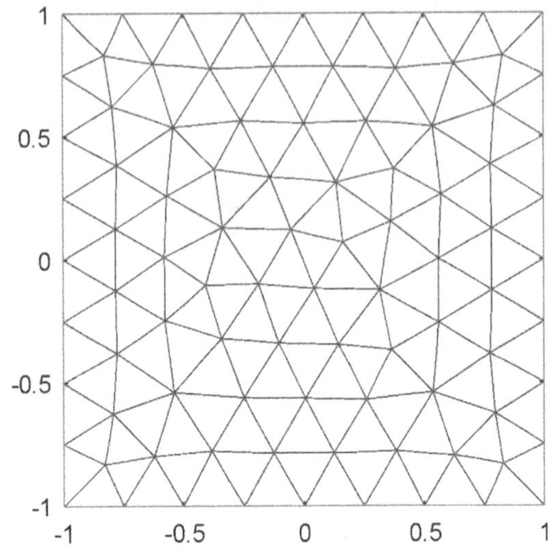

FIGURE 6.8 Generated triangular mesh.

The two last commands assign the defined eigenvalues and membrane displacements to the l and u variables, respectively. The defined values are originally located in the Eigenvalues matrix and Eigenvector vector of the res object. The *u* values defined for the mesh nodes are contained together with the node coordinates in the NodalSolution matrix of the res object.

7. As described in Step 7 above, the required two graphs for the last two eigenvalues can be obtained using the following commands:

```
pdeplot(model,'XYData',u(:,length(l)-1),...
    'ZData',u(:,length(l)-1),'ColorMap','jet')
title(['\lambda=',num2str(l(end-1))])
pdeplot(model,'XYData',u(:,length(l)),...
    'ZData',u(:,length(l)),'ColorMap','jet')
title(['\lambda=',num2str(l(end))])
```

The resulted graphs are presented in Figure 6.9 a and b.

The complete live script along with the defined λ values and generated graphs is presented in Figure 6.10.

In this problem, only three eigenvalues (see the l vector in the right part of Figure 6.10) were defined in the required range –inf …15.

6.5 APPLICATION EXAMPLES

For the purposes of the book, the following are applications for some simple geometric shapes.

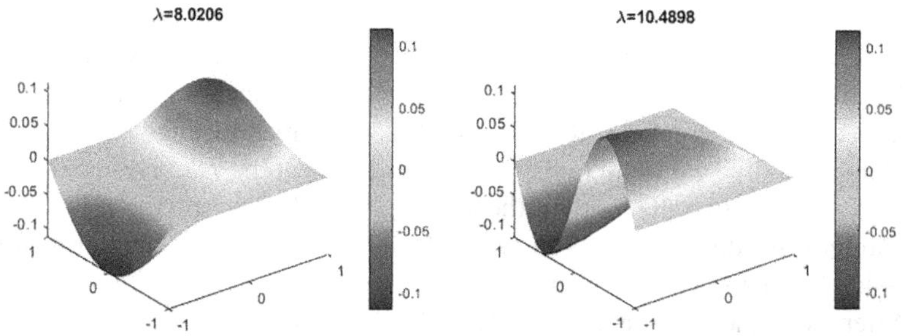

FIGURE 6.9 Solution of the eigenvalue PDE for square membrane for two of the found eigenvalues: (a) $\lambda = 8.0206$ and (b) $\lambda = 10.4898$.

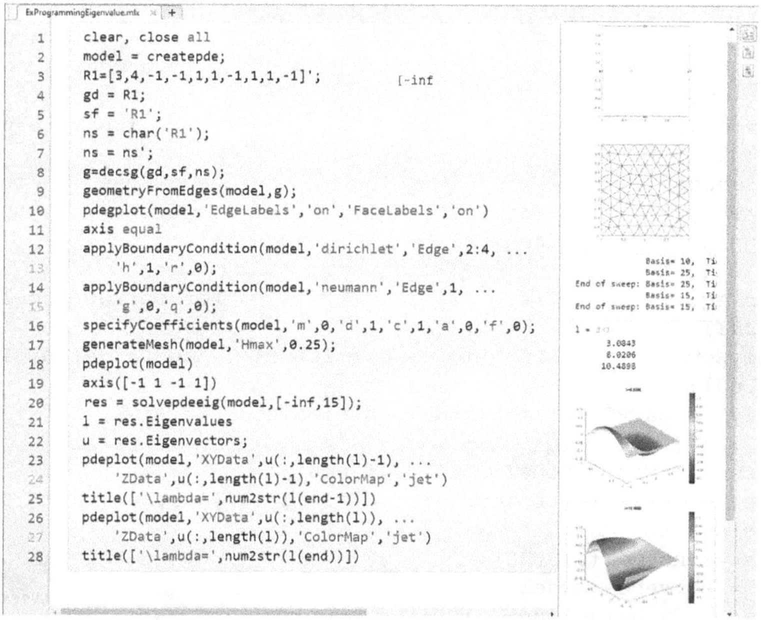

FIGURE 6.10 Live script with the commands and results for solving the eigenvalue PDE.

6.5.1 Heating a Small Metallic Plate: Transient Heat Transfer

The 2D PDE for the unsteady heat transfer has the form:

$$\frac{\partial T}{\partial t} = \alpha \Delta T$$

where T – temperature, K; t – time, sec; α – coefficient of diffusivity, mm²/s; $\Delta T = \dfrac{\partial^2 T}{\partial x^2} + \dfrac{\partial^2 T}{\partial y^2}$ and x and y are coordinates, mm.

Problem: Write a live script with the name ApExample_6_1 that solves the above equation for a small rectangular 8 × 6 mm aluminum plate with a diffusivity 9.9 mm²/s. The left edge of the rectangle has a temperature of 333 K while the rest of the edges are 283K. The initial plate temperature is 283K. Solve the problem at 20 time values in range 0 … 5 s and present the *T*(*x*,*y*) results for times 0 and 5 s on the two 3D plots in the same figure. Use the mirror view. Display the numeric results for $t = 5$ s in the table according to a rectangular grid with $x = -4:2:4$ mm and $y = -3:3$ mm.

After using the “Export to Word …” option of the Live Editor (located in the Save popup menu list), the ApExample_6_1 a live script solving this problem looks like:

```
clear, close all
model = createpde;
R1 = [3,4,-4,-4,4,4,-3,3,3,-3]';                              % square
gd = R1;                                              % geometry matrix
sf = 'R1';                                                % formula set
ns = char('R1');                                          % square name
ns = ns';
g = decsg(gd,sf,ns);                            % decomposed geometry
geometryFromEdges(model,g);                       % geometry to model
pdegplot(model,'EdgeLabels','on','FaceLabels','on')
% plotting
axis equal                                      % for correct scaling
u0=283;                                                           % K
applyBoundaryCondition(model,'dirichlet','Edge',2:4,'h',1,
'r',u0);
applyBoundaryCondition(model,'dirichlet','Edge',1, 'h',1,
'r',333);
setInitialConditions(model,u0,'Face',1);
specifyCoefficients(model,'m',0,'d',1,'c',9.9,'a',0,'f',0);
generateMesh(model,'Hmax',0.25);
tlist=linspace(0,5,20);
res = solvepde(model,tlist);
u=res.NodalSolution;
figure
subplot(2,1,1)
pdeplot(model,'XYData',u(:,5),'Zdata',u(:,1),'Colormap',
'jet');
view(37.5,30),title('t=0')
subplot(2,1,2)
pdeplot(model,'XYData',u(:,5),'Zdata',u(:,20),'Colormap','
jet');
view(37.5,30),title('t=5 s')
[x,y]=meshgrid(-4:2:4,-3:3);
uintrp=interpolateSolution(res,x,y,tlist(end));
Temperature = reshape(uintrp,size(x))
```

The commands operate as follows:

- The first two commands clear the workspace, close all Figure windows, and generate a model object;
- The next nine commands enter the coordinates of the rectangle edges, form the rectangle with its name, simplify the geometry of the model, and draw a rectangle with boundaries and facial labels;
- The boundary and initial conditions were specified by the commands in lines 12 … 15. The Dirichlet boundary conditions h=1 and r=283 are specified for the second, third, and fourth rectangle boundaries while, for the first boundary, the Dirichlet condition was specified as $h = 1$ and $r = 333$. The setInitialCondition command was used with the u0 starting temperature value but without the starting temperature change rate ut0 since the heat equation does not have the $\frac{\partial^2 T}{\partial t^2}$ part (corresponds to $d = 0$ in standard PDE form – Table 6.1);
- The sixteenth command specifies the coefficients of the heat transfer equation; comparing our PDE with the standard PDE, we set the following coefficients: $m = 0$, $d = 1$, $c = 9.9$, $a = 0$, and $f = 0$ (see Table 6.1);
- A triangular mesh was created with a maximum triangle edge length of 0.25 – line 17;
- The solution was obtained for the required time values, with the results assigned to variable u by the commands in lines 18 … 20. The solvepde command is used in a non-stationary form with the tlist vector containing the required time values;
- The commands in lines 21 … 27 provide a graphical representation of the results for times 0 and 5 s on two 3D plots of the same figure; the view commands provide a mirror view of the $T(x,y)$ surfaces;
- The numerical data for the required rectangular grid are displayed for the final time by the commands on lines 28 … 30.

The rectangle geometry, temperature distribution plots, and temperature table that appear after running the live script are presented in Figure 6.11a–c.

6.5.2 Drumhead Vibrations

The wave equation describing a drumhead vibration reads

$$\frac{\partial^2 u}{\partial t^2} = c^2 \Delta u$$

where u is the displacement, t – time, c – velocity of wave propagation; $\Delta u = \frac{\partial^2 u}{\partial x^2} + \frac{\partial^2 u}{\partial y^2}$, x and y are the coordinates.

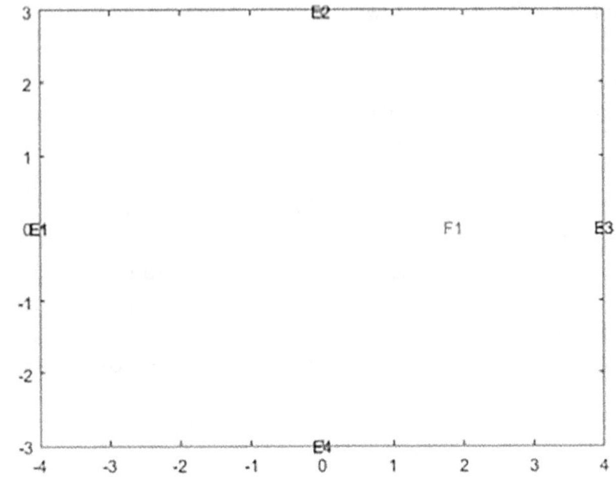

a)

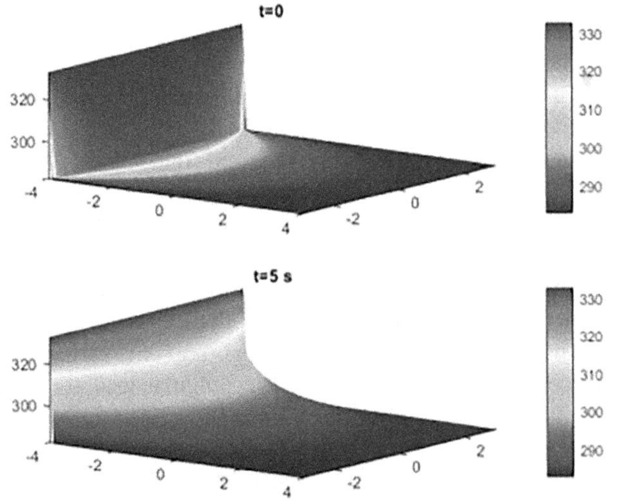

b)

Temperature = 7×5

333.0000	283.0000	283.0000	283.0000	283.0000
333.0000	295.0231	286.8036	284.1439	283.0000
333.0000	302.1312	289.5194	284.9785	283.0000
333.0000	304.2980	290.4889	285.2828	283.0000
333.0000	302.1311	289.5194	284.9785	283.0000
333.0000	295.0232	286.8036	284.1439	283.0000
333.0000	283.0000	283.0000	283.0000	283.0000

c)

FIGURE 6.11 Rectangle with labeled boundaries and face (a), initial temperature distribution and the temperature distributions at $t = 5$ s (b), and final temperature table (c).

Problem:

Write a live script with the name ApExample_6_2 that calculates displacements u, mm, of a circular drumhead with a radius of 0.3 m. Consider coefficient c = 1.73 mm/s and assume the initial velocity applied to the drumhead is $ut0 = x/0.2$. Solve the problem for 51 time-values in the range 0 … 10 s and present the $u(x,y)$ results for times 3, 6, 9, and 10 s on the four 3D plots in the same figure.

An ApExample_6_2 live script solving this problem (obtained with the "Export to Word …" option of the Live Editor) is

```
clear, close all
model = createpde;
r=0.3;
C1=[1,0,0,r]';                                          % circle
gd = C1;                                       % geometry matrix
sf = 'C1';                                         % formula set
ns = char('C1');                                   % circle name
ns = ns';
g=decsg(gd,sf,ns);                         % decomposed geometry
geometryFromEdges(model,g);                  % geometry to model
pdegplot(model,'EdgeLabels','on','FaceLabels','on') % plotting
axis equal                                 % for correct scaling
u0=0;
applyBoundaryCondition(model,'dirichlet','Edge',1:4,'h',1,
'r',0);
ut0=@(location)location.x/0.2;
setInitialConditions(model,u0,ut0,'Face',1);
specifyCoefficients(model,'m',1,'d',0,'c',1.73^2,'a',0,
'f',0);
generateMesh(model,'Hmax',0.05);
tlist=linspace(0,10,51);
res = solvepde(model,tlist);
u=res.NodalSolution;
figure
subplot(2,2,1)
pdeplot(model,'XYData',u(:,16),'Zdata',u(:,16),...
    'Colormap','hot','ColorBar','off');
title(['t=',num2str(tlist(16))])
subplot(2,2,2)
pdeplot(model,'XYData',u(:,31),'Zdata',u(:,31),...
    'Colormap','jet','ColorBar','off');
title(['t=',num2str(tlist(31))])
subplot(2,2,3)
pdeplot(model,'XYData',u(:,46),'Zdata',u(:,46),...
    'Colormap','hot','ColorBar','off');
title(['t=',num2str(tlist(46))])
subplot(2,2,4)
pdeplot(model,'XYData',u(:,51),'Zdata',u(:,51),...
    'Colormap','jet','ColorBar','off');
title(['t=',num2str(tlist(51))])
```

The commands perform the following operations:

- The first two commands clear the workspace, close all Figure windows, and generate a model object;
- The next ten commands enter the circle parameters, form the circle with its name, simplify the model geometry, and draw a circle with the boundaries and facial labels;
- The boundary and initial conditions were specified by the commands in lines 13 … 16; the starting velocity applied to the drumhead was specified using the ut0 anonymous function with the location argument and using the location.x variable as x in the function expression (required by the setInitialConditions commands);
- The coefficients of the wave equation were converted to the standardized PDE form and specified using the command on line 17 as follows: $m = 1$, $d = 0$, $c = 1.73^2$, $a = 0$, and $f = 0$;
- A triangular mesh was created with a maximum triangle edge length of 0.05 – line 18;
- A solution was obtained for the required time values; the results are assigned to the variable u by the commands in lines 18 … 21. The solvepde command is used in a non-stationary form with the tlist vector containing the required time values;
- The graphical representation of the results for times 3, 6, 9, and 10 s is shown in four 3D plots at the same figure using the commands in lines 21 … 34.

The drumhead plot with the boundaries and facial labels and four vibration shapes at times 3, 6, 9, and 10 s that appear after running the live script are presented in Figure 6.12a and b.

6.5.3 Elliptic Membrane Eigenvalue Problem

Problem: Consider an eigenvalue PDE $-\Delta u = \lambda u$ for an elliptic membrane with semi-axes 0.4 and 0.6 and angle $\pi/2$ between the main axis of the membrane and the x axis; u = 0 at all membrane boundaries. Compose a script that solves the problem in the 0 …150 λ-range, displays the defined eigenvalues, and plots in three different figure windows: (a) the membrane geometry, (b) mesh plot, and (c) 2D resulting plots with contour lines for the second and the last of the defined eigenvalues, both in the third Window.

A ApExample_6_3 script that solves this problem looks like this:

```
clear, close all
model = createpde;
r1=0.6;r2=0.4;
Ellipse1=[4,0,0,r1,r2,0]';                                   % circle
gd = Ellipse1;                                       % geometry matrix
```

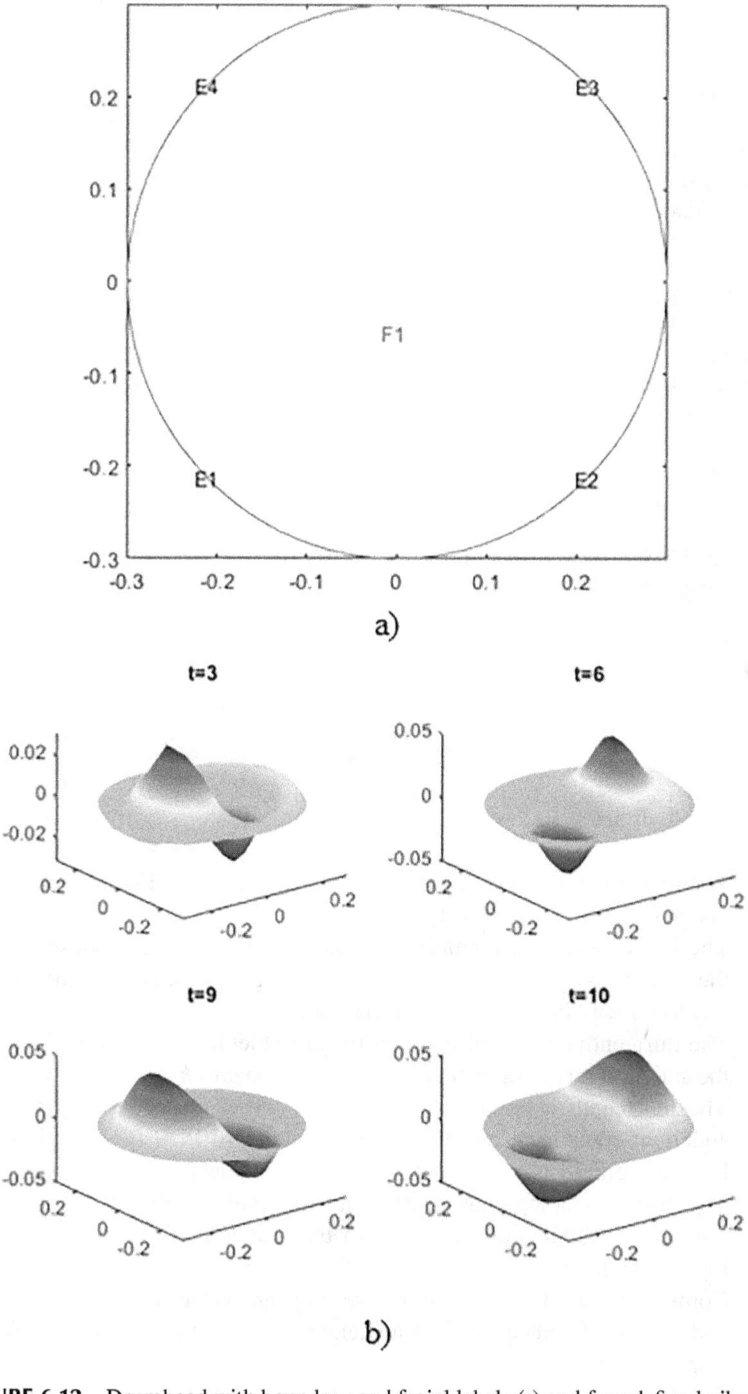

FIGURE 6.12 Drumhead with boundary and facial labels (a) and four defined vibration shapes at $t=3$, 6, 9, and 10 s (b).

```
sf = 'Ellipse1';                                    % formula set
ns = char('Ellipse1');                              % circle name
ns = ns';
g=decsg(gd,sf,ns);                         % decomposed geometry
geometryFromEdges(model,g);                 % geometry to model
pdegplot(model,'EdgeLabels','on','FaceLabels','on') %
plotting
axis equal                                 % for correct scaling
applyBoundaryCondition(model,'dirichlet','Edge',1:4,'h',1,
'r',0);
specifyCoefficients(model,'m',0,'d',1,'c',1,'a',0,'f',0);
generateMesh(model,'Hmax',0.1);
figure
pdeplot(model)
axis([-r2 r2 -r1 r1])
res = solvepdeeig(model,[0,150]);
l = res.Eigenvalues
u = res.Eigenvectors;
figure
subplot(2,1,1)
pdeplot(model,'XYData',u(:,2),'contour','on','ColorMap','
jet')
title(['\lambda=',num2str(l(1))])
subplot(2,1,2)
pdeplot(model,'XYData',u(:,length(l)),...
'contour','on','ColorMap','jet')
title(['\lambda=',num2str(l(end))])
```

The commands act as follows:

- The first two commands clear the workspace, close all Figure windows, and generate a model object;
- The ten following commands enter the parameters of the ellipse, form the ellipse and its name, simplify the geometry of the model, and draw an ellipse with boundaries and facial labels;
- The thirteenth command specifies the Dirichlet boundary conditions in the standard form (Table 6.2). In our case, it means $h = 1$ and $r = 0$;
- The fourteenth command specifies the coefficients of the membrane equation; comparing our PDE with the standard PDE, we set the following coefficients: $m = 0$, $d = 1$, $c = 1$, $a = 0$, and $f = 0$ (see Table 6.1);
- The next four commands generate a mesh with a value of 0.1 for the maximum triangle edge and draw a triangulated ellipse in a separate Figure window;
- Commands 19–21 solve the model setting eigenvalues in range 0 … 150 and assign defined eigenvalues and eigenvectors to the l and u variables, respectively;
- The last commands draw two 2D requiring plots in a separate Figure window.

After running this program, we receive the eigenvalue vector together with some statistics (outputted automatically) in the Command window with the required plots appearing in three Figure windows – Figure 6.13a–c.

```
>> ApExample_6_3
                Basis= 10,  Time=  0.08,  New conv eig=  1
                Basis= 24,  Time=  0.13,  New conv eig=  5
                Basis= 38,  Time=  0.17,  New conv eig= 11
End of sweep:   Basis= 38,  Time=  0.17,  New conv eig= 11
                Basis= 21,  Time=  0.22,  New conv eig=  0
End of sweep:   Basis= 21,  Time=  0.22,  New conv eig=  0
l =
   25.9887
   53.1430
   78.5891
   92.2762
  117.7581
  143.6076
```

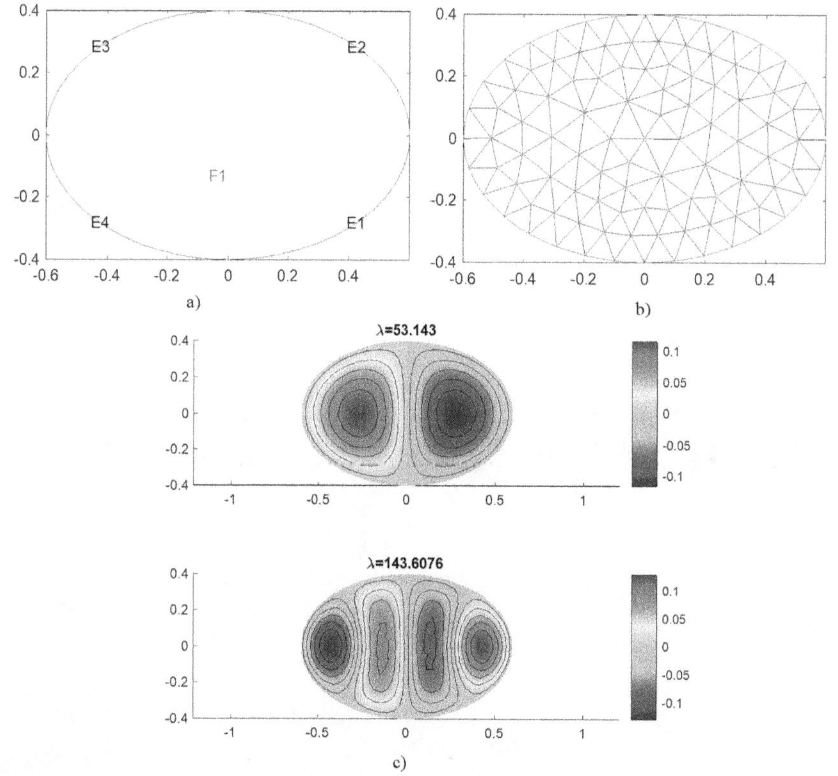

FIGURE 6.13 Graphical output of the ExApplication_6_3 script: model geometry (a), triangulated ellipse (b), and the resulting +2D graphs for the second and the last eigenvalues (c).

7 Solving Two-Dimensional Partial Differential Equations with PDE Modeler

7.1 INTRODUCTION

The commands of the Partial Differential Equations Toolbox™ described in the preceding chapter belong to the programmatic tool. Additionally, this toolbox has a graphical interface that allows the user to solve two-dimensional PDEs interactively without writing commands. This can be performed with a special interface called PDE Modeler (formerly known as PDE Tool)[1]. This chapter describes the steps to be applied to solve a 2D PDE and, for some engineering examples and applications, presents solutions to various PDEs under various boundary and initial conditions. Among the considered problems:

- Poisson type equation adopted for the axial velocity of a laminar flow in a horizontal pipe;
- Reynolds equation describing hydrodynamic lubrication between surfaces covered with hemispherical pores;
- the unsteady thermal conduction equation with material property coefficient changes with temperature;
- PDE Modeler application "Structural Mechanics, Plane Stress" used to determine stresses acting in a thin plate with an elliptical hole;
- the eigenvalue equation adopted to the oscillations of the T-shaped membrane.

7.2 PDE TOOLBOX INTERFACE

The PDE Modeler represents an interface for solving elliptic, parabolic, hyperbolic, and eigenvalue types of PDEs with Dirichlet or Neumann boundary conditions. The general standard forms of the PDE types and BCs that can be solved are the same as described in the preceding chapter (Section 6.2 and Tables 6.1,6.2).

[1] Some text and table materials from Burstein, 2021a (Sections 8.2, 8.4) are used in the chapter; with permissions from IGI Global.

DOI: 10.1201/9781003200352-7

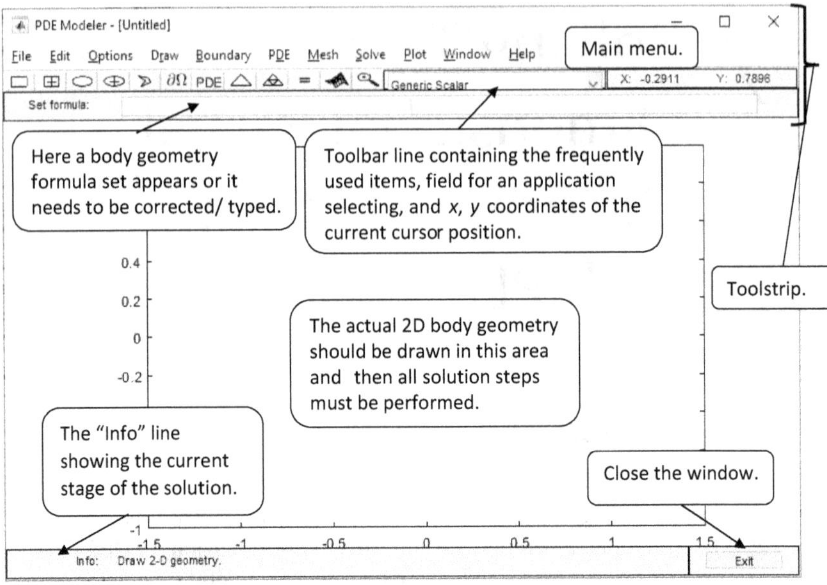

FIGURE 7.1 The PDE Modeler window with coordinate axes bounding place where the solution steps will be performed.

To launch the PDE Modeler interface, enter the pdeModeler command in the Command Window:

```
>> pdeModeler
```

After entering this command, the PDE Modeler window appears – Figure 7.1. The toolstrip of this window contains the menu, toolbar line with the buttons presenting the frequently used options, and the dialog field line where the formula set describing the required body shape should be entered/corrected. The large empty field bounded by the x and y coordinate axes with ticks is the place where the 2D geometry of the actual body must be drawn at the first step of the PDE solution. The menu of this window contains a line with the File, Edit, Options, Draw, Boundary, PDE, Mesh, Solve, Plot, Window, and Help options, each of which opens a popup menu with a list of additional options. Below, we describe all the necessary toolstrip options along with the steps for a PDE solution.

7.2.1 Solution Steps in the PDE Modeler

To represent the solution steps with PDE Modeler, we solve the 2D Poisson equation that describes, for example, the axial velocity of laminar flow in a horizontal pipe:

$$-\Delta u = f$$

The equation is solved with Dirichlet BCs, $u = 0$, at the inner wall of the pipe $Y^2 + Z^2 = 1$.

In these equations: $\Delta = \frac{\partial^2 u}{\partial Y^2} + \frac{\partial^2 u}{\partial Z^2}$; u – axial fluid velocity in the Y and Z directions, m/s; $Y = y/r$, $Z = z/r$ dimensionless coordinates that change in the range −1 … 1; y and z are dimensional Cartesian coordinates of the pipe cross-section; $f = \frac{r^2}{\mu}\frac{dP}{dX}$ where the pressure gradient $\frac{dP}{dx}$, the dynamic viscosity of the fluid μ, and pipe radius r are constants. In the equations being solved, u and f are dimensional, while X, Y, and the pipe radius are dimensionless.

Problem: Solve the above PDE assuming $\mu = 0.87$ N·s/m², $\frac{dP}{dX} = 0.9 \cdot 10^8$ Pa/m, and $r = 8$ mm, accordingly $f = (0.008^2/0.87) \cdot 0.9 \cdot 10^8 =$ m/s. The coordinate center (0,0) is placed at the pipe cross-section center; dimensionless diameter is equal to 1. Present results in 2D and 3D plots. Save automatically generated program in file named Example_Modeler.

Step 1. At this step, the applied model mode corresponds to the problem being solved. For this, the appropriate Application line should be marked in the popup menu of the Options button located on the main menu – Figure 7.2.

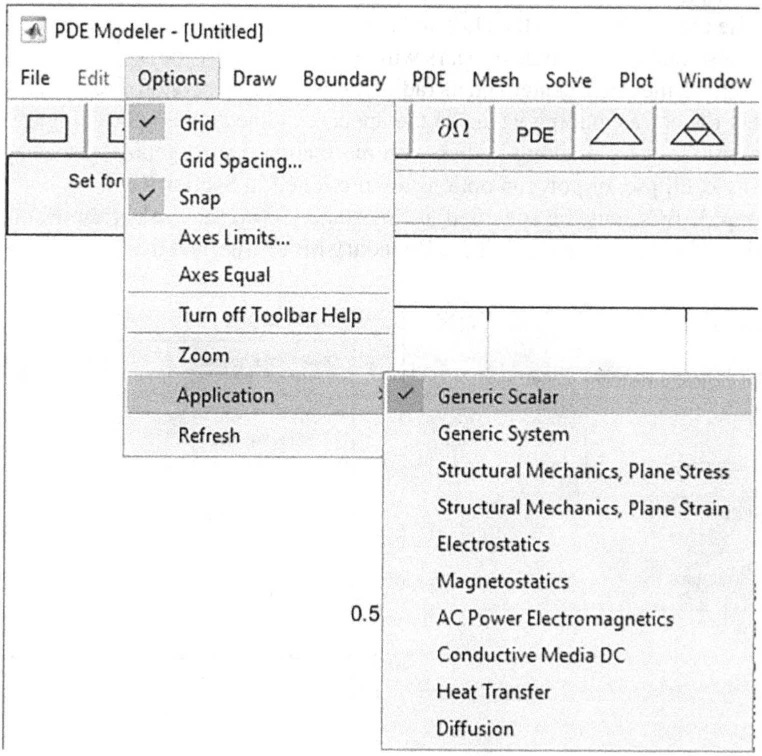

FIGURE 7.2 The first step in the PDE solution: selecting the Application option.

As can be seen, there is no flow dynamics option. In this case, for further solution, the Generic Scalar options (default) should be chosen.

Step 2. At this stage, the 2D body geometry should be drawn. To do this, the Grid and Snap options (see Figure 7.2) must be marked. Then the limits of the axes must be settled correspondingly to the sizes of the studied body. To perform this, click the Axes Limits or Grid Spacing option. In the appeared dialog boxes, type the needed values in the appropriate fields. For our pipe with a cross-section diameter of 1, the x and y limits of −1.25 and 1.25 were entered (Figure 7.3a). The default grid spacing was selected for x-axis and 0.5 for y-axis (Figure 7.3b). Additionally, the Axes equal line (see Figure 7.2) was clicked to generate the non-deformed circle view at the drawing and other steps. The selected limits and gridlines appear in the graph field of the PDE Modeler window.

After this, we can start to draw the necessary geometry. In general, the drawing is done by combining shapes, which can be selected from the popup menu of the Draw menu button (see Figure 7.4).

The same shapes – rectangle, square, ellipse, circle, or polygon – can be selected as it was presented in the programmatic tool (Section 6.2). But here we can just click the appropriate button in the line of frequently used items. Thus, to draw the circle (pipe cross-section shape), click the Ellipse/Circle (centered) button, press then the mouse + pointer at the (0,0) point, and drag the mouse pointer with the pressed button to the (1,0) and then to the (0,1) grid points; after releasing the mouse button, the circle appears with its name (Figure 7.5).

To check the circle center and its radius, click on the generated circle and check/correct the corresponding values in the appeared Object Dialog box (Figure 7.6).

Instructions for drawing bodies with more complicated geometries using the rectangle, ellipse, or polygon options are presented in Section 7.3.4.

Step 3. BCs must be specified at this stage. To do this, select the Boundary button in the main menu, select the Boundary Mode line, and then mark the Show

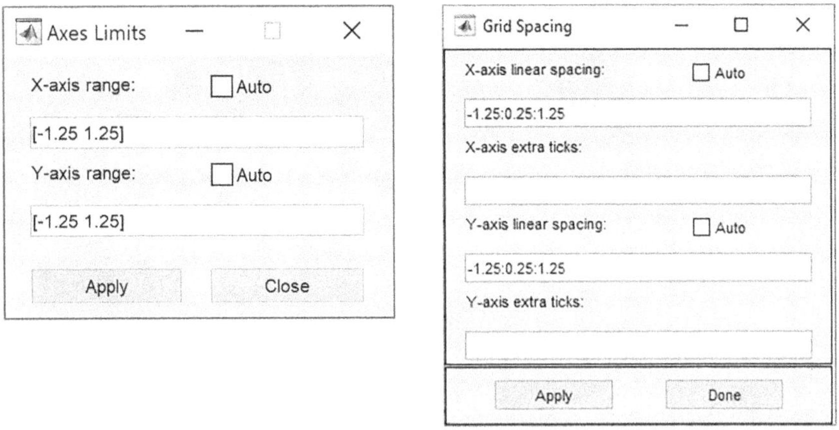

FIGURE 7.3 The Axes Limits (a) and Grid Spacing (b) dialog boxes completed according to the problem.

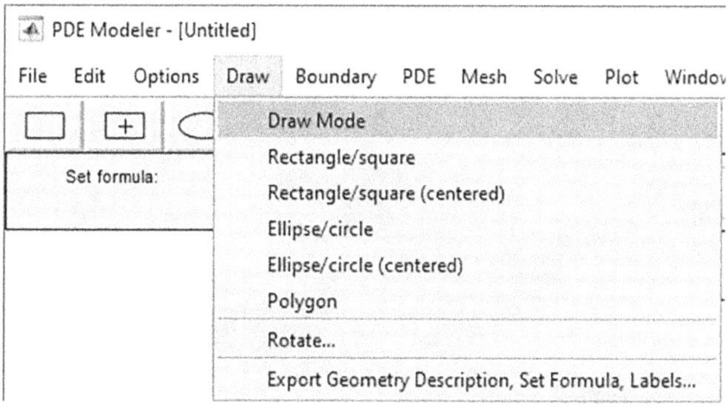

FIGURE 7.4 The second step in the PDE solution: selecting the Draw Mode.

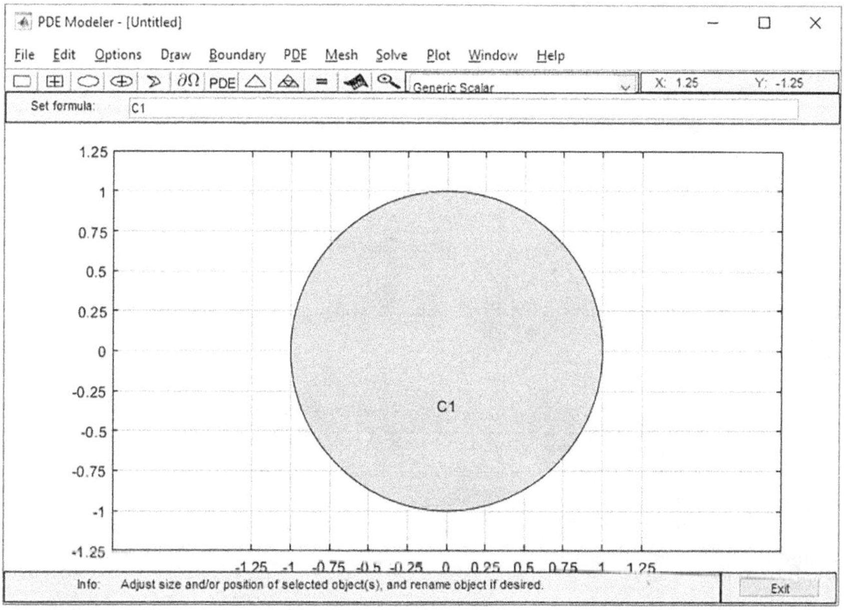

FIGURE 7.5 A circle (cross-section of a pipe) drawn in the PDE Modeler window.

Edge Labels line to see the segments of the circle into which it was automatically divided and where the BC should be inputted. BC can be entered by double click on the mouse button when the mouse pointer is placed on the segment. The Boundary Condition box is opened (Figure 7.7). In this panel, the corresponding type of the BC should be marked and the necessary boundary equation parameters should be entered in the appropriate fields. In our case, the default BC is the Dirichlet condition therefore the following coefficients should be entered: $h = 1$ and $r = 1$ at each of circle boundaries (Figure 7.8).

This procedure must be performed for each segment of the drawn body.

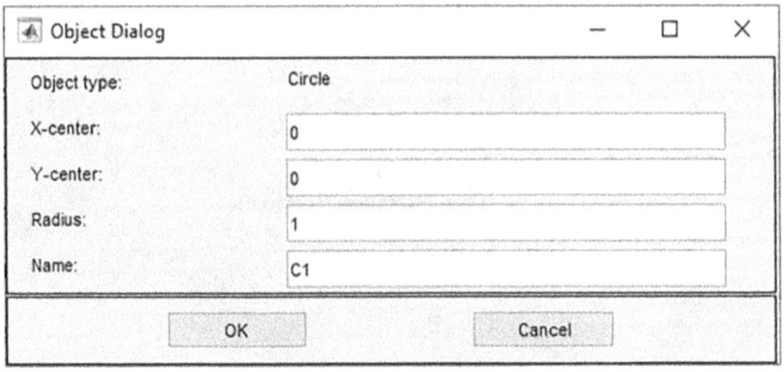

FIGURE 7.6 Object Dialog box for checking/correcting the coordinates of the circle center, its radius, and name.

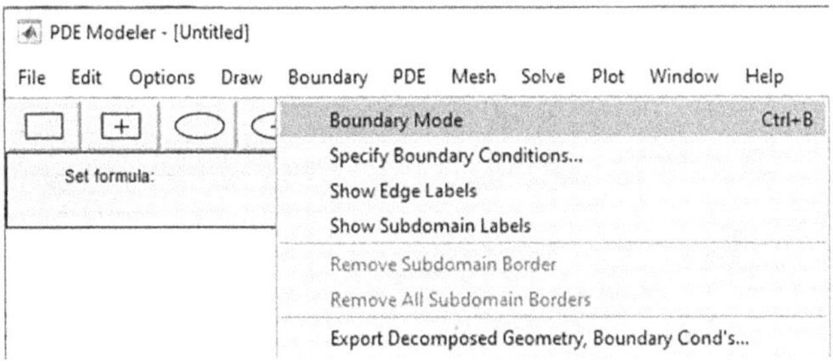

FIGURE 7.7 The third step in the PDE solution: setting the Boundary Mode.

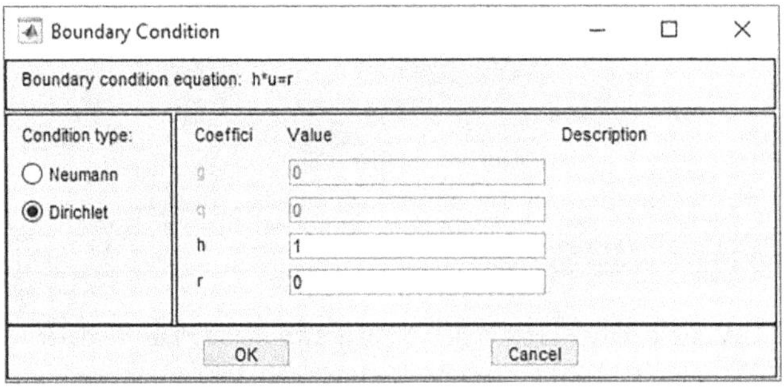

FIGURE 7.8 Boundary Condition dialog box filled for Dirichlet type boundary for one circle segment of the example problem.

Note that *r* here is the standard notation adopted by the PDE Modeler for the Dirichlet boundary type (and not the pipe radius).

Dirichlet boundaries are set by default and do not need to be inputted; however, it is recommended to check them for each segment, especially in the case of complicated geometry, and also to avoid confusion due to inaccurate user actions during the solution steps.

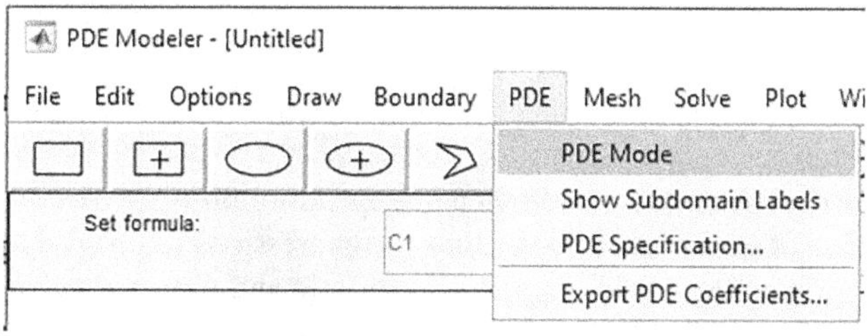

FIGURE 7.9 The fourth step in the PDE solution: selecting the PDE Mode.

<u>Step 4.</u> Now we must specify the PDE. Select for this the PDE Mode line of the PDE menu button and then the "PDE Specification …" line (Figure 7.9). Alternatively, after signing the PDE Mode, just click within the drawn shape (circle, in our case).

After the PDE Specification dialog panel appears (Figure 7.9), we should determine the type of PDE to be solved. Comparison of the equation being solve with the standard equation (Table 6.1) shows that in our case we have an elliptic equation, and thus the Elliptic option (default) of the dialog panel should be marked. Further, to enter PDE coefficients into the appropriate fields of this panel, we need to determine them, firstly, by comparing our equation and the equation written at the top of the panel. The equation Equation: -div(c*grad(u))+a*u=f appearing in the panel is written in a general vector form via the divergence, *div*, and gradient, *grad*; for scalar case, it can be rewritten as $-\nabla(c\nabla(u))+au=f$. Therefore, the coefficients c, a, and f should be inputted as $c = 1$, $a = 0$ and $f = 7.8900 \cdot 10^6$ (Figure 7.10) to match our equation.

Note:

- The set of parameters of the PDE Specification panel varies depending on the selected PDE types.

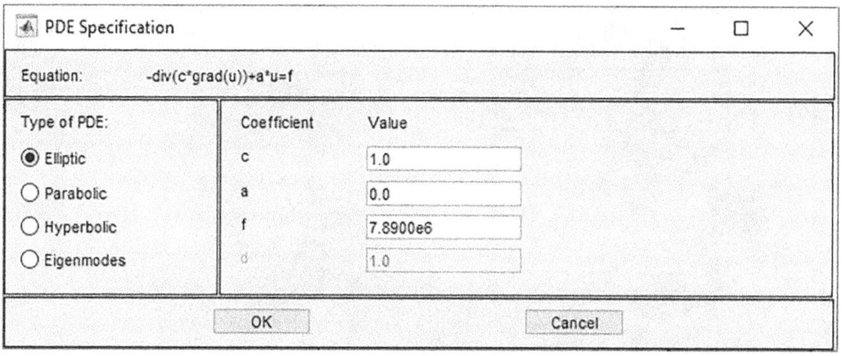

FIGURE 7.10 PDE Specification dialog panel completed for the example problem.

- When the actual body geometry includes two or more domains, the "Show Subdomain Labels" line of the PDE menu button should be clicked to see each subdomain number and to specify the PDE for each subdomain.
- If the *c*, *a*, *f*, or other coefficients (depending on the PDE type) are a function of coordinates and/or time the expressions can be written in the appropriate panel field via coordinates *x*, y and time *t* (these notations are mandatory), for example, if $c = 0.5sin\left(\sqrt{x^2 + y^2}\right)$ in the c-field we should write 0.5 * sin (sqrt(x.^2+y.^2)).

Step 5. Now, a mesh should be created to obtain a solution at the mesh nodes. For this, the Mesh Mode line in the popup menu of the Mesh menu button should be selected. To see the node numbers, the Show Node Labels line can be additionally marked (Figure 7.11).

The mesh can be refined by selecting the Refine Mesh or Jiggle Mesh line and returned to the starting view by selecting the Initialize Mesh line. In our case, we click twice the "Refine Mesh" line for better circle matching and the higher accuracy result. The circle with mesh is presented in Figure 7.12. The mesh is presented here without node numbers for a better image.

Step 6. At this stage, the Solve PDE line of the popup menu of the PDE menu button should be selected (Figure 7.13).

Parameters for PDE solution are set by default, but when non-default solver parameters are required, the "Parameters …" option can be selected. In such case, the Solve Parameters panel is opened. The default view of this panel for a parabolic PDE is presented in Figure 7.14a.

Note, in case of parabolic equation, the Solve Parameters panel has another view adopted to the actual PDE type (Figure 7.14b). For example, for a parabolic equation with a non-default initial condition (u(t0) = 0, default), the actual initial condition should be typed in the u(t0) field as a string line – in single quotes,

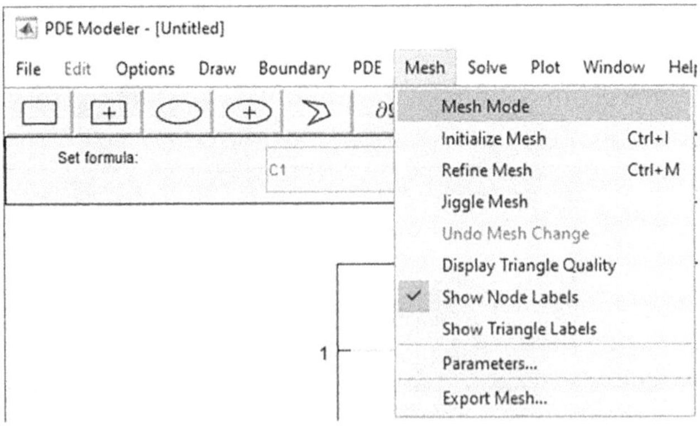

FIGURE 7.11 The fifth step in the PDE solution: selecting the Mesh Mode; the Show Node Labels line is marked.

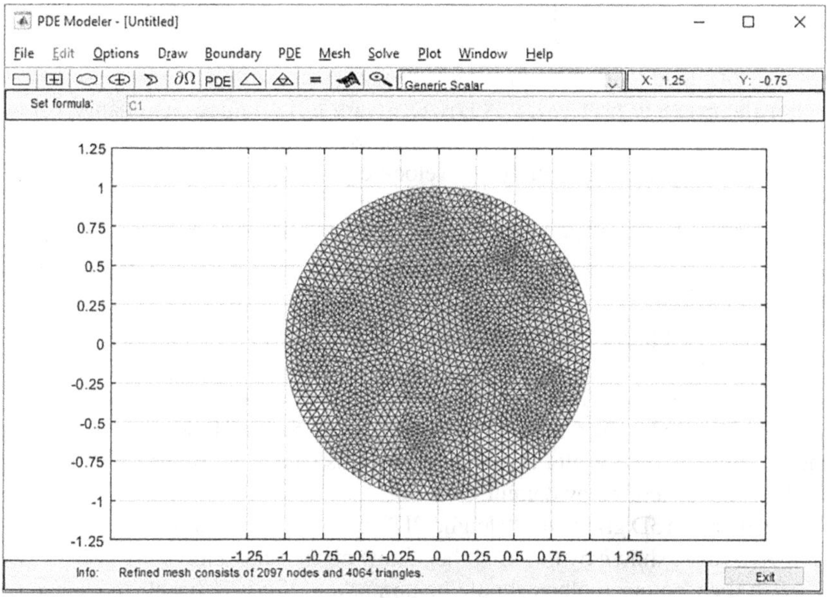

FIGURE 7.12 The Mesh Mode: circle with a generated mesh.

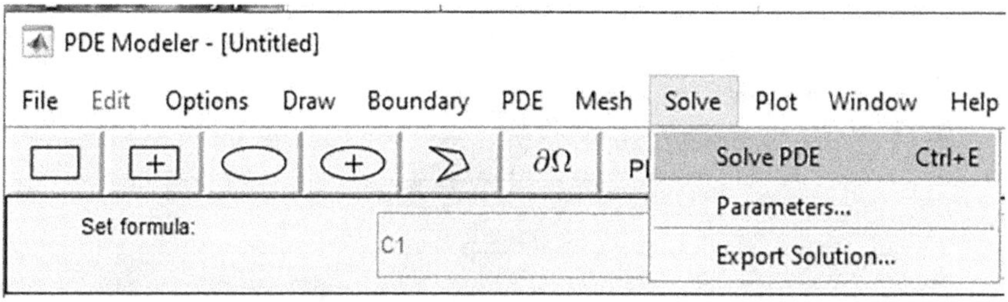

FIGURE 7.13 The sixth stage in the PDE solution: selecting the Solve PDE mode.

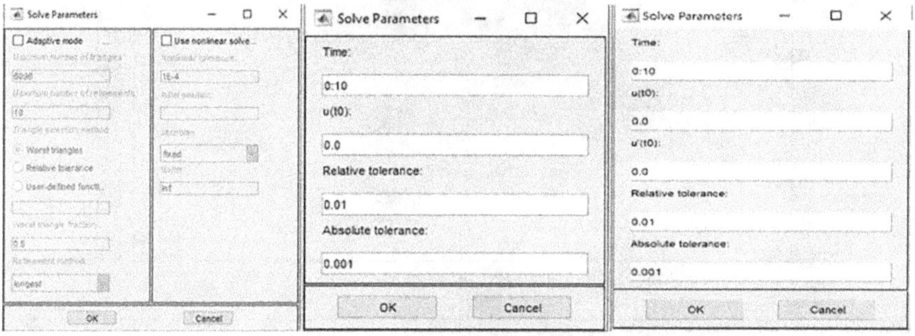

FIGURE 7.14 The Solve Parameters panel for the elliptic (a), parabolic (b) hyperbolic (c) PDEs; default views.

e.g. 'sqrt(x.^2+y.^2)'. In case of parabolic equation, the Solve Parameters panel has view presented in Figure 7.14c. For the eigenmode PDE type, this panel has only one field – “Eigenvalue search range”.

For the example problem, it is not necessary to introduce any changes in the Solve Parameters panel. Therefore, after selecting the Solve PDE mode, the solution is executed and the result (flow velocities) is presented automatically in the 2D plot (Figure 7.15).

Step 7 (optional). To perform some changes in the resulting plot or generate 3D graph, the “Parameters …” line in the popup menu of the Plot button should be selected (Figure 7.16).

After selecting the “Parameters …” line, the Plot Selection window appears (Figure 7.17).

To add, for example, ten contour lines (iso-velocity lines, in our case) to the early produced plot the Contour box should be marked and the value 10 must be entered in the Contour plot levels field; result is shown in Figure 7.18. Note: 20 *iso*-lines are generated by default.

To generate a 3D graph, the “Height (3D)” box should be marked. The rectangular mesh can be shown by marking the “Show mesh” and “Plot in x-y grid” boxes, wherein the Contour toolbox should be unmarked. If it is desirable to change the color map, for example to “hot”, you can select it from the “Colormap:” popup menu. The default colormap is “cool”.

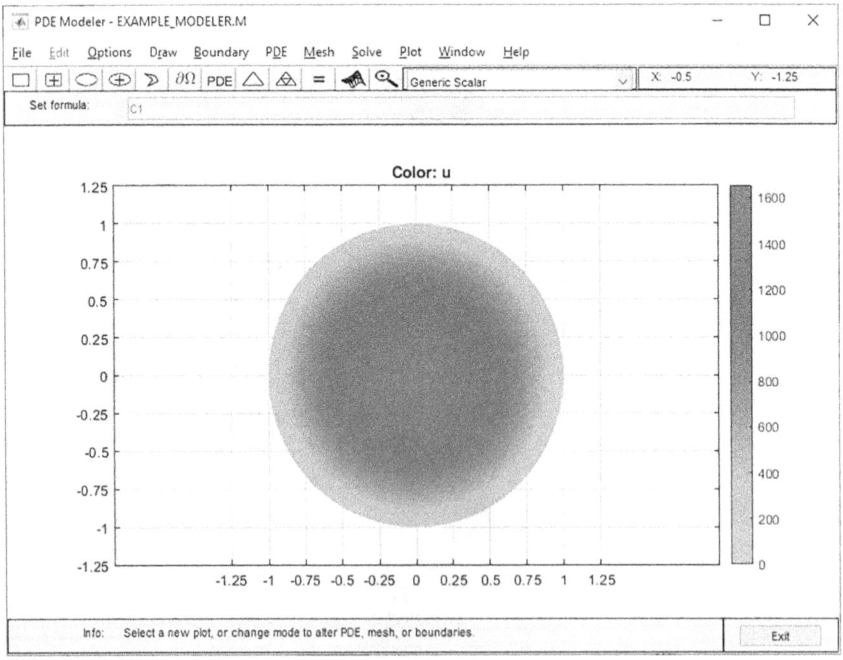

FIGURE 7.15 2D view of the example PDE solution in the PDE Modeler window.

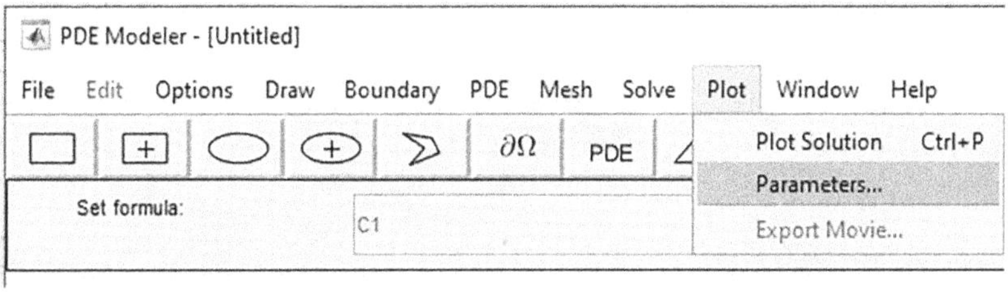

FIGURE 7.16 The seventh step in the PDE solution: selecting the "Parameters …" line in the Plot mode.

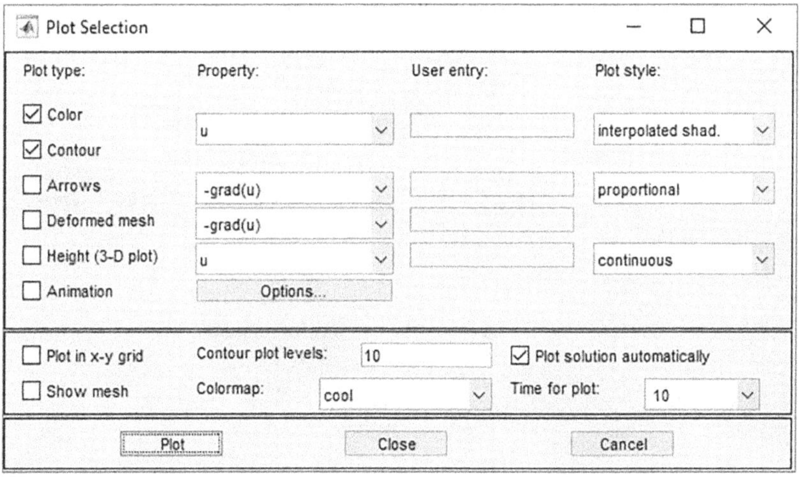

FIGURE 7.17 The Plot Selection window completed to add ten contour lines to the 2D solution plot.

Finally, after selecting all the desired parameters of the graph, click the Plot button of the Plot Selection window. The resulting plot is presented in Figure 7.19.

Step 8 (optional). The PDE Modeler generates a program containing commands that perform the all above steps. To save this program in a file, select the "Save As …" line of the popup File options of the main menu and type the desired file name in the "File name" field of the "Save As" dialog panel that appears. The saved file can be opened and run for any future calculations.

Thus, to save the program generated for the studied example, select the "Save As …" line of the popup menu of the main menu File option and type the name Example_Modeler in the File name field of the panel that appears.

7.2.2 Export Defined Solution and Generated Mesh to the Workspace

PDE Modeler outputs the results of the PDE solution as 2D and 3D graphs with a color bar showing the values in a color scale. However, numerical values are often required in engineering calculations. This can be accomplished by passing

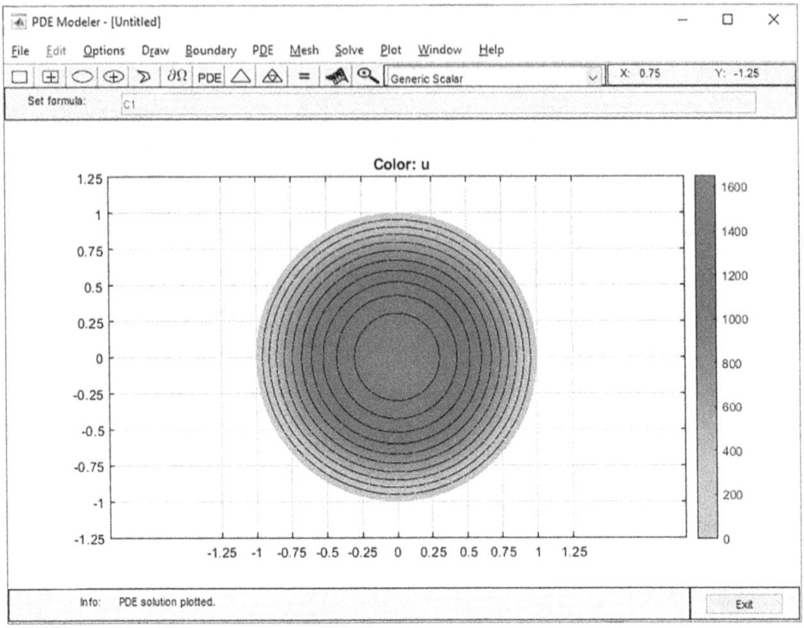

FIGURE 7.18 The resulting 2D plot with the contour lines added using the Plot Selection window.

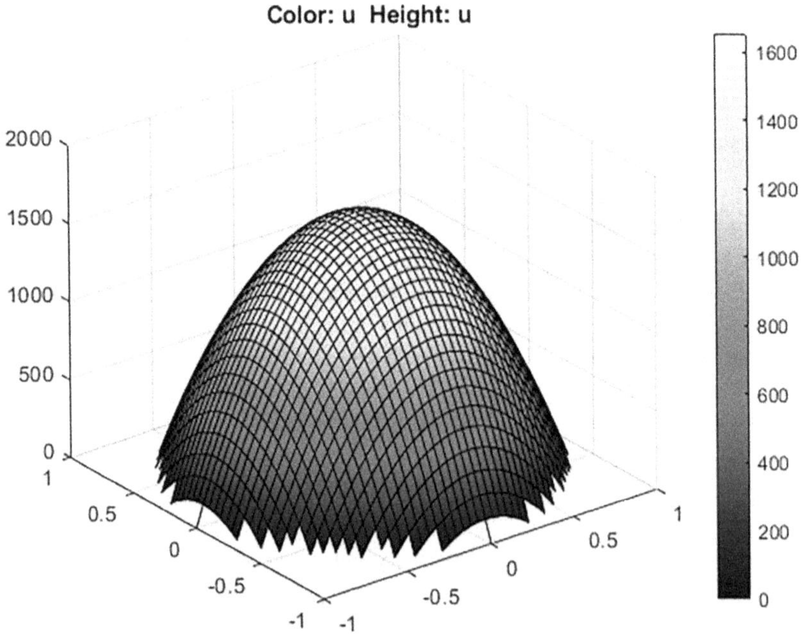

FIGURE 7.19 3D solution generated using the Plot Selection window.

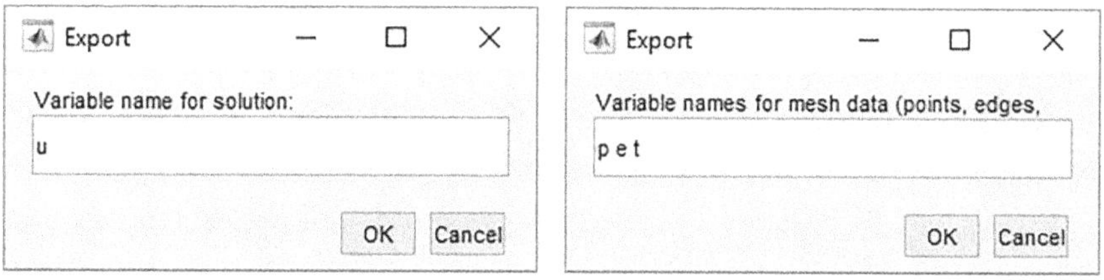

FIGURE 7.20 Boxes intended to export (a) the *u* values of the solution and (b) *x*,*y* node coordinates (p-matrix) together with some other specific mesh parameters (e and t-matrices).

defined *u*-values and the node coordinates (at which the *u*-values were obtained) to the MATLAB® workspace. To export the solution, select the "Export Solution ..." line in the popup menu of the Solve main menu button. In the appearing small box (Figure 7.20a), enter a new name or leave the default variable name and click the OK. The variable with the selecting name appears in the Workspace window of the MATLAB® desktop. For the non-stationary PDEs, *u* is the matrix with columns corresponding to the current time each, but in stationary cases, *u* is the one-column vector (contains the 2097 *u*-values obtained in our example).

Solution is obtained at the certain *x*,*y* mesh nodes, to export this data, select the "Export Mesh ..." line in the popup menu of the Mesh main menu button. In the small box that opens (Figure 7.20b), enter new names or leave the default names for mesh data and click OK. Three variables *p*, *e*, and *t* (or with the names that you entered) appear in the Workspace window of the MATLAB® desktop. The *p* (derived from the *point* word) matrix has two rows with node coordinates; the first row is the *x*-coordinates of the mesh nodes while the second row – *y*-coordinates. The *e* (derived from the *edge* word) matrix has seven rows that contain the ending point indices, starting and ending parameter values, edge segment, and subdomain numbers. The *t* (derived from the *triangle* word) matrix has four rows that contain the indices of the corner points and the subdomain numbers. For practical purposes, only the *p* data is important, but some commands of the PDE Toolbox use *e* and *t* parameters, and thus it is recommended to move all three parameters to the workspace.

After the solution and node coordinates were exported, their values can be displayed. For the example problem, to display in the shortg format (best of fixed or floating point, five-digit format) the coordinates *y* and *x* together with *u*-values at, for example, the 1:200:2001 points and at the final point, the following commands can be entered in the Command Window:

```
>>node = [1:200:2001 length(u)];format shortg
>>disp('x  y  u' ),disp([p(:,node)' u(node)])
     x            y              u
        -1   -1.2246e-16         0
   0.65376     -0.50638     522.92
   0.22928      0.12644     1541.5
  -0.67156      0.74095          0
```

```
 0.073497  -0.65106      944.53
-0.92568   -0.24895      134.24
 0.61163    0.49238      634.53
-0.19285   -0.041658     1590.7
-0.010092   0.36331      1436.4
 0.2487     0.25474        1445
 0.77148   -0.14849      633.46
 0.44292    0.55504      819.98
```

7.2.3 From the Triangular to Rectangular Grid

Each obtained *u*-value corresponds to a certain node of the triangular mesh. Nevertheless, a rectangular grid are often of practical engineering interest. To convert obtaining solution from a triangle mesh to a rectangular grid, the tri2grid command can be used. The simplest form of this command is

```
U= tri2grid(p,t,u,x,y)
```

where p, t, and u are, respectively, the exported node coordinates, triangle parameters, and solution values; x and y are the vectors of the rectangular coordinate nodes at which we want to determine *u*-values; and U is the determined *u*-values computed using the linear interpolation of the solution.

For our example, assuming rectangular grid in range 0 … 1 with seven points for each coordinate, we should enter the following command to convert *u*-data from the triangular grid to rectangular:

```
>> x=linspace(-1,1,7);y=x;
>> U=tri2grid(p,t,u,x,y)
U =
  1.0e+03 *
NaN    NaN          NaN   -0.0000     NaN       NaN         NaN
NaN   0.1830   0.7345    0.9188   0.7347   0.1832     NaN
NaN   0.7346   1.2863    1.4705   1.2863   0.7351     NaN
 -0   0.9190   1.4710    1.6538   1.4707   0.9184       0
NaN   0.7347   1.2865    1.4702   1.2863   0.7348     NaN
NaN   0.1826   0.7346    0.9185   0.7344   0.1827     NaN
NaN    NaN          NaN   -0.0000     NaN          NaN      NaN
```

Note that in case non-rectangular shapes, circles, ellipses, as well as cavities, scale-different elements, etc., rectangular grid nodes may appear in some places outside the body; such points are denoted as NaN (not a number), as in the above example.

7.3 DRAWING TWO-DIMENSIONAL OBJECTS

To draw geometry of a real object, the PDE Modeler provides some basic shapes and appropriate buttons on the toolbar line or in the Draw menu. These shapes are an ellipse/circle, rectangle/square, and polygon; additionally, the same shapes can be generated using appropriate commands by entering them in the Command Window. Available button icons, their purposes, alternative commands, and examples with generated shapes with formula fields are presented in Table 7.1.

Various shapes of complicated geometry can be drawn with a combination of the basic shapes presented in Table 7.1 with additional correction (if necessary) of the expression appearing in the Set formula field. Examples of drawing some complex shapes are presented in Table 7.2.

TABLE 7.1[a]

Drawing Basic Shapes Using the PDE Modeler Interface and Alternative Commands.

Icon, Button Name, and Description	Alternative Command	Example of the Button and Command Implementation
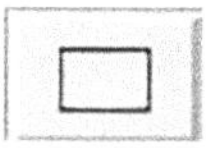 - rectangle/ square button: draws a rectangle or square starting at a left bottom corner and dragging the + (mouse pointer) to the diagonal corner; in the example, the + pointer is shown at the point where you should end to draw the square.	`pderect([xmin xmax ymin ymax])` where xmin, xmax, ymin, and ymax are the *x* and *y* coordinates of the diagonal corners of the rectangle	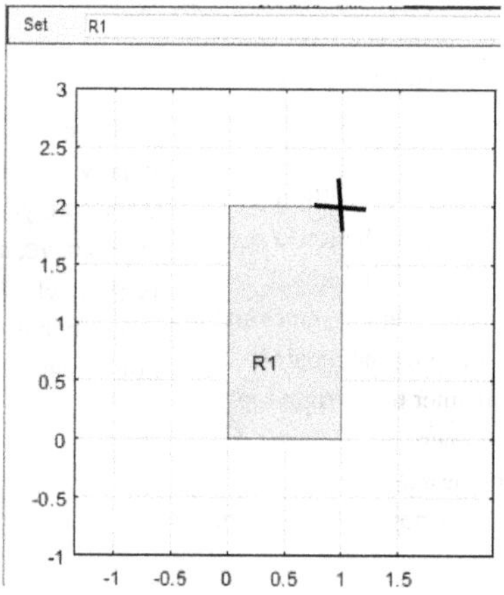 `>>pderect([0 1 0 2])`
 - rectangle or square (centered) button: draws a rectangle or square starting at the coordinates of the rectangle center and dragging mouth pointer to the right top corner of the rectangle. In the example, the + pointer is shown at the point where you should end to draw the square	The same command as above	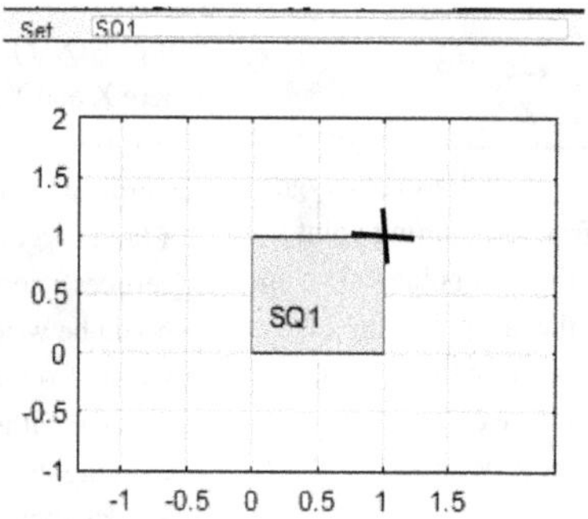 `>>pderect([0 1 0 1])`

(*Continued*)

TABLE 7.1[a] (*Continued*)
Drawing Basic Shapes Using the PDE Modeler Interface and Alternative Commands

Icon, Button Name, and Description	Alternative Command	Example of the Button and Command Implementation
 - ellipse or circle button: draws an ellipse or circle starting at the left bottom point of the perimeter and dragging mouth pointer to the diagonal corner. In the example, the + pointer is shown at the point where you should end to draw the ellipse	pdeellip(xc,yc,a,b, phi) where xc, yc, are the ellipse center coordinates, a and b are the ellipse semi-axes, phi – rotation angle in radian	Ellipse: 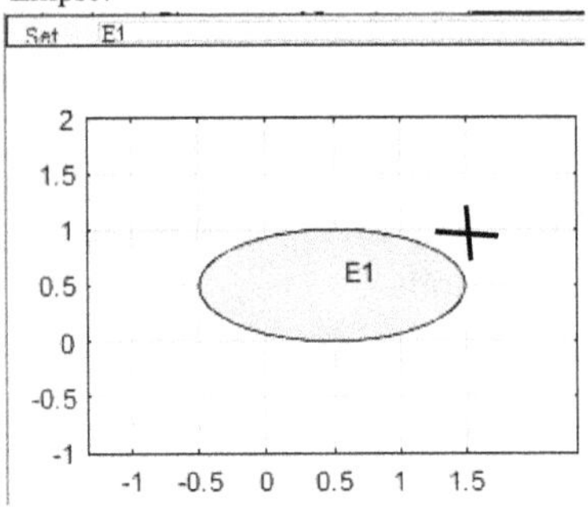 >>pdeellip(0.5,0.5,1,0.5,0)
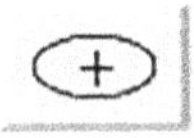 - Ellipse/circle (centered) button: draws an ellipse or circle starting at the center and dragging the mouth pointer to the right top corner. In the example, the + pointer is shown at the point where you should end to draw the ellipse	The same command as in previous case or: pdecirc(xc,yc,r) where xc, yc, and r are respectively the *x* and *y* coordinates of the circle center and radius	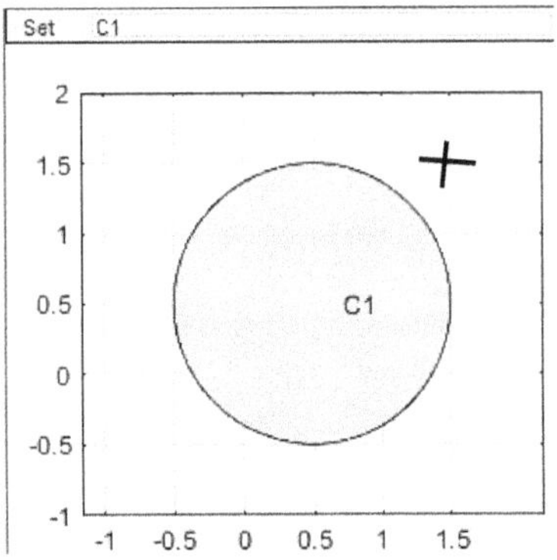 >>pdecirc(0.5,0.5,1)
 - Polygon button: draws a polygon from the starting point to the next point and so on; the last point should be the starting point. In the example, the + pointer is shown at the point where you should start and end to draw the polygon	pdepoly(X,Y) where X and Y are the vectors with coordinates of the each of the polygon corners (vertices); coordinates should be written starting from any point and then without changing direction (e.g., clockwise) to the one point before starting	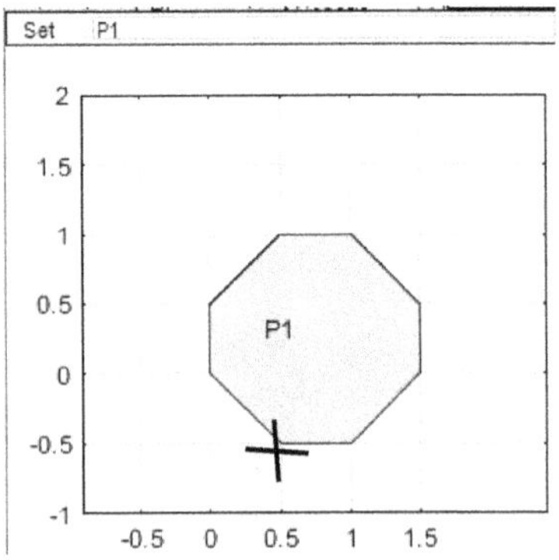 >>pdepoly([0.5 0 0 0.5 1 1.5 1.5 1],[-0.5 0 0.5 1 1 0.5 0-0.5])

[a] Based on Table 8.2 in Burstein 2021a

TABLE 7.2b

Examples of Some Complex Shapes Drawing Sing the Basic PDE Modeler Shapes

What to Draw	Steps to Draw	Drawn Shape, Set Formula, and Alternatively Commands
A 2 × 2 square plate centered at (0,0) and rounded corners with a radius of 0.2 each.	• Mark the Grid and Snap options and adjust the Axes Limits, and Grid spacing as follows: x-axis to `-1.5:0.1:1.5` and y-axis to `-1.5:0.1:1.5` • Click the Rectangle (centered) button. Place the mouse pointer on point (0,0) and drag to (1,1) • Draw four squares with the adge 0.2, one in each corner using for this the Rectangle (centered) button; centers of squares are (–0.9, –0.9), (–0.9, 0.9), (0.9, 0.9), (0.9, –0.9) and end points for drawing are (–0.8, –1), (–0.8, 1), (0.8, 1), (0.8, –1) • Draw four circles with the radius 0.2 and the centers at (–0.8, –0.3), (–0.8, 0.3), (0.8, –0.3), and (0.8, 0.3) – with the Ellipse/Circle button each • Enter the following expression R1-(SQ1 + SQ2 + SQ3 + SQ4) + C1 + C2 + C3 + C4 in the Set Formula field • Select Boundary mode line in the Boundary menu option and then select the "Remove All Subdomain Borders" line	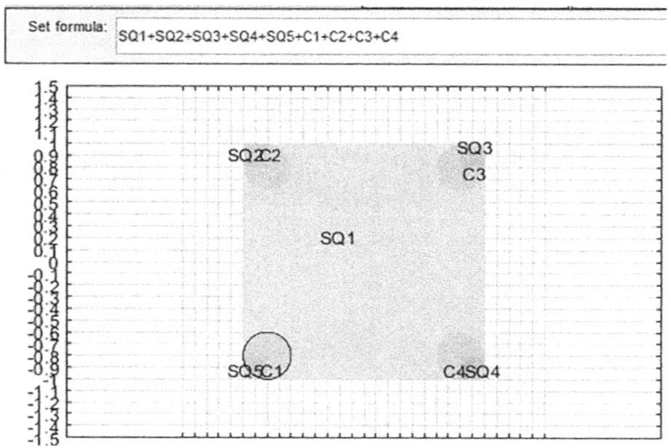

(*Continued*)

TABLE 7.2[b] (*Continued*)

Examples of Some Complex Shapes Drawing Sing the Basic PDE Modeler Shapes

What to Draw	Steps to Draw	Drawn Shape, Set Formula, and Alternatively Commands
		This view shows the plate in the Draw Mode

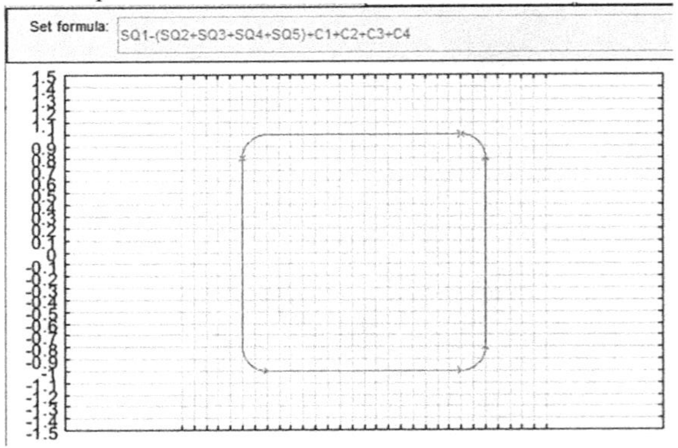

This view shows the plate in the Boundary mode after selecting the Remove All Subdomain Borders option.

Alternatively:

```
>>pderect([-1 1-1 1])
>>pderect([-1-0.8-1-0.8])
>>pderect([-1-0.8 0.8 1])
>>pderect([0.8 1 0.8 1])
>>pderect([0.8 1-0.8-1])
>>pdecirc(-.8,-.8,.2)
>>pdecirc(-0.8,0.8,0.2)
>>pdecirc(0.8,0.8,0.2)
>>pdecirc(0.8,-0.8,0.2)
```

(*Continued*)

TABLE 7.2[b] (*Continued*)

Examples of Some Complex Shapes Drawing Sing the Basic PDE Modeler Shapes

What to Draw	Steps to Draw	Drawn Shape, Set Formula, and Alternatively Commands
		After this enter R1–(SQ1 + SQ2 + SQ3 + SQ4) + C1 + C2 + C3 + C4 in the Set formula field, activate the Boundary Mode and select the Remove All Subdomain Borders option
A shape composed by two basic figures and has more than one sub-domains: centered rectangle 0.4 × 4 and circle with center at the (0,1) point and radius $r = 0.5$.	• Mark the Grid and Snap options; adjust the Axes Limits as follows: x in the –2 … 2 range, and y in the –2.4 … 2.4 range, set the Grid Spacing with step suitable to draw the rectangle and circle • Click the Rectangle (centered) button. Place the mouse pointer on point (–0.2, –2) and drag to (0.4, 2) • Click the Ellipse/Circle (centered) button. Place the mouse pointer on point (0,1) and drag to (0.5,0.5) with the right mouse button pressed. Check the circle by clicking on the circle and then completing the Object Dialog box • Check the expression in the Set Formula field - it should be R1 + E1	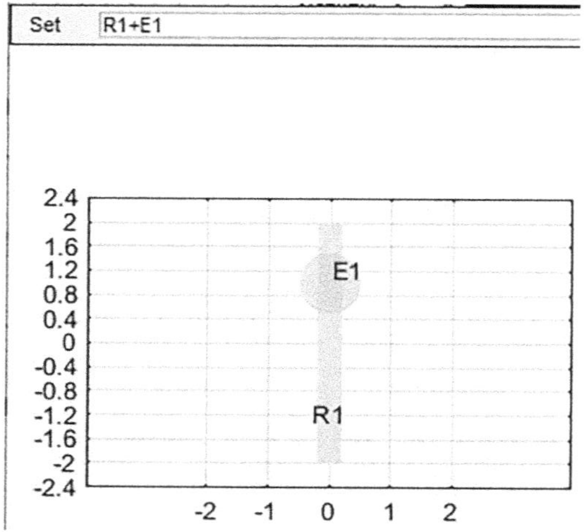 The view is given in the Draw Mode. Alternatively: >>pderect([-.2.2-2 2]) >>pdecirc(0,1,.5)

(*Continued*)

TABLE 7.2[b] **(*Continued*)**

Examples of Some Complex Shapes Drawing Sing the Basic PDE Modeler Shapes

What to Draw	Steps to Draw	Drawn Shape, Set Formula, and Alternatively Commands

A centered square plate 2 × 2 with centered rectangular slot 0.2 × 1.2	• Mark the Grid and Snap options and adjust the Axes Limits, and Grid spacing as follows: each of the *x*- and *y*-axis to –0.5…2.5 with spacing 0.1
	• Click the Rectangle/Square (centered) button. Place the mouse pointer to point (1.1) and drag to (2,2)
	• Click the Rectangle/Square (centered) button again. Place the mouse pointer on point (1,1) and drag to (1.6,0.1)
	• Enter the following expression in the Set Formula field: SQ1–R1
	• Select the Boundary Mode line in the Boundary menu
		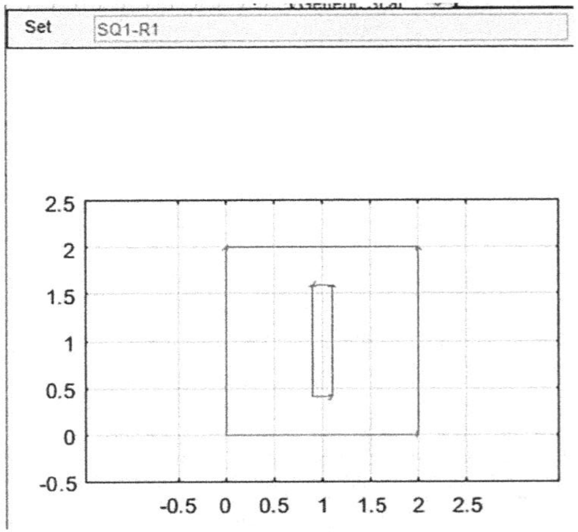
		This plot is a view after activating the Boundary Mode.
		Alternatively:
		`>>pderect([0 2 0 2])`
		`>>pderect([0.9 1.1 0.4 1.6])`
		After this enter SQ1–R1 in the Set formula field and activate the Boundary Mode

(Continued)

TABLE 7.2[b] (*Continued*)

Examples of Some Complex Shapes Drawing Sing the Basic PDE Modeler Shapes

What to Draw	Steps to Draw	Drawn Shape, Set Formula, and Alternatively Commands
T-shaped plate with two rounded corners. The sizes as per the second column.	• Mark the Grid and Snap options and adjust the Axis limits and Grid spacing as follows: ***x***-axis to –1.5:0.2:1.5 and ***y***-axis to–1:0.2:1 • Click on the polygon button and draw the polygon with the following (*x*,*y*) corner coordinates: (–0.4, –0.9), (–0.4,0.5), (–0.8,0.5), (–0.8,0.9), (0.8,0.9), (0.8.0.5), (0.4,0.5) and (–0.4, –0.9) to close this shape • With the rectangle/square (centered) button, draw two squares, the first – from the center (–0.6,0.3) drag the pointer to the point (–0.4,0.5) and the second – from the center (0.6,0.3) drag the pointer to the point (0.4,0.5) • With the Ellipse/Circle button, draw two circles with the radius 0.2 each, first centered at point (–0.6,0.3) and the second – at point (0.6,0.3) • Enter the following expression in the Set Formula field: P1 + (SQ1 + SQ2)–(C1 + C2) • Activate the Boundary Mode line of the Boundary menu and then select the Remove All Subdomain Borders line	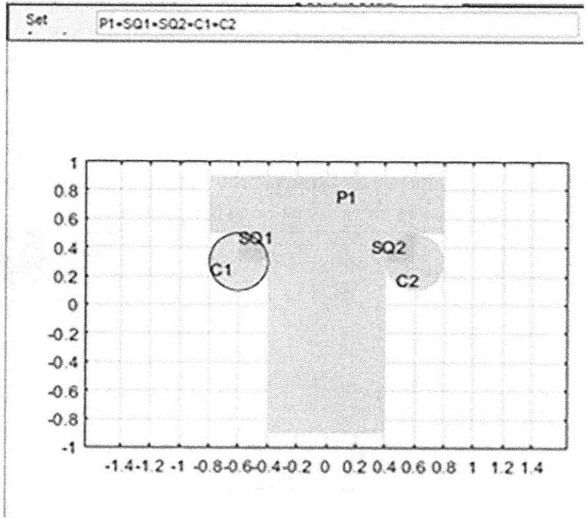

(*Continued*)

TABLE 7.2[b] ***(Continued)***

Examples of Some Complex Shapes Drawing Sing the Basic PDE Modeler Shapes

What to Draw	Steps to Draw	Drawn Shape, Set Formula, and Alternatively Commands

		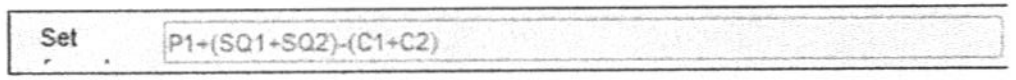 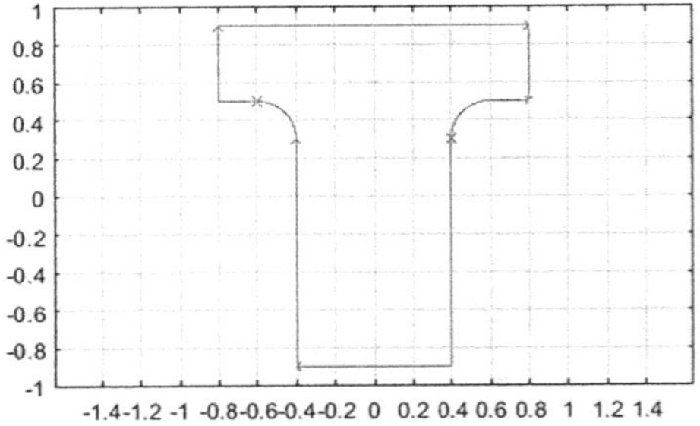 These two plots show the T-shaped plate view before and after activating the Boundary Mode. The lower plot represents the L-shaped plate with rounded corners after selecting the Remove All Subdomain Borders option.
		Alternatively:
		`>>pdepoly([-0.4-0.4-0.8-0.8 0.8 0.8 0.4 0.4],[-0.9 0.5 0.5 0.9 0.9 0.5 0.5-0.9])`
		`>>pderect([0.4 0.6 0.3 0.5])`
		`>>pderect([-0.6-0.4 0.3 0.5])`
		`>>pdecirc(-0.6,0.3,.2)`
		`>>pdecirc(0.6,0.3,.2)`
		After this enter P1 + (SQ1 + SQ2) − (C1 + C2) in the Set formula field, activate the Boundary Mode and select the Remove All Subdomain Borders option

[2] Based on Table 8.3 in Burstein, 2021a.

Remember:

- to draw a square or circle using the Rectangle/Square (centered) or Ellipse/Circle (centered) buttons, place the mouse pointer to the center point, and drag the mouse keeping the right button down to any desired diagonal point of the square or the circle radius;
- it is recommended to check/correct each drawn basic shape; to do this, place the mouse pointer within the shape and click, in the appeared window you can see and correct the required coordinates of the shape; in the case of a polygon, the same action opens a window showing a popup list with coordinates of the vertices.

7.4 APPLICATION EXAMPLES

The following application examples contain some PDEs and a number of PDE Modeler elements that were not previously encountered; however, the user should be familiar with them to use the PDE Modeler effectively.

7.4.1 Momentary Pressure Distribution in a Lubricating Film between Two Surfaces Covered by Hemispherical Pores

The 2D Reynolds equation describes the hydrodynamic behavior of a thin lubricating film between two surfaces covered by hemispherical pores, when one of the surfaces is stationary, and the second is moving (Burstein, 2021b). The equation has the form

$$\frac{\partial}{\partial X^2}\left(H^3\frac{\partial P}{\partial X}\right)+\frac{\partial}{\partial Z^2}\left(H^3\frac{\partial P}{\partial Z}\right)=\frac{dH}{dX}+2\frac{dH}{d\underline{t}}$$

where P is hydrodynamic pressure arising in the film, X – coordinate, H – gap between the upper and lower surfaces, $\underline{t}$ – time; all variables in this equation are dimensionless.

The difference between the position of the upper H_u and lower H_l surfaces forms a gap.

For upper surface position, the gap and its derivatives $\frac{dH_u(X)}{dX}+2\frac{dH_u(X)}{d\underline{t}}=-\frac{dH_u(X)}{dX}$ are given with the following equations for a hemispherical pore and adjacent surface parts (called pore cell)

$$H_u=\begin{cases}1+\psi\sqrt{1-((X-\underline{t}-\xi)^2+Z^2)}, & (X-\underline{t}-\xi)^2+Z^2<1\\ 1, & otherwise\end{cases}$$

$$\frac{dH_u}{dX}=\begin{cases}\frac{\psi(X-\underline{t}-\xi)}{\sqrt{1-((X-\underline{t}-\xi)^2+Z^2)}}, & (X-\underline{t}-\xi^2+Z^2)<1\\ 0, & otherwise\end{cases}$$

And for lower surface position, the gap and derivatives $\frac{dH_l(X)}{dX}+2\frac{dH_l(X)}{d\underline{t}}=\frac{dH_l(X)}{dX}$:

$$H_l=\begin{cases}1+\psi\sqrt{1-((X-\xi)^2+Z^2)}, & (X-\xi)^2+Z^2<11, \ otherwise\end{cases}$$

$$\frac{dH_l}{dX}=\begin{cases}\frac{\psi(X-\xi)}{\sqrt{1-((X-\xi)^2+Z^2)}}, & (X-\xi^2+Z^2)<10, \ otherwise\end{cases}$$

Where in the above expressions, ψ is the pore radius to the gap ratio and ξ is the ratio of the cell dimension to the pore radius.

When parts of upper and lower pores are against each other, the gap and its derivatives are

$$H=H_u-H_l$$

$$\frac{dH}{dX}+2\frac{dH}{d\underline{t}}=-\frac{dH_u}{dX}-\frac{dH_l}{dX}$$

The hydrodynamic pressures at boundaries of a rectangular pore cell, X∈[–ξ,ξ] and Z∈[–ξ,ξ], are assumed to be zeros. In other words, the boundary conditions are: P = 0 at X, Z = ±ξ.

Problem: Solve the Reynolds equation for a rectangular pore cell X€[–0,2ξ] and Z€[–ξ,ξ] with Dirichlet (default of the PDE Modeler) boundaries: P = 0 along four pore cell boundaries. Assume $\underline{t}=\frac{\xi}{2}$, $\xi=2$, and $\psi=8$. Represent the pressure distribution $P(X,Z)$ in a 3D plot. Transfer defined P-values and mesh parameters to the workspace and display the P(X,Z) table for the orthogonal grid with seven equally spaced values of X in the range 0 … 2ξ and Z in the range – ξ … ξ. Find the maximum and minimum pressure values. Save the automatically created program in a file named ApExample_7_1.

Determine firstly the type of the solving equation by matching this equation with the required standard form (Table 8.1); it is easy to see that the Reynolds equation at the specified time is an elliptic type of the PDE as in the standard form reads

$$\nabla\left(H^3\nabla P\right)=\frac{dH}{dX}+2\frac{dH}{d\underline{t}}$$

The equation is identical to the standard PDE form when u = P, Z = Y, m = 0, d = 0, c = H^3, a = 0, and $f=\frac{dH}{dX}+2\frac{dH}{d\underline{t}}$. Thus, further we must remember that *y* and *u* of the PDE Modeler notations are coordinate *Z* and pressure *P* of the Reynolds equation, respectively.

Activate now the PDE Modeler with the command

```
>>pdeModeler
```

Mark the Grid and Snap lines in the popup menu of the Options button of the PDE Modeler menu, and type the *x* and *y* limits as [–0.5 4.5] and [–2.5 2.5] in the Axis Limits dialog box. Use the general Application option – Generic Scalar (default).

Then, activate the Draw Mode and draw a rectangle with the 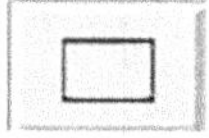 rectangle button. Now place the mouse arrow on point (0, –2) and drag to point (4,2) of the plot. Check the rectangle parameters in the Object Dialog panel that appears after clicking within the rectangle. Draw now a lower surface circle domain clicking for this the 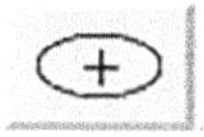 Ellipse/Circle (centered) button. Place the mouse at the (2,0) point and, keeping the right mouth button, drag the mouse diagonally to the point (4,1). Check the circle geometry parameters with the Object Dialog panel that appears after clicking within the circle. Draw now an upper surface circle domain clicking for this the 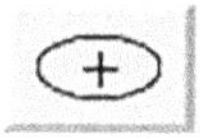 Ellipse/Circle (centered) button. Place the mouse at the (3,0) point and, keeping the right mouth button, drag the mouse diagonally to the point (4,1). Check the circle geometry parameters with the Object Dialog panel that appears after clicking within the circle.

Activate the Boundary button and, in the appeared "Boundary Conditions" panel, select the Dirichlet option and enter 1 in the h field and 0 in the r field; do this for each side of the rectangle (these actions are optional since the Dirichlet boundaries and the corresponding h and r values are set by default).

Now select the PDE Mode from the popup menu of the PDE Modeler main menu button. The studied body has four subdomains – a rectangle, two non-overlapping circle parts on opposite surfaces, and overlapping circle subdomain; subdomain numbers can be shown when the "Show domain labels" options of the popup menu are marked (see Figure 7.21).

Move the mouse pointer on subdomain 1 and click the mouse button. In the opened PDE Specification panel, check/mark the Elliptic type of PDE and enter 1, 0, and 0 into the fields c, a, and f, respectively.

Place now the mouse pointer on subdomain 2 and press the mouth button. In the appeared PDE Specification panel, check the Elliptic type of PDE (default) and type:

(1+8*sqrt(1- (x-3).^2-y.^2)).^3 in the *c* field; 0 in the *a* field; and 8 * (x-3)./ sqrt(1- (x-3).^2-y.^2) in the *f* field.

In subdomain 3, after the same procedure as for previous domain, check the Elliptic type of PDE (default) and type: -(8*sqrt(1-(x-2).^2-y.^2)).^3 in the *c* field; 0 in the *a* field; and (8*(x-2)./sqrt(1- (x-2).^2-y.^2) in the *f* field.

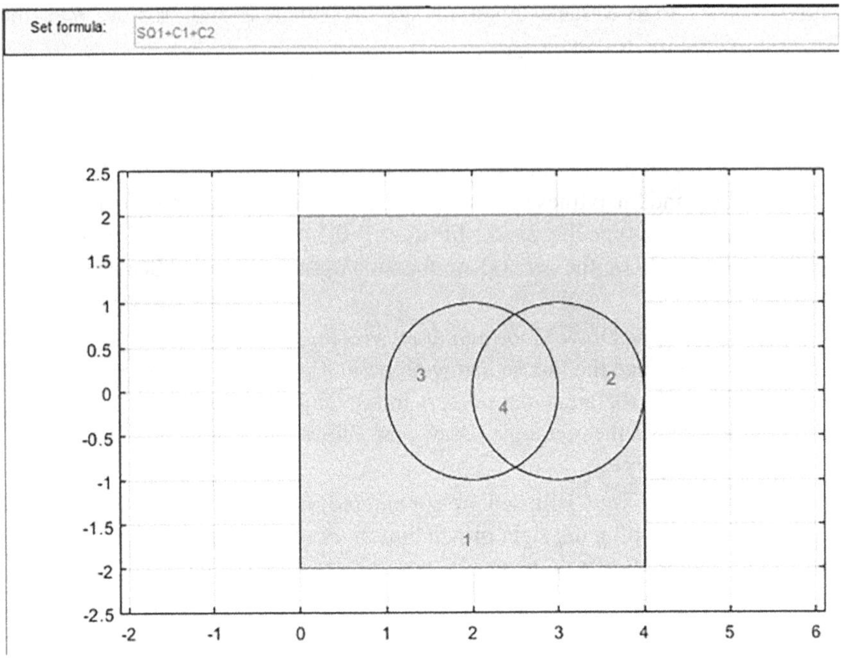

FIGURE 7.21 The PDE Mode; pore cell with four subdomains.

Finally, in subdomain 4, check the Elliptic type of PDE (default) and type: (1+8*(sqrt(1− (x−3).^2−y.^2)+sqrt(1− (x−2).^2−y.^2))).^3 in the *c* field; 0 in the *a* field; and −8*((x−3)./sqrt(1− (x−3).^2−y.^2)+(x−2)./sqrt(1− (x−2).^2−y.^2)) in the *f* field.

Activate the Mesh Mode and initialize the triangle mesh and then refine mesh two times.

Now we can solve the Reynolds equation, for this select the Solve PDE line in the popup list of the main menu Solve button. The 2D solution with colored bar appears as below (Figure 7.22).

The graph shown in this figure was obtained with the Colormap option selected as "jet" in the appropriate field of the "Parameter …" panel, which appears after selecting the appropriate line from the popup list of the main menu "Plot" button. Additionally, after marking the boxes "Height(3D plot) " and "Show mesh", the following 3D plot with the solution appears in the separate Figure window (Figure 7.23):

To transfer the solution and mesh parameters to the MATLAB® workspace, select the lines Export Mesh and Export Solution in the popup menus of the corresponding Solve and Mesh buttons of the main menu. If you have not changed the variable names, then the solution is in the *u* matrix, and the mesh parameters are in the *p*, *e*, and *t* matrices. To obtain now the *u*-values at the required orthogonal grid points, the following commands should be entered in the Command Window to display the resulting table.

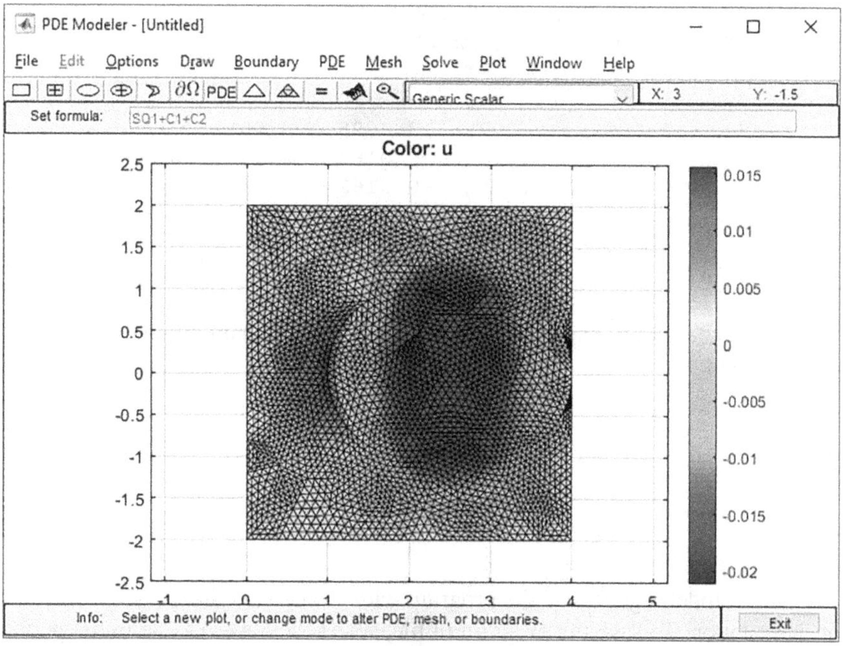

FIGURE 7.22 Momentary pressure distribution in the 2D PDE Modeler representation.

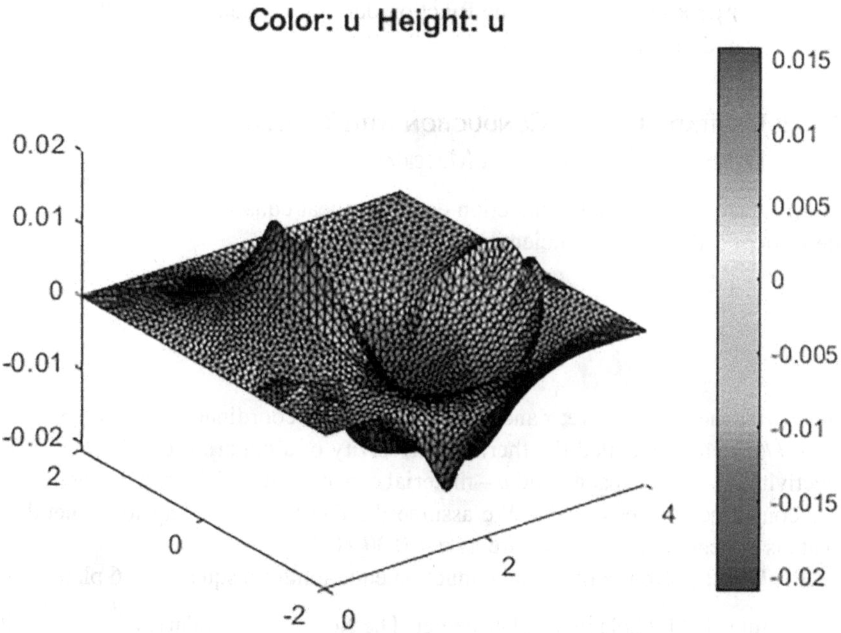

FIGURE 7.23 Momentary pressure distribution in the 3D PDE Modeler representation.

```
>> x=linspace(0,4,7);y=linspace(-2,2,7);
>> P=tri2grid(p,t,u(:,end),x,y)
P =
0        0         0         0         0         0  0
0   0.0011   -0.0009   -0.0079   -0.0098   -0.0053  0
0   0.0048    0.0016   -0.0147   -0.0163   -0.0099  0
0   0.0082   -0.0038   -0.0182   -0.0162   -0.0109  0
0   0.0048    0.0019   -0.0147   -0.0163   -0.0099  0
0   0.0011   -0.0010   -0.0080   -0.0098   -0.0054  0
0        0         0         0         0         0  0
```

To find the maximum and minimum values of hydrodynamic pressure, enter in the Command Window

```
>> max(u),min(u)
ans =
0.0157
ans =
-0.0211
```

The PDE Modeler generates the program with commands that perform all steps of the solution. To save the program in file, select the Save As line in the popup menu of the File button of the main menu and type the name ApExample_7_1 in the File name field of the panel that appears. To check the generated file, close the PDE Modeler window, then open the saved file in the MATLAB® Editor window and type ApExample_7_1 in the function definition line instead of the default name located there.

7.4.2 Unsteady Thermal Conduction with Temperature-Dependent Property of a Material

An unsteady 2D thermal conduction equation (heat equation) with temperature-dependent diffusivity coefficient has the form

$$\frac{\partial \theta}{\partial t} = \frac{\partial}{\partial x}\left(\alpha(\theta)\frac{\partial \theta}{\partial x}\right) + \frac{\partial}{\partial y}\left(\alpha(\theta)\frac{\partial \theta}{\partial y}\right)$$

where θ is the temperature; x and y are the Cartesian coordinates; t – time; α is the ratio $k/(c_p\rho)$ and is called the thermal diffusivity of a material, k – thermal conductivity, c_p – heat capacity and ρ – material density. All variables in this equation are considered dimensionless. We assume the thermal diffusivity for a metallic plate is dependent on temperature as $\alpha = 0.3\theta + 0.4$.

Problem: Solve the thermal conduction equation for a square 6×6 plate with rectangular 1×1.2 hole in the plate center. The Neumann boundaries $\frac{\partial \theta}{\partial x} = \frac{\partial \theta}{\partial y} = 0$ are on the outer plate sides while the Dirichlet boundaries $\theta = 0$ (default) are on the borders of the hole. The initial conditions are $\theta = 1$ in small area of the plate with coordinates $-0.5 \le x \le 0.5$ and $-2.5 \le y \le -1.5$. Set times in range 0 …0.1 with step 0.01. Chose the refined options for the triangle mesh. Represent the final u-values in the 2D and 3D plots, which should show the four contour levels and "jet" color map. Save the automatically created program in a file named ApExample_7_2. Export to the MATLAB® workspace the resulting u and triangle p, e, t- values; transform u at final $t = 0.1$ at the orthogonal mesh with the tri2grid command, and display results for $x = -0.5:0.25:0.5$ and $y = -0.6:0.25: -0.6$.

The thermal conduction equation to be solved is a parabolic type. To match this equation with the required standard form (see Table 6.1), present the thermal conduction equation as

$$\frac{\partial \theta}{\partial t} - \nabla\left(\alpha(\theta)\nabla\theta\right) = 0$$

This equation is identical to the standard form when $u = \theta$, $\mathrm{m} = 0$, $d = 1$, $c = \alpha(\theta) = 0.3\theta + 0.4$ and $a = f = 0$.

To solve the problem, activate the PDE Modeler by typing in the Command Window

```
>> pdeModeler
```

Mark the Grid and Snap lines in the popup menu of the Options menu button, and type limits [-3.5 3.5] for both the x and y axis (after selecting the Axis Limits line in the same popup menu). Use the Generic Scalar in the Application option.

Activate now the Draw Mode and draw a square and a rectangle using the

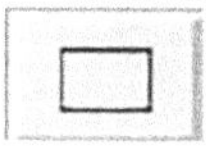

Rectangle/Square button. The first shape – by placing the mouse arrow at point (–3.5, –3.5), and dragging the mouse to the point (3.5,3.5), and the second shape – by placing the mouse arrow at point (–0.6, –0.5), and dragging the mouse to the point (0.5,0.6). Click inside the each of shapes to check the rectangle geometry with the Object Dialog panel. Enter R1–R2 in the Set Formula field.

Move now to the Boundary mode by selecting the Boundary Mode line of the popup menu of the main menu Boundary button. Click sequentially on each outer edge of the plate, mark the Neumann box, and enter 0 in the g and q fields of the Boundary Conditions panel that appears. Make sure the boundaries of the hole meet the required Dirichlet (default) conditions.

Select now the PDE Mode of the popup menu of the PDE main menu button, mark the Parabolic type of PDE, and enter: 0.3 + 0.4 *u within the c field; 0 within the a and f fields and 1 within the d field.

Initialize the triangle mesh in the Mesh Mode (a line in the popup menu of the main menu Mesh button) and click twice on the Refine line.

Select the "Parameters …" line in the popup menu of the Solve main menu button and in the Solve Parameters panel enter 0:0.01:0.1 in the Time field and the initial conditions '(x>−0.5 & x<0.5) & (y>−2.5 & y<−1.5)' (in the single

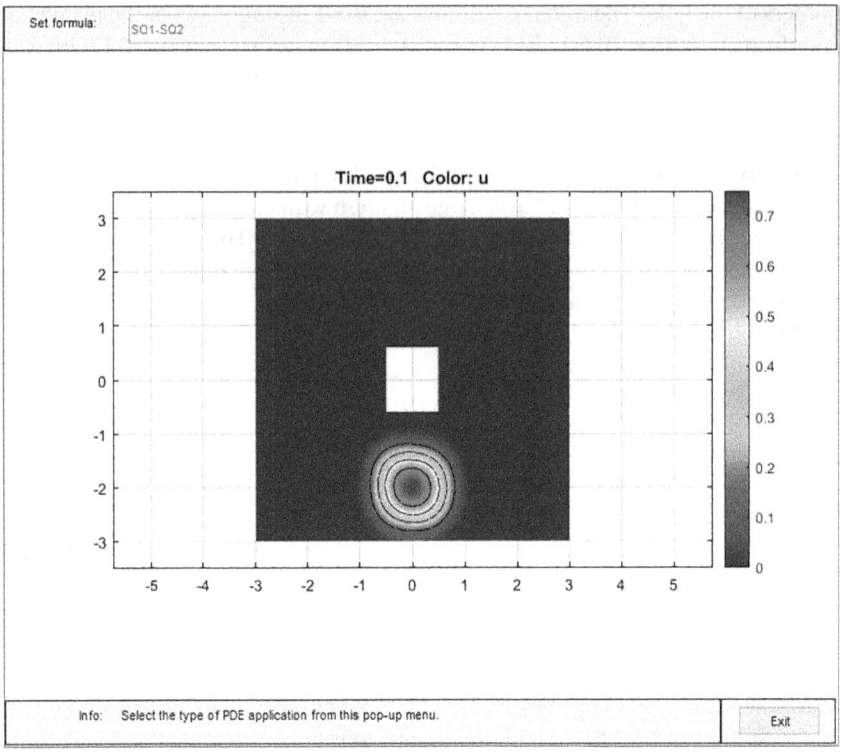

FIGURE 7.24 2D plot with dimensionless temperature distribution at the final time with four contour lines.

quotes) in the u(t0) field. Here we used the logical relation property to generate 1 (initial condition of the problem) when the condition is true and 0 when it is false. After that, click the Solve PDE line of the popup menu of the main menu Solve button. A 2D solution with a colored bar appears. Figure 7.24 represents this solution after selecting the "Parameters …" option of the main menu Plot button and entering/marking the following plot parameters in the appeared "Plot Selection" window: "jet" in the Colormap box, Contour box, and value 4 in "Contour plot levels"(instead of the default).

Additionally, mark the "Height(3D plot)" option in the Plot Selection window to generate a 3D plot with the solution- Figure 7.25.

To transfer the solution and mesh parameters to the MATLAB® workspace, select the Export Mesh and Export Solution lines within the popup menus of the respective Solve and Mesh menu buttons. If the variable names are not changed, the solution in the *u* matrix (each matrix column corresponds to the given times, i.e., *u*(:,1) corresponds to $t = 1$ and *u*(:,11) – to $t = 0.08$) while mesh parameters are within the *p*, *e*, and *t* matrices. After this, to obtain *u*-values at the required orthogonal grid points, the following commands should be entered with the resulting table displayed in the Command Window.

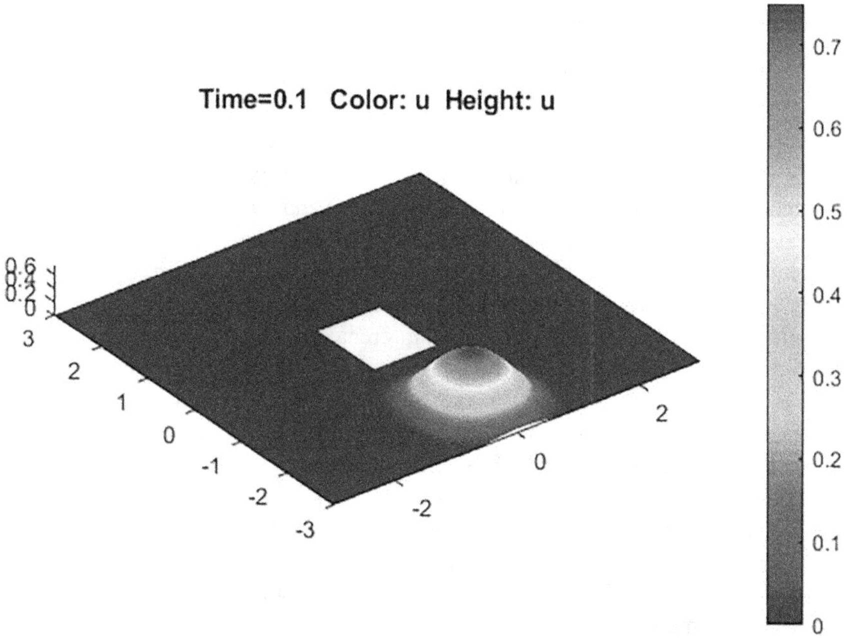

FIGURE 7.25 3D plot with dimensionless temperature distribution at final time.

```
>> x = -.5:0.25:0.5;y = -2.5:0.25:-1.25;
>> T_table = tri2grid(p,t,u(:,end),x,y)
T_table =
0.2685   0.4159   0.4726   0.4242   0.2797
0.4105   0.6043   0.6742   0.6072   0.4185
0.4674   0.6721   0.7489   0.6685   0.4634
0.4246   0.6102   0.6668   0.5943   0.4028
0.2904   0.4248   0.4603   0.3972   0.2548
0.1163   0.1820   0.1957   0.1582   0.0906
```

The PDE Modeler automatically produces a file with commands that follow all the steps in the solution that you complete. To save the program, select the "Save As …" line in the popup File menu and type the name ApExample_7_2 in the "File name" field of the "Save As" panel that appears. After this, to use the file for recalculations, open the saved file in the MATLAB® Editor window and write ApExample_7_2 in the function definition line instead of the default name assigned automatically by the PDE Modeler.

7.4.3 "Structural Mechanics, Plane Stress" Application Example

PDE Modeler has several applications that adapt the generic PDE nomenclature to the form of equations and terminology used in the relevant field of engineering and science. Here we show how the "Structural mechanics, plane stress" application can be used to determine the stress distribution and maximum stress in a rectangular plate with an elliptical hole.

Problem: Using the "Structural Mechanics, Plane Stress" application determine the maximum stress for a 16 × 8 (cm) rectangle plate with an elliptic hole having a semiaxis $a = 2.5$ and a semiaxes $b = 1.25$ (cm). The vertical boundaries of the rectangle and the boundaries of the hole are stress-free. Stresses σ act on two horizontal boundaries and are equal to 100 and −100 (N/cm^2) for positive and negative y-axis directions, respectively. Assume Young's modulus E is $2{\cdot}10^7$ (N/cm^2). The "Refine mesh" option should be used twice. The elliptic type of the PDE is solved. Represent the final y-stress values in the 2D plot with contour lines and the "jet" colormap. Save the program created by the PDE Modeler in a file named ApExample_7_3.

The "Structural Mechanics, Plane Stress" application is designed for thin plates, so the user does not need to formulate a PDE and then compare it to a standard form, just know the PDE type.

There are two main types of boundary conditions in this PDE Modeler application – displacements and surface stresses, which are classified as Dirichlet and Neumann, respectively. In the studied problem, all boundary conditions are Neumann conditions. In the general case, two stress components can act on each boundary segment – in the x and in y directions, denoted by g_1 and g_2. Each of these components typically includes normal and shear stresses. In our problem, there are stresses acting only in y-directions, therefore $g1 = 0$ and $g2 = \pm 100$.

To solve the problem, activate the PDE Modeler with the command

```
>> pdeModeler
```

In the popup menu of the Options menu button, mark the Grid and Snap lines and type the x and y limits as [−9.5 9.5] and [−4.5 4.5], respectively in the appropriate fields of the Axis Limits box (opening from the same Options menu). Set the grid spacing to 1 in the Grid Spacing box, which opens with the "Grid Spacing ..." line of the current menu. Select the Application line and mark the "Structural Mechanics, plane stress" line in the popup menu that appears.

Now draw the studied body, go to Draw Mode and draw a rectangle selecting the 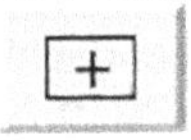 Rectangle/Square (centered) button and placing the mouse pointer at point (0,0) and dragging it, holding down the mouse button, to point (8,4). Check the rectangle geometry with the Object Dialog panel that appears after clicking inside the rectangle. Now let's draw an ellipse by clicking the 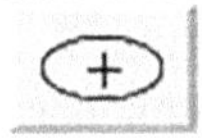 Ellipse/Circle (centered) button. Place the mouse and press its button at the (0,0) point and, keeping the mouse button, drag the mouse in direction of point (2.5,1.25). Click within the ellipse to check the ellipse center and geometry parameters in the appeared Object Dialog panel, and correct the ellipse (if necessary) typing 2.5 and 1.25 in the A- and B-semiaxes fields, respectively. In the Set Formula field, change the expression R1 + E1 to R1 − E1 to "cut" the elliptical hole.

Click the Boundary button of the main menu and select the Boundary Mode line. Place the mouse pointer on the left rectangle edge and click the mouse button in the appeared Boundary Conditions panel, select the Neumann option, and make sure that the g1 and g2 fields contain zeros; do this for the opposite side of the rectangle and for the four boundaries of the elliptical hole. Click on the top rectangle edge and in the appeared Boundary Condition panel select the Neumann option and type 100 in the g2 field; in our problem, the g1 field should contain 0 (default value). Perform the same actions for the bottom rectangle edge, but type now –100 in the g2 field.

Now select PDE Mode from the popup menu of the PDE main menu option, then select the "PDE Specification ..." and mark the Elliptic box on the appeared PDE Specification panel. Enter the 2e7 value in the E field and accept the default Poisson ratio (*nu* = 0.3) and the default mass density (*rho* = 1, not used in the steady-state problem).

Initialize the triangle mesh in Mesh Mode (by clicking the appropriate line in the popup menu of the main menu Mesh button) and then refine the mesh twice by selecting the Refine Mesh line.

Now select the Solve PDE line in the popup menu of the main menu Solve button. PDE Modeler then performs a 2D solution for the *x* displacement (denoted as *u* by default).

Many different results can be best displayed using the Plot Selection window that opens by selecting the "Parameters ..." line of the main menu Plot button. For example, to display the *y*-stress results (σ_{yy}), select *y*-stress line from the popup menu of the top box in the Property column; mark the Contour box and select the "jet" color from the Colormap popup menu. Click then the Plot button. The displayed result is shown in Figure 7.26.

Various data can be obtained by exporting the solution, mesh parameters, and PDE coefficients to the workspace. For example, to obtain the maximum stress σ_{yy} we export solution by selecting the lines "Export Solution ...", "Export Mesh ...", and "Export PDE Coefficients ...", using respectively the main menu buttons Solution, Mesh, and PDE. The panels that open offer and allow you to change the variable names for the exported data. We leave these names unchanged. To obtain maximal stress, the special pdesmech command should be used. This command, entered in the Command Window, looks like this:

```
>> StressY = pdesmech(p,t,c,u,'tensor','syy');
```

Now the maximal stress value σ_{yy} can be determined as:

```
>> syy = max(StressY)
syy =
605.1317
```

Note that all the obtained stresses have a dimension of N/cm^2.

The automatically generated program can be saved by selecting the "Save As ..." line from the popup menu of the main menu Save button. Enter the name

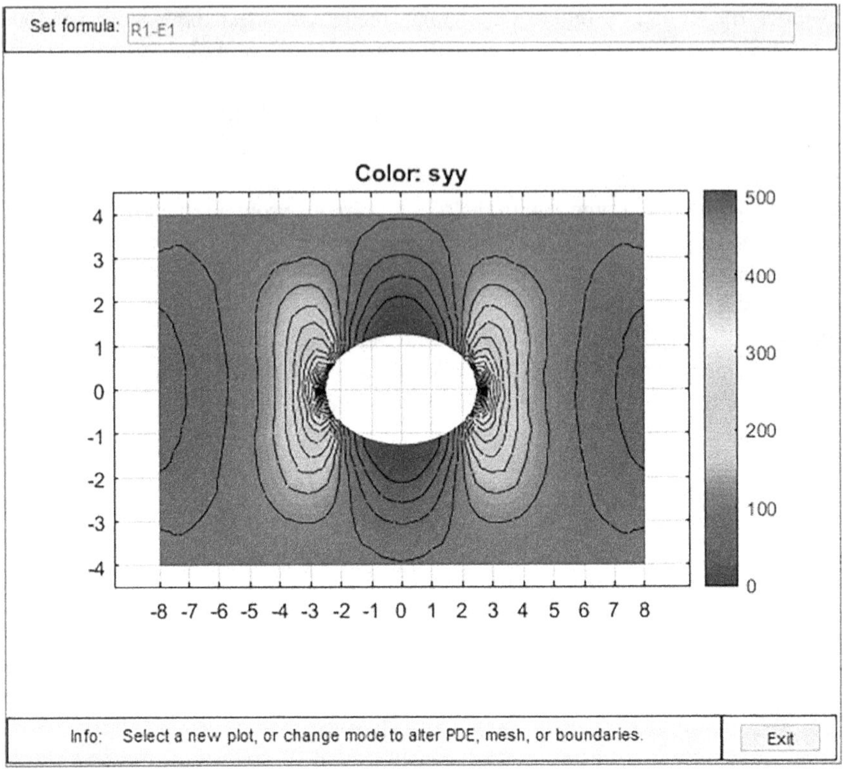

FIGURE 7.26 2D Stress σ_{yy} distribution with contour lines.

ApExample_7_3 in the "File name:" field of the panel that appears. You can open the saved file in the Editor window and write ApExample_7_3 in the function definition line instead of the default auto-generated name.

7.4.4 T-Shaped Membrane

The eigenvalue equation $-\Delta u = \lambda u$, where u is the displacement, λ is the eigenvalue, and $\Delta = \frac{\partial}{\partial x} + \frac{\partial}{\partial y}$ can be solved for oscillating membranes of various shapes (see for example programmatic solutions in Section 6.4.3 and 6.5.3). Here we describe the solution for a T-shaped membrane with two rounded corners.

Problem: Use PDE Modeler to solve an eigenvalue problem for membrane constructed using the Polygon drawing option with the following vertices (–1, 1), (1, 1), (0.5, 0), (0.5, –1), (–0.5, 0), (–1, 0). Two corners at the junction of the vertical and horizontal T-forming parts of the membrane are rounded with a radius of 0.5. Assume u = 0 (Dirichlet conditions) on all boundaries. Set the maximum triangle edge size value to 0.05. Find all solutions for positive eigenvalues λ smaller than 100. Plot a 2D solution with contour lines for the first defined eigenvalue and two 3D solutions for the first and last of the defined eigenvalues. Export the resulting displacements along with the eigenvalues to the workspace and display the exported eigenvalues. Save the auto-generated program in a file named ApExample_7_4.

To solve the problem, open the PDE Modeler window with the command

```
>> pdeModeler
```

In the popup menu of the Options button of the main menu, mark the Grid and Snap lines and type the limits [−1.5 1.5] for x axis and [−1.25 1.25] for the y-axis (in the Axis Limits box that appears after selecting the "Axis Limits …" line). Set the Grid spacing with step 0.25 in the Grid Spacing box (after selecting the "Grid Spacing …" line). Check the Generic Scalar option in the Application popup menu.

To draw the studied shape, click the polygon button , place the mouse arrow at the start point (−1,1), press mouse button, and then drag the mouse sequentially to points (1, 1), (0.5, 0), (0.5, −1), (−0.5, 0), (−1, 0), and (1,1). Check the geometry with the Object Dialog panel that appears after clicking within the polygon.

Activate the Boundary Mode line from the popup menu of the main menu Boundary option and select the Remove All Subdomain Borders line. By default, all boundaries have Dirichlet conditions with h=1 and r=0. You can verify this by clicking each boundary line and checking the Boundary Conditions panel that appears.

Now select the PDE Mode of the popup menu of the PDE main menu option and select the PDE Specification and the Eigenmodes type of PDE found there; enter 1 in the c and d fields, and 0 in the a field.

Initialize the triangle mesh in the Mesh Mode (a line in the popup menu of the Mesh button) and select the "Parameters …" line. In the appearing Mesh Parameters panel, enter 0.05 in the "Maximum edge size" field.

Select the "Parameters …" line in the popup menu of the Solve main menu option and type the eigenvalue range [0 100] in the appeared small box "Eigenvalue search range". Now, click the Solve PDE line of the Solve button. The 2D solution for the first defined eigenvalue (default) has appeared.

To show the contour line and better color representation, mark the Contour box and select the "jet" color in the Plot Selection panel (opened with selecting the "Parameters …" line of the popup menu of the Plot main menu option). The resulting 2D plot is presented in Figure 7.27.

Check the Height(3D) box and uncheck the Contour box to generate the required 3D plot for the first eigenvalue (Figure 7.28a), then click the Plot button. To plot results with the last of the defined eigenvalues, select the last value in the Eigenvalue dropdown list and click the Plot button. The generated graph is shown in Figure 7.28b.

To transfer the solution and eigenvalues to the MATLAB® workspace, select the Export Solution line from the popup menu of the main menu Solve button.

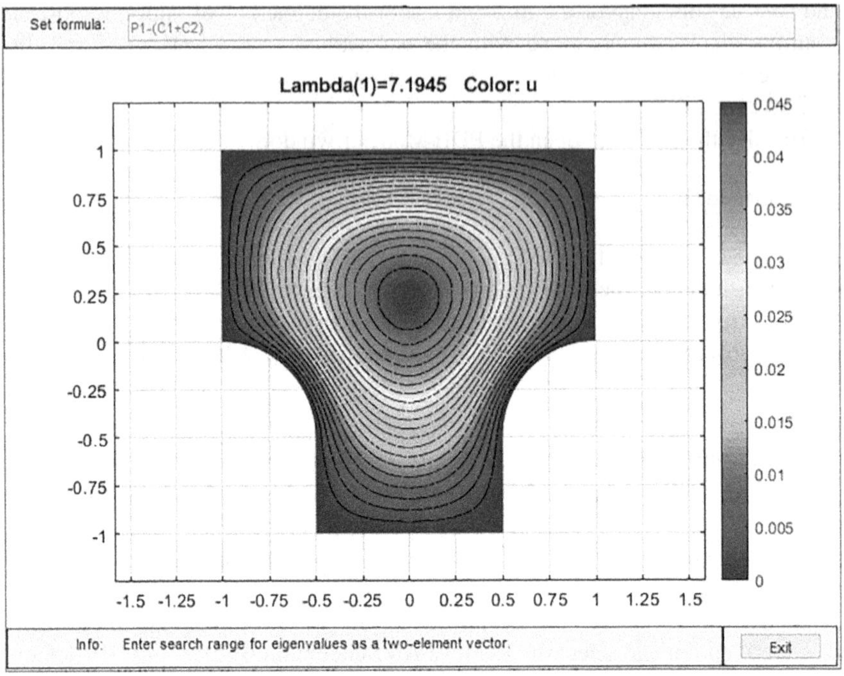

FIGURE 7.27 2D plot of the solution for the first eigenvalue with contour lines.

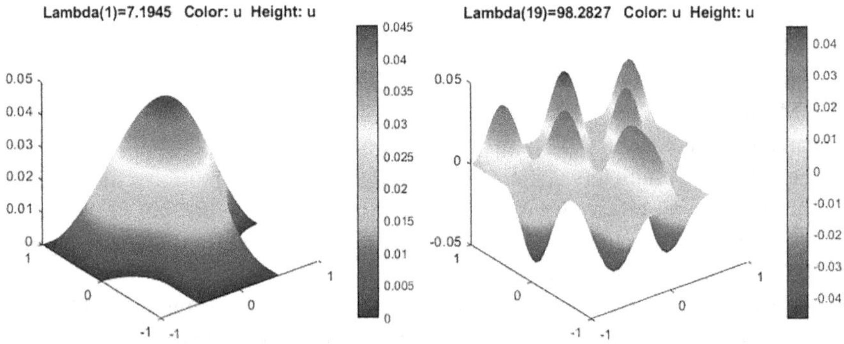

FIGURE 7.28 3D plots of the solution for the first (a) and last (b) eigenvalues.

The Export box that appears offers two names *u* and *l* for a displacement matrix and an eigenvalue vector. If the variable names have not been changed, the solution is in *u* matrix (each matrix column corresponds to the serial number of the eigenvalue, for example, in our problem *u*(:,1) corresponds to the *l*(1) = 7.1945, and *u*(:,end) to the *l*(19) = 98.2827). To display all eigenvalues, type in the Command Window:

```
>> l
l =
7.1945
16.1294
16.6835
27.8105
30.3489
33.3511
44.3008
46.5649
48.7736
53.1827
62.2143
66.1365
70.1533
74.8495
79.1663
83.5748
92.4967
97.2894
98.2827
```

To save a program created by PDE Modeler, select the “Save As …” line of the File popup menu and type the name ApExample_7_4 in the “File name:” field of the panel that appears. After that, open the saved file in the MATLAB® Editor and write ApExample_7_4 in the function definition line instead of the generated default name.

8 Solving One-Dimensional Partial Differential Equations

8.1 INTRODUCTION

In the partial differential equations (PDEs) discussed in the preceding two chapters, an unknown function depends on two spatial variables. However, in many real cases, it is sufficient to solve one-dimensional (1D) PDEs, in which the unknown function depends on one spatial and one transient variable. Many real processes and phenomena in technology and physics are described with one PDEs, such as, for example, dynamics, heat and mass transfer, lubrication hydrodynamics, stress and strain analysis, engineering, machine parts, and many others. Here we present the pdepe solver with some sample engineering applications that provide solutions of relevant spatially 1D partial differential equations[1]. In particular, the transient diffusion equation with Neumann boundaries and piecewise initial conditions coupled PDEs in an action potential model incorporating a diffusion term, pipe flow, and Bateman-Burgers PDE.

8.2 STANDARD FORMS FOR 1D PDE AND INITIAL AND BOUNDARY CONDITIONS

For solving 1D PDEs, MATLAB® provides a special solver that can be used to solve one or set of PDEs under certain initial and boundary conditions. The solution can be realized when the solving equation, as well as the initial and boundary conditions, corresponds to some general (here called "standard") form, presented below.

8.2.1 ONE-DIMENSIONAL PDE IN STANDARD FORM

To solve 1D PDEs, the pdepe command should be used. The command is designed to solve a PDE matching the following standard form

$$c\left(x,t,u,\frac{\partial u}{\partial t}\right)\frac{\partial u}{\partial t}=x^{-m}\frac{\partial}{\partial x}\left(x^{m}f\left(x,t,u,\frac{\partial u}{\partial x}\right)\right)+s\left(x,t,u,\frac{\partial u}{\partial x}\right)$$

[1] Some text and table materials from Burstein, 2021a (Section 7.3) and Burstein, 2020 (Sections 6.2.3, 6.4.2) are used in the chapter; with permissions from IGI Global and Elsevier respectively

DOI: 10.1201/9781003200352-8

where:

u is the function to be determined as a result of the PDE solution;

t is the time, the range of which varies from the initial t_0 to the final t_f value;

x is the coordinate, the range of which varies from $x = a$ to $x = b$;

m is an integer equal to 0, 1, or 2 (called the symmetry constant), which corresponds to a Cartesian, cylindrical, or spherical coordinate system, respectively;

$c\left(x,t,u,\frac{\partial u}{\partial t}\right)$ is a coefficient that can be function of the x, t, u, and $\frac{\partial u}{\partial x}$, either of each of these variables or any combinations of them;

$f\left(x,t,u,\frac{\partial u}{\partial x}\right)$ is called a flux term that can be function of the x, t, u, and $\frac{\partial u}{\partial x}$, either of each of these variables or any combinations of them;

$s\left(x,t,u,\frac{\partial u}{\partial x}\right)$ is the source term that can be function of the x, t, u, and $\frac{\partial u}{\partial x}$, either of each of these variables or any combinations of them.

The general form of this equation solution is u as a function of coordinate and time – $u(x,t)$.

According to the standard equation form, the **pdepe** command is designed to solve PDEs of the first or second order with respect to the x coordinate. In the accepted PDE type classification (Chapter 6), the **pdepe** command resolves elliptic or parabolic PDE types. Examples of regular PDEs and their adaptation to the standard equation are presented in Table 8.1.

8.2.2 Proper Form of the Initial Conditions

Solution of the transient PDE should be accompanied by certain initial conditions setting the starting (t_0) u-values for all coordinate points. Initial values should be presented as:

$$u(x,t_0)=u_0(x)$$

where u_0 denotes the values of the function u at starting time t_0; for example, if $u(x,t_0) = 2$ (constant u value at all x-points at the initial time t_0), thus $u_0 = 2$; or if $u(x,t_0) = 5x–1$ (coordinate-dependent u value at the initial time t_0), thus $u_0 = 5x–1$.

8.2.3 Proper Form of the Boundary Conditions

The boundaries represent two extreme points of the x-interval in the 1D problem. The values of u and/or $\frac{\partial u}{\partial x}$ for each of the both $x = a$ and $x = b$ boundaries (a and b are the x-interval ends) should be set in the following general form

$$p(x,t,u)+q(x,t)f\left(x,t,u,\frac{\partial u}{\partial x}\right)=0$$

where f is the same function as in the above standard form of PDE; $p(x,t,u)$ is a function of the x, t, and u or any their combination; and q is a function of x and/or t.

TABLE 8.1
PDEs in Regular and pdepe-Adopted Forms

Regular PDE	Adopted PDE	x	m	c	f	s
		Relevant Coordinate and PDE Terms				
$\frac{\partial u}{\partial t}-\frac{1}{x}\frac{\partial}{\partial x}\left(x\frac{\partial u}{\partial x}\right)=0$	$\frac{\partial u}{\partial t}=x^{-1}\frac{\partial}{\partial x}\left(x^{1}\frac{\partial u}{\partial x}\right)$	x	1	1	$\frac{\partial u}{\partial x}$	0
$\frac{1}{x}\left(\frac{\partial u}{\partial t}\right)-\frac{\partial}{\partial x}\left(\frac{1}{t}u\right)=0$	$\frac{1}{x}\left(\frac{\partial u}{\partial t}\right)=\frac{\partial}{\partial x}\left(\frac{1}{t}u\right)$	x	0	$\frac{1}{x}$	$\frac{1}{t}u$	0
$\left(\frac{\partial u}{\partial t}\right)+u\frac{\partial u}{\partial x}=\nu\frac{\partial^2 u}{\partial x^2}$	$\left(\frac{\partial u}{\partial t}\right)=\frac{\partial u}{\partial x}\left(\nu\frac{\partial u}{\partial x}\right)-u\frac{\partial u}{\partial x}$	x	0	1	$\nu\frac{\partial u}{\partial x}$	$-u\frac{\partial u}{\partial x}$
$\frac{\partial T}{\partial t}=\frac{k}{\rho c_p}\frac{\partial^2 T}{\partial x^2}+\frac{q}{\rho c_p}$	$\frac{\partial u}{\partial t}=\frac{\partial}{\partial x}\left(\frac{k}{\rho c_p}\frac{\partial u}{\partial x}\right)+\frac{q}{\rho c_p}$	x	0	1	$\frac{k}{\rho c_p}\frac{\partial T}{\partial x}$	$\frac{q}{\rho c_p}$
$\rho C_p\frac{\partial T}{\partial t}=\frac{1}{r}\frac{\partial}{\partial r}\left(kr\frac{\partial T}{\partial r}\right)$	$\rho C_p\frac{\partial T}{\partial t}=x^{-1}\frac{\partial}{\partial x}\left(x^{1}\left(k\frac{\partial T}{\partial x}\right)\right)$	r	1	ρC_p	$k\frac{\partial T}{\partial r}$	0
$\frac{\partial \varphi}{\partial t}=D\frac{\partial^2 \varphi}{\partial x^2}$	$\frac{\partial u}{\partial t}=\frac{\partial}{\partial x}\left(D\frac{\partial \varphi}{\partial x}\right)$	x	0	1	$D\frac{\partial \varphi}{\partial x}$	0
$\pi^2\frac{\partial u}{\partial t}=\frac{\partial^2 u}{\partial x^2}$	$\pi^2\frac{\partial u}{\partial t}=\frac{\partial}{\partial x}\left(\frac{\partial u}{\partial x}\right)$	x	0	π^2	$\frac{\partial u}{\partial x}$	0
$\frac{\partial u}{\partial t}=a\frac{\partial^2 u}{\partial x^2}-Fu$	$\frac{\partial u}{\partial t}=\frac{\partial}{\partial x}\left(a\frac{\partial u}{\partial x}\right)+(-Fu)$	x	0	1	$a\frac{\partial u}{\partial x}$	$-Fu$
$\frac{\partial u}{\partial t}=\frac{5}{r^2}\frac{\partial u}{\partial r}\left(r^2\frac{\partial u}{\partial r}\right)+q$	$\frac{\partial u}{\partial t}=x^{-2}\frac{\partial}{\partial x}\left(x^2\left(5\frac{\partial u}{\partial x}\right)\right)$	r	2	1	$5\frac{\partial u}{\partial x}$	q

The certain boundary conditions should be rewritten to match the standard form $p + qf = 0$. As the f function is determined at PDE; therefore, we need to assign values for two variables only $-p$ and q – to set the actual boundary conditions. For example, if $u = 5$ at a boundary point thus matching this equation, $u - 5 = 0$, and the standard boundary equation we need to require $p = u - 5$ and $q = 0$. In case $du/dx = 0$ at a boundary point and $f = D * du/dx$ in our PDE, the matching with the standard boundary form gives $p = 0$ and $q = 1$. Some possible boundary conditions and corresponding p and q values are summarized in Table 8.2.

In this table:

- It is assumed that the function f is equal to $k(du/dx)$, in general it is the same as the PDE being solved; therefore, to match the term f in BC and PDE, both sides of the BC equation are multiplied by k.
- The g is a constant or a function of the x and t while the h is a constant or a function of the x, t, and u.

TABLE 8.2
Boundary Conditions (BC) and Corresponding *p* and *q* That Should Be Specified

BC and *f* of the PDE	BC Rewritten in Standard Form	Required *p* and *q*
$u = 0$ and any f	$u = 0$	$p = u$, $q = 0$
$u = 0$ and any f, cylindrical or spherical coordinates	$u = 0$	At the one boundary $p = u$, $q = 0$ and at the second boundary ($r = 0$) $p = 0$, $q = 0$
$u = 4.1$ and any f	$U{-}4.1 = 0$	$p = u{-}4.1$, $q = 0$
$\frac{\partial u}{\partial x} = 0$ for PDE containing $f = k\frac{\partial u}{\partial x}$	$k\frac{\partial u}{\partial x} = 0$	$p = 0$, $q = 1$
$\frac{\partial u}{\partial x} = 5.1$ for PDE containing $f = k\frac{\partial u}{\partial x}$	$-5.1k + k\frac{\partial u}{\partial x} = 0$	$p = -5.1k$, $q = 1$
$\frac{\partial u}{\partial x} = 2.5{-}u$ for PDE containing $f = k\frac{\partial u}{\partial x}$	$-2.5k + uk + k\frac{\partial u}{\partial x} = 0$	$p = -2.5k + uk$, $q = 1$
$g\cdot\frac{\partial u}{\partial x} = h$ for PDE containing $f = k\frac{\partial u}{\partial x}$	$-\frac{h}{g}k + k\frac{\partial u}{\partial x} = 0$	$p = \frac{h}{g}k$ $q = 1$

Note, in case of the cylindrical or spherical PDE ($m = 1$ or 2), p and q must be set equal to zero at the a-boundary ($r = 0$).

The above conditions, except for the case of cylindrical and spherical coordinates, shall apply to each of the boundaries a and b.

8.2.4 About Finite Difference Methods for Solving PDEs

In Chapter 6, the finite element methods were explained in a simplified manner; here we introduce a simple explanation of the finite difference methods used in numerical solutions of the ODEs (ordinary differential equations) and 1D PDEs. The numerical solution of PDE is performed by replacing the derivatives with the finite difference. In the case of a PDE, there are spatial time-dependent differences. Consider the 1D second-order ODE describing diffusion/adhesion-like processes that in its dimensionless has the form $\frac{\partial u}{\partial t} = D\frac{\partial^2 u}{\partial x^2}$, where D is a constant. Discretizing the space interval in $N + 1$ (from 0 to N) and the time interval in $M + 1$ (from 0 to M) evenly spaced points, this equation can be written as

$$\frac{u_i^{k+1}-u_i^k}{\Delta t}=D\frac{u_{i+1}^k-2u_i^k+u_{i-1}^k}{\Delta x^2}$$

where i and k are the current numbers of the spatial and time point, respectively, while $\Delta x = x_{i+1}^k - x_i^k$ and $\Delta t = t^{k+1} - t^k$ are the x and t differences between adjacent points that are assumed to be constants.

Setting all the u-values at starting time t_s (i.e. $k = 0$) and at the boundary point $x = 0$ (i.e. $i = 0$), we can calculate u at time $k = 1$ for each of the coordinate points, so for the $i = 1$-point:

$$u_1^1=u_1^0+D\frac{u_2^0-2u_1^0+u_0^0}{\Delta x^2}\Delta t=r(u_2^0+u_0^0)+(1-2r)u_1^0$$

where $r=D\frac{\Delta t}{\Delta x^2}$,

for $i = 2$,

$$u_2^1=u_2^0+D\frac{u_3^0-2u_2^0+u_1^0}{\Delta x^2}\Delta t=r(u_3^0+u_1^0)+(1-2r)u_2^0$$

and so on until $i = N-1$

$$u_{N-1}^1=u_{N-1}^0+D\frac{u_N^0-2u_{N-1}^0+u_{N-2}^0}{\Delta x^2}\Delta t=r\left(u_N^0+u_{N-2}^0\right)+(1-2r)u_{N-1}^0$$

At this point, the u value at the boundary $i = N$ is used, which should be given. Solving these equations we can determine all u-values for $k = 1$ time point, the next time point $k = 2$ can be calculated in the same way. This process can be continued until a given end point in time $k = M$.

The script realizing this scheme for mentioned diffusion/adhesion PDE is

```
D = .1;
N = 10;M = 12;
ua = 10;ub = 20;
xs = 0;ts = 0;xf = 0.5;tf = 0.1;
x = linspace(xs,xf,N);t = linspace(ts,tf,M);
dx = x(2)-x(1);dt=t(2)-t(1);
u = ua*ones(N-1,M);
u = [u; ub*ones(1,M)];
r = D*dt/dx^2;
for k = 1:M-1
    for i = 2:N-1
        u(i,k+1) = r*(u(i+1,k)+u(i-1,k))+(1-2*r)*u(i,k);
    end
end
```

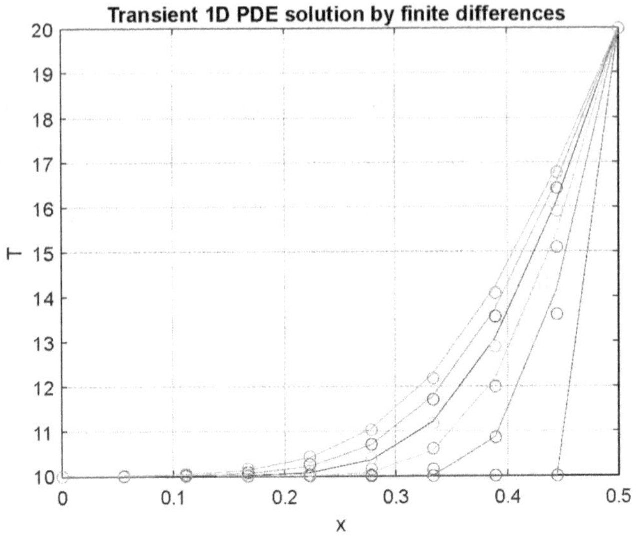

FIGURE 8.1 Solution of the transient 1D PDE using a simple finite difference scheme (solid lines) and using the pdepe command (o-points).

Results of calculations by this script were compared with the further studied pdepe command – Figure 8.1. Here the above finite differences scheme is presented with solid lines, and the most accurate pdepe calculations are marked with o. The curves correspond to the time values 0, 0.182, 0.364, 0.545, 0.727, and 0.0909. As it can be seen, there is a sufficiently good coincidence.

The described explicit scheme, with some improvements and complications, is applicable to all finite difference methods used to solve 1D PDEs. More sophisticated methods are taught in special courses on numerical methods, and their explanation is beyond the scope of this book.

8.3 THE PDEPE COMMAND, ITS INPUT AND OUTPUT

A single or a set of spatially one-dimensional PDEs can be solved using the pdepe command, which has the following form:

```
sol = pdepe(m,@PDE,@IC,@BC,x_mesh,t_span)
```

- the input parameters of this command:
  - m is a symmetry constant equal to 0, 1, or 2 for Cartesian, cylindrical, or spherical coordinate system, respectively (the corresponding Laplacians have the forms $\frac{\partial^2 u}{\partial x^2}$, $\frac{1}{\rho}\frac{\partial}{\partial \rho}\left(\rho\frac{\partial u}{\partial \rho}\right)+\frac{1}{\rho^2}\frac{\partial^2 u}{\partial \varphi^2}$, and $\frac{1}{r^2}\frac{\partial}{\partial r}\left(r^2\frac{\partial u}{\partial \rho}\right)+\frac{1}{r^2\sin\theta}\frac{\partial}{\partial \theta}\left(\sin\theta\frac{\partial u}{\partial \theta}\right)+\frac{1}{r^2\sin^2\theta}\frac{\partial^2 u}{\partial \varphi^2}$).

- PDE is the name of an user-defined function that must contain the terms of the equation to be solved; definition line of this function reads

```
function [c,f,s] = PDE(x,t,u,DuDx)
```

  the input and output variables x, t, u, DuDx, c, f, and s being the same x, t, u, $\frac{du}{dx}$, c, f, and s as in the standard form of the PDE expression (Section 8.2.1); the DuDx denotes the derivative $\frac{du}{dx}$ of the standard PDE expression. In the case of two or more PDEs, the c, f, and s terms in the standard equation should be written as column vectors, and therefore, elementwise operations should be applied to these vectors.
- IC is the name of a user-defined function where the initial conditions should be written. The definition line of this function reads

```
function u0 = IC(x)
```

- u0 being the vector with the $u(x)$ value/s at t=0;
  - BC is the name of a user-defined function that should contain the boundary conditions for each of both $x = a$ and $x = b$ (a and b – x-interval ends). The definition line of the BC function reads

```
function [pa,qa,pb,qb] = BC(xa,ua,xb,ub,t)
```

    xa=a and xb=b being the coordinates of the boundary points, ua and ub are the u values at these points; pa, qa, pb, and qb represent the p and q values of the standard boundary condition form (see Section 8.2.3) given at the a and b boundary points at time t.
  - x_mesh is a vector of x-coordinates at which a solution is sought for each time value contained in t_span; it is required to write the x_mesh values in ascending order from a to b;
  - t_span is a vector with the values of the time points, which should be written in ascending order;
- the output parameter of this command:

sol denotes the found solution represented as a 3D array comprising k (number of solving PDEs) 2D arrays with M (number of time points) rows and N (number of coordinate points) columns each. Elements in the 3D arrays are numbered similarly to 2D ones (see Section 2.3). For example, sol(2,4,2) denotes a term located in the second row and fourth column of the second array (termed sometimes as matrix page) while sol(2,:,1) denotes in the first array (page) all columns of the second row. In general, the sol(i,j,k) notates defined u-values for k-th PDE (in case more than one PDE) at the t_i time and x_j coordinate points. Therefore, for example, sol (:,:,1) is the 2D array containing the solution for a single or first PDE; each row of this array corresponds to a certain moment in time and contains u values calculated for the x coordinate points.

For more information, enter command >>doc pdepe or use the MATLAB® Help window.

8.3.1 The Steps of Solution by Example

Consider, for example, the simple thermo-diffusion equation without a heat source. Solution of this equation describes the temporal and spatial distributions of temperature u, for example, for some mechanical part (such as an insulated wire or thin rod):

$$\frac{\partial u}{\partial t} = k\frac{\partial^2 u}{\partial x^2}$$

where t and x are the time and coordinate, respectively, while k is the diffusivity coefficient. All variables in this equation are given in dimensionless units; we assume k is 1, x is in the range from 0 to 1 and t is in the range from 0 to 0.4.

This equation is taken with initial and boundary conditions as below.

The initial conditions:

$$u(x,0)=0.4+0.6x$$

The boundary conditions:

$$u(0,t)=u_a=0.4,\ u(1,t)=u_b=0.4$$

Problem: Arrange a live program named ExStepsPDEPE that solves the above equation for 20 x and 20t points with given initial and boundary conditions. The program should have a function ExSteps that presents the result on two graphs located on the same page: one as a 3D graph of $u(x,t)$ and the other as a 2D graph of $u(x)$ at three time values t = 0.001, 0.01, and 0.1 (using the contour command).

The solution of this PDE goes through the following steps:

1. First, the equation must be presented in the standard form. The heat-diffusion equation rewritten in the required form reads

$$\frac{\partial u}{\partial t} = \frac{\partial}{\partial x}\left(k\frac{\partial u}{\partial x}\right)$$

Matching this and the standard PDE equations, we obtain their identity when:

$$m = 0,$$

$$c\left(x,t,u,\frac{\partial u}{\partial t}\right)=1,$$

$$f\left(x,t,u,\frac{\partial u}{\partial x}\right)=k\frac{\partial u}{\partial x},$$

$$s\left(x,t,u,\frac{\partial u}{\partial x}\right)=0$$

Appropriate PDE function, written in the Live Editor, should look as:

```
function [c,f,s] = PDE(x,t,u,DuDx) % PDE terms for solution
c = 1;
k = 1;
f = k*DuDx;
s = 0;
end
```

2. The initial and boundary conditions should also be presented in the standard form. Accordingly, the initial condition:

$$u_0 = 0.4+0.6x$$

Thus the IC function is

```
function u0 = IC(x) % initial conditions
u0 = 0.4+0.6*x;
end
```

and the boundary conditions at the point boundaries a ($x = x_a = 0$) and b ($x = x_b = 1$) are the first-type boundary conditions (termed frequently Dirichlet boundary conditions):

$$u_a = 0.4 \ \ u_b = 0.4$$

Comparing these equations with the $p + qf = 0$ standard form (Table 8.2, row 2), we can conclude that for the boundary points:

$$p_a = u_a - 0.4, \ q_a = 0$$

$$p_b = u_b - 0.4, \ q_b = 0$$

Therefore, the BC function is

```
function [pa,qa,pb,qb] = BC(xa,ua,xb,ub,t) % boundary
conditions
pa = ua-0.4;pb = ub-0.4;
qa = 0;qb = 0;
end
```

3. At this stage, the pdepe command should be written together with the m, x_mesh and t_span parameters of this command.

```
m = 0;
x_mesh = linspace(0,1,n_x);
t_span = linspace(0,0.4,n_t);
u = pdepe(m,@myPDE,@i_c,@b_c,x_mesh,t_span);   % pdepe
command
```

Here n_x and n_t are the desired x coordinate and time numbers, respectively.

Now we represent a live user-defined program named ExStepsPDEPE that contains the ExSteps function and running command solving the above equation by the described steps and showing the results in two graphs on the same page:

the first – 3D plot showing the obtained temperatures along the coordinate and at each of given time values (Figure 8.2a),
the second – 2D plot represented temperatures at three times – 0.001, 0.01, and 0.1 (Figure 8.2b).

```
ExSteps (25,25)
function ExSteps(n_x,n_t)
%          solves an one-dimensional heat-diffusion PDE with
Dirichlet BC
%                                  To run: >>ExSteps(25,25)
m=0;
x_mesh=linspace(0,1,n_x);
t_span= linspace(0,0.4,n_t);
u=pdepe(m,@PDE,@IC,@BC,x_mesh,t_span);
[X,T]=meshgrid(x_mesh,t_span);
subplot(1,2,1)
mesh(X,T,u)                % mesh plot with the X - T domain
xlabel('Coordinate'),ylabel('Time'), zlabel('Temperature')
title('1D heat-diffusion with the pdepe')
subplot(1,2,2)
c=contour(X,u,T,[0.001 0.005 0.05]);      % 3 iso-time lines
clabel(c)     % labels for the iso-lines in the contour plot
xlabel('Coordinate, nondim'), ylabel('Temperature, nondim')
title('Temperatures at three times')
grid
end
function [c,f,s]=PDE(~,~,~,DuDx)            % PDE for solution
c=1;
k=1;
f=k*DuDx;
s=0;
end
function u0=IC(x)                          % initial conditions
u0=0.4+0.6*x;
```

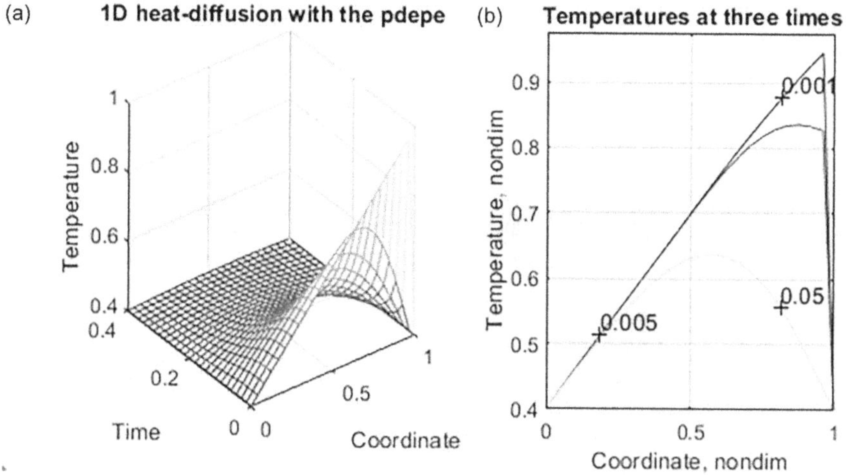

FIGURE 8.2 The solution with the live user-defined function: coordinate – time-temperature 3D mesh plot (a) and coordinate-temperature 2D contour plot with three iso-time lines (b).

```
end
function [pa,qa,pb,qb]=BC(~,ua,~,ub,~) % boundary conditions
pa=ua-0.4;pb=ub-0.4;
qa=0;qb=0;
end
```

The ExSteps function is written without output arguments. Its input arguments n_x and n_t are the serial numbers of the *x*- and *t* points that should be specified for the solution. The pdepe function calls here three sub-functions: PDE comprising the terms of the heat-diffusion equation, IC defining the initial condition, and BC defining the boundary conditions. The xmesh and tspan vectors are generated using two linspace commands, in which n_x and n_t must be entered when the ExStepsPDEPE run. The numerical results are stored in the u array, which in our case is 2D, and used in subsequent graphical commands. The mesh command is used to generate a 3D mesh subplot $u(x,t)$ while the contour and clabel commands are used to draw a 2D contour subplot $u(x)$ showing three labeled iso-time curves (0.001, 0.01, and 0.1).

Note: The input parameters that are not used inside sub-functions can be replaced with the ~ tilde sign (usually recommended by the Code Analyzer).

8.3.2 Passing Additional Parameters to the PDE, IC, and PC Functions

Besides dependent and independent variables, the PDE can include some additional parameters (e.g., constant *k* in the above example), the pdepe command can be written in the following form to pass these parameters to its sub-functions – PDE, IC, and BC:

```
sol = pdepe(m,myPDE,myIC,myBC,x_mesh,t_span)
```

where:

myPDE, myIC, and myBC are names of anonymous functions passing parameters to the PDE, IC, and BC functions, respectively. They should look like

```
myPDE = @(x,t,u,DuDx)PDE(x,t,u,DuDx,prm1, prm2,…)
myIC = @(x)IC(x, prm1, prm2,…)
myBC = @(xa,ua,xb,ub,t)BC(xa,ua,xb,ub,t,prm1, prm2,…)
```

where prm1, prm2,… are parameters that should be passed; other parameters are the same as described in Section 8.3.

For example, the ExStepsPDEPE live program can be modified in the following way to introduce the *k* coefficient in the PDE function of the pdepe commands:

```
ExStepsWithPrms(25,25,1)
function ExStepsWithPrms(n_x,n_t,k)
%            solves an one-dimensional heat-diffusion PDE
with Dirichlet BC
%                            To run: >>ExStepsWithPrms(25,25,1)
m=0;
xmesh=linspace(0,1,n_x);
tspan= linspace(0,0.4,n_t);
myPDE=@(x,t,u,DuDx)PDE(x,t,u,DuDx,k);         % anonymous fnct
passing k
u=pdepe(m,myPDE,@IC,@BC,xmesh,tspan);
[X,T]=meshgrid(xmesh,tspan);
subplot(1,2,1)
mesh(X,T,u)                  % mesh plot with the X - T domain
xlabel('Coordinate'),ylabel('Time'), zlabel('Temperature')
title('1D heat-diffusion with the pdepe')
subplot(1,2,2)
c=contour(X,u,T,[0.001 0.005 0.05]);       % 3 iso-time lines
clabel(c)                    % labels for the iso-lines in the
contour plot
xlabel('Coordinate, nondim'), ylabel('Temperature, nondim')
title('Temperatures at three times')
grid
end
function [c,f,s]=PDE(~,~,~,DuDx,k)          % PDE for solution
c=1;
f=k*DuDx;
s=0;
end
```

```
function u0=IC(x)                                % initial conditions
u0=0.4+0.6*x;
end
function [pa,qa,pb,qb]=BC(~,ua,~,ub,~) % boundary conditions
pa=ua-0.4;pb=ub-0.4;
qa=0;qb=0;
end
```

To keep the previous program, this function can be saved in a file with a new name, for example, ExStepsPDEPE_Params. The results of its run are the same as above (Section 8.3.2).

The advantage of the presented pdepe command form is its greater versatility, since many additional parameters can be passed and used in the functions of this command.

8.4 APPLICATION EXAMPLES

The following engineering-oriented applications continue acquaintance with the pdepe command by real examples, which include some additional materials useful for solving 1D PDE, in particular, the problem of solving a pair of PDEs, Neumann bounds, piecewise initial conditions, and some others.

8.4.1 Transient 1D Diffusion Equation with Neumann Boundaries and Piecewise IC

Consider a simple diffusion equation describing the temporal and spatial changes in substance concentration u:

$$\frac{\partial u}{\partial t} = D\frac{\partial^2 u}{\partial x^2}$$

The u here is a substance concentration, and other variables are the same as in Section 8.3. Notwithstanding, we intend to solve this equation with the Neumann boundary conditions and piecewise IC, instead of the Dirichlet boundaries and continuous IC studied above. Assumed BC and IC are more realistic as the final concentrations at boundaries are not known in advance, and there is a section of high initial concentration.

Specify IC as:

$$u(x,0)=\{1,\ 0.4\leq x\leq 0.6\ \ 0,\ elsewhere$$

and the BC:

$$D\frac{\partial u(0,t)}{\partial x}=0,\ D\frac{\partial u(1,t)}{\partial x}=0$$

Problem: Compose a user-defined function with name ApExample_8_1 that calculates and generates a graph of the concentrations u along coordinate x from 0 to 1 and at t from 0 to 1. Use the point numbers of x and t as input parameters, and 2D matrix as an output parameter; each row of this matrix must represent u-values along the x coordinate at a certain t. To better view, present plot with azimuth angle 230° and elevation angle 30°. Take the x and t point numbers each equal to 20. To simplify the output, set the outputting u-matrix so that it has only five lines (No 1, 5, 10, 15, and 20) and six columns (No 1, 5, 10, 11, 15, and 20).

To solve the problem, the PDE, IC, and BC must be represented in standard form.

Rewriting the diffusion equation

$$\frac{\partial u}{\partial t} = \frac{\partial}{\partial x}\left(D\frac{\partial u}{\partial x}\right)$$

and comparing with the standard form equations (presented early in Section 8.2), we obtain that they are identical when:

$$m = 0,$$

$$c\left(x,t,u,\frac{\partial u}{\partial t}\right) = 1,$$

$$f\left(x,t,u,\frac{\partial u}{\partial x}\right) = D\frac{\partial u}{\partial x},$$

$$s\left(x,t,u,\frac{\partial u}{\partial x}\right) = 0$$

The IC in standard form is

$$u_0 = \{1,\ 0.45 \le x \le 0.55\ \ 0,\ \ elsewhere$$

The BCs in standard form (p + gf = 0) are

$$0 + 1\cdot D\frac{\partial u}{\partial x} = 0 \text{ at } x = x_a = 0, \quad \text{and} \quad 0 + 1\cdot D\frac{\partial u}{\partial x} = 0 \quad \text{at} \quad x = x_b =$$

and in standard notations:

$$p(0,t,u) = 0,\ p(1,t,u) = 0$$

$$q(0,t) = q_a = 1,\ q(1,t) = q_b = 1,$$

$$f\left(x,t,u,\frac{\partial u}{\partial x}\right) = D\frac{\partial u}{\partial x}$$

The commands for solving this problem are:

```
function u_tab=ApExample_8_1(n_x,n_t)
% Diffusion with Neumann boundaries and piecewise initial
cond.
%                       To run: >> u_tab=ApExample_8_1(20,20)
x_mesh=linspace(0,1,n_x);t_span= linspace(0,1,n_t);
m=0;
u=pdepe(m,@PDE,@IC,@BC,x_mesh,t_span);
u_tab= u([1 5:5:n_t],[1 5 10 11 15 n_x]);
[X,T]=meshgrid(x_mesh,t_span);
surf(X,T,u)
xlabel('Coordinate'),ylabel('Time'), zlabel('Concentration')
title({'Transient 1D diffusion';'Neumann BC, piecewise IC'})
view(230,30)
axis square

function [c,f,s]=PDE(~,~,~,DuDx)                            % PDE
c=1;
D=0.1;
f= D*DuDx;
s=0;

function u0=IC(x)                              % initial condition
u0=(x>=0.4&&x<=0.6)

function [pa,qa,pb,qb]=BC(~,~,~,~,~)    % boundary conditions
pa=0;pb=0;
qa=1;qb=1;
```

Note: Here the IC function produces 1 in the range x 0.4 …0.6 and 0 out of range due to the logical operator's property to produce 1 if the results are true, or 0 if they are false.

After saving this program in the ApExample_8_1 file and entering the running command in the Command Window, the following table and graph (Figure 8.3) appear.

```
>>u_table=ApExample_8_1(20,20)
u_table =
     0        0   1.0000   1.0000        0        0
0.0507   0.1571   0.3929   0.3929   0.2121   0.0507
0.1492   0.1952   0.2716   0.2716   0.2153   0.1492
0.1885   0.2051   0.2323   0.2323   0.2123   0.1885
0.2026   0.2086   0.2184   0.2184   0.2112   0.2026
```

Each row in the resulting table presents concentrations along coordinate at a certain time.

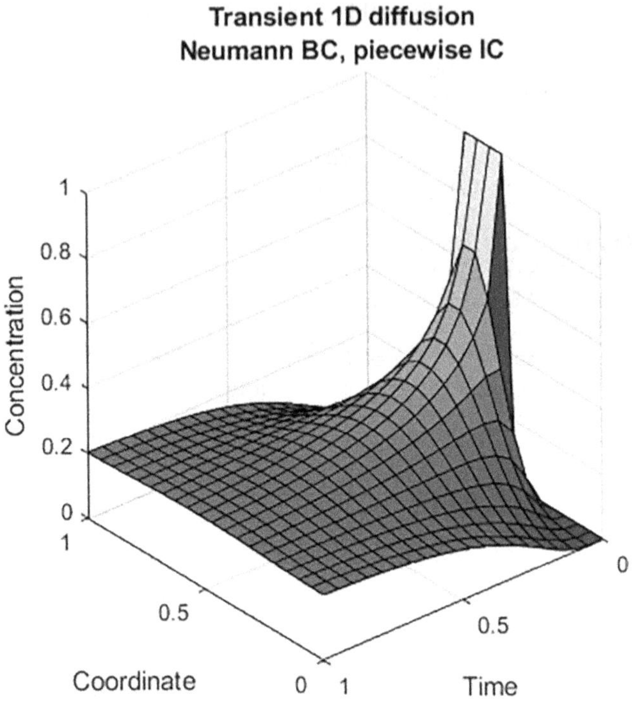

FIGURE 8.3 Resulting concentration changes along coordinate and over time.

8.4.2 Action Potential Model with Diffusion Term

Many models in various fields, such as image processing, wave dynamics, nerve spike propagation, and others, are described by FitzHugh-Nagumo-type PDE (abbreviated as the FHN equation) with the diffusion term:

$$\frac{\partial u}{\partial t} = D\frac{\partial^2 u}{\partial x^2} + u(a-u)(u-1) - w$$

$$\frac{\partial w}{\partial t} = bu - cw$$

where u and w denote the action potential (e.g., electrical) and the recovery processes, respectively; D is the diffusion coefficient; a, b, and c are parameters of the state of rest and dynamics of the system.

The ICs can vary depending on the form of actual starting pulse and are simulated here as

$$u(x,0) = e^{-x^2}$$

$$w(x,0) = e^{-(x+2)^2}$$

The BCs at both ends, l and r (left and right), of the solving x-range are:

$$\frac{\partial u(l,t)}{\partial x} = \frac{\partial u(r,t)}{\partial x} = 0$$

$$\frac{\partial w(l,t)}{\partial x} = \frac{\partial w(r,t)}{\partial x} = 0$$

Note: Here we use l and r notations for boundary points instead of a and b to avoid confusion with the a and b coefficients of the FHN equation.

Problem: Write a live user-defined function with name ApExample_8_2 that calculates u and w and generates $u(x,t)$ graph with coordinates x from $x_s = 0$ to $x_f = 10$ and at t from $t_s = 0$ to $t_f = 200$. Use the form pdepe with anonymous function to pass D, a, b, and c constants to the PDE sub-function. Consider the following values of the constants $D = 0.05$, a = 0.1, b = 0.01, c = 0.01 and the number of points in time and coordinate $n_t = 21$ and $n_x = 401$. Set x_s, x_f, t_s, t_f, n_t, n_x, D, a, b, and c as input parameters of the live function, and one output parameter that displays the table of u values with only five evenly spaced time lines and five evenly spaced coordinate columns.

To solve the FHN equations, each of them must be rewritten in the standard form:

$$\frac{\partial u}{\partial t} = \frac{\partial}{\partial x}\left(D\frac{\partial u}{\partial x}\right) + u(a-u)(u-1) - w$$

$$\frac{\partial w}{\partial t} = \frac{\partial}{\partial x}\left(0\frac{\partial w}{\partial x}\right) + bu - cw$$

Comparing the solving equations with the standard form equation, we obtain their identity when:

$m = 1$ (Cartesian coordinates) and:

$c\left(x,t,u,\frac{\partial u}{\partial t}\right) = 1$ for each equation,

$f\left(x,t,u,\frac{\partial u}{\partial x}\right) = D\frac{\partial u}{\partial x}$ for the first equation,

$f\left(x,t,w,\frac{\partial w}{\partial x}\right) = 0\frac{\partial w}{\partial x}$ for the second equation,

$s\left(x,t,u,\frac{\partial u}{\partial x}\right) = u(a-u)(u-1) - w$ for the first equations,

$s\left(x,t,w,\frac{\partial w}{\partial x}\right) = bu - cw$ for the second equation.

Alternatively, in the matrix form:

$$\begin{bmatrix}1 & 1\end{bmatrix}.*\frac{\partial}{\partial t}\begin{bmatrix}u_1 & u_2\end{bmatrix} = \frac{\partial}{\partial x}\begin{bmatrix}D\frac{\partial u_1}{\partial x} & 0\frac{\partial u_2}{\partial x}\end{bmatrix} + \begin{bmatrix}u_1(a-u_1)(u_1-1) - u_2 & bu_1 - cu_2\end{bmatrix}$$

where u_1 and u_2 denote u and w, respectively.

The ICs in standard form are

$$\left[u_1(x,0)\ u_2(x,0)\right]=\left[e^{-x^2}\ e^{-(x+2)^2}\right]$$

The BCs rewritten in the standard $p + qf$ form are

$$[0\ 0\ ].*\left[u_1(l,t)\ u_2(l,t)\right]+[1\ 1\ ].*\left[D\frac{\partial u_1(l,t)}{\partial x}\ 0\frac{\partial u_2(l,t)}{\partial x}\right]=[0\ 0\ ]$$

$$[0\ 0\ ].*\left[u_1(r,t)\ u_2(r,t)\right]+[1\ 1\ ].*\left[D\frac{\partial u_1(r,t)}{\partial x}\ \alpha\frac{\partial u_2(r,t)}{\partial x}\right]=[0\ 0\ ]$$

In the MATLAB® command notations:

$$c=[1;1]$$
$$f=[0.01;0]*DuDx$$
$$s=\left[u(1)*(a-u(1)).*(u(1)-1)-u(2);b*u(1)-c*u(2)\right]$$
$$u_0=\left[\exp\ (-x.\wedge 2);0.2*exp(-(x-2).\wedge 2)\right]$$
$$pl=0;pr=0;$$
$$ql=1;gr=1$$

The commands for solving this problem are (Figure 8.4):

After saving this function in the ApExample_8_3 file and running it from the Command Window, the following resulting table and graph (Figure 8.5) appear:

```
>> v_table=ApExample_8_2(0,10,0,200,20,400,.05,.1,.01,.01)
v_table =
     1.0000     0.0019     0.0000     0.0000     0.0000
    -0.2350    -0.2544     0.7560     0.0237     0.0000
    -0.0440    -0.0595    -0.1218    -0.2157     0.3099
     0.0038     0.0045    -0.0082    -0.0629    -0.1298
    -0.0001    -0.0003    -0.0001     0.0037    -0.0019
```

The ApExample_8_2 function includes:

- Definition line with required input arguments xs, xf, ts, tf, nt, nx, D, a, b, c1 (c1 designates here the c-coefficient of the FHN equations to avoid confusion with c-term used in the ODE subfunction) and an output argument u_table for displaying the required table.
- Vectors x_mesh and t_span are created by two linspace commands.
- The pdepe function that invokes three sub-functions: PDE with the terms of the FHN equation, IC with the initial condition, and BC with the boundary conditions. The numerical results are stored in the sol

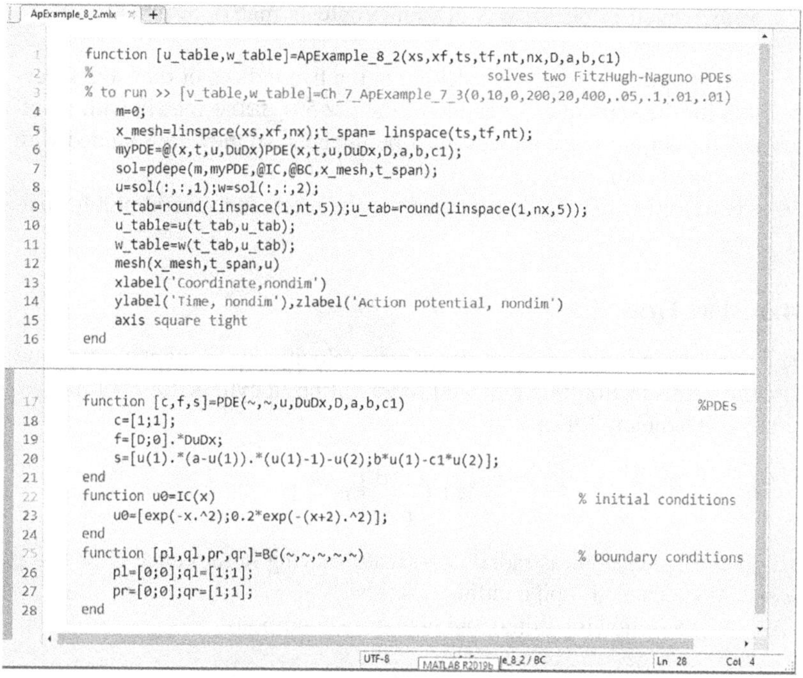

```
function [u_table,w_table]=ApExample_8_2(xs,xf,ts,tf,nt,nx,D,a,b,c1)
%                                                    solves two FitzHugh-Naguno PDEs
% to run >> [v_table,w_table]=Ch_7_ApExample_7_3(0,10,0,200,20,400,.05,.1,.01,.01)
    m=0;
    x_mesh=linspace(xs,xf,nx);t_span= linspace(ts,tf,nt);
    myPDE=@(x,t,u,DuDx)PDE(x,t,u,DuDx,D,a,b,c1);
    sol=pdepe(m,myPDE,@IC,@BC,x_mesh,t_span);
    u=sol(:,:,1);w=sol(:,:,2);
    t_tab=round(linspace(1,nt,5));u_tab=round(linspace(1,nx,5));
    u_table=u(t_tab,u_tab);
    w_table=w(t_tab,u_tab);
    mesh(x_mesh,t_span,u)
    xlabel('Coordinate,nondim')
    ylabel('Time, nondim'),zlabel('Action potential, nondim')
    axis square tight
end

function [c,f,s]=PDE(~,~,u,DuDx,D,a,b,c1)                                  %PDEs
    c=[1;1];
    f=[D;0].*DuDx;
    s=[u(1).*(a-u(1)).*(u(1)-1)-u(2);b*u(1)-c1*u(2)];
end
function u0=IC(x)                                             % initial conditions
    u0=[exp(-x.^2);0.2*exp(-(x+2).^2)];
end
function [pl,ql,pr,qr]=BC(~,~,~,~,~)                          % boundary conditions
    pl=[0;0];ql=[1;1];
    pr=[0;0];qr=[1;1];
end
```

FIGURE 8.4 Live program solving the FHN-diffusion equation.

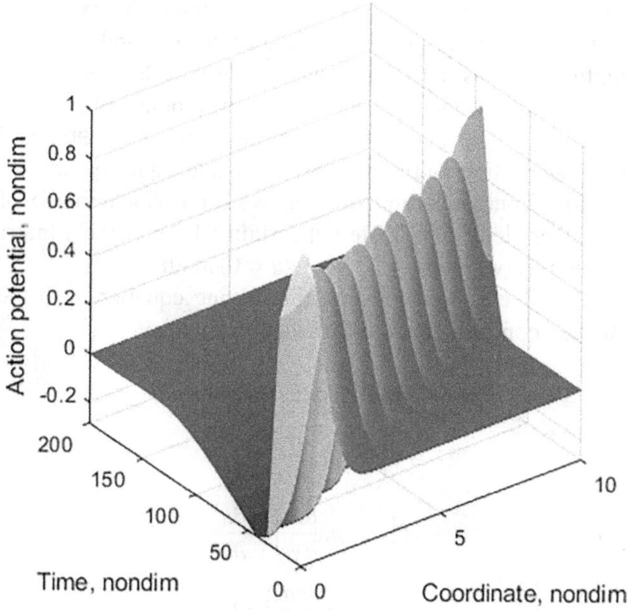

FIGURE 8.5 Solution of the FHN-diffusion equation.

array, which in our case is 3D, and contents matrix of u-values in the sol(:,:,1) page.

- Two linspace commands calculating the five indices of the t and x vectors for the subsequent creation of the 5×5 u_table matrix with u values for output; these indices must be integers, so they are rounded with the round command.
- The resulting 3D plot is generated by the mesh and several plot formatting commands.

8.4.3 Pipe Flow

The partial differential equation describing velocity *u* of fluid at startup flow in a circular cross-section of a pipe and representing in cylindrical coordinates are (based on Batchelor, 2012):

$$\frac{\partial u}{\partial t} = 1 + \frac{1}{r}\frac{\partial}{\partial r}\left(r\frac{\partial u}{\partial r}\right)$$

here *r* is the dimensionless radial coordinate varying in the range 0…1 (0 – pipe cross-section center, 1 – pipe radius), *t* – time.

Assume the following initial and boundary conditions:

$$u(x,0) = 0$$
$$u(1,t) = 0$$

Problem: Compose a user-defined function with name ApExample_8_3 that calculates flow velocity as a function of radial coordinate and time and generates the following three graphs in the same window: *u(r)* at different *t*, *u(t)* for center of the pipe ($r = 0$), and surface plot *u(r,t)*. Take the 8 points of *r* and 6 points of *t* in ranges from 0 to 1 each. The input parameters of the ApExample_8_3 function should be the starting and final points for *r* and *t* and the amounts of the *x* and *t* points, and the output parameter – the two-dimensional matrix of each row of which is *u*-values along coordinate *r* at certain *t*. Display the *u*-matrix with for seven first *r*-columns (without $r = 1$, where $u = 0$) at all *t*.

To solve the problem, compare first the solving equation with the standard PDE form; we can conclude that they are identical when:

$$m = 1 \quad \text{(cylindrical coordinates)}$$

$$c\left(x,t,u,\frac{\partial u}{\partial t}\right) = 1$$

$$f\left(x,t,u,\frac{\partial u}{\partial x}\right) = \frac{\partial u}{\partial x}$$

$$s\left(x,t,u,\frac{\partial u}{\partial x}\right) = 1$$

The standard forms of the initial and boundary conditions are:

$$u_0 = 0$$

$$p_a = 0,\ q_a = 0 \quad \text{at} \quad r = r_a = 0 \text{ (since } m = 1)$$

$$p_b = u_b,\ \ q_b = 0\ at r = r_b = 1$$

The script that solves this problem is:

```
function u_table=ApExample_8_3(xs,xf,ts,tf,nt,nx)
%   solves flow in pipe equation, in cylindrical coordinates
%                 to run >> u_table=ApExample_8_3(0,1,0,1,6,8)
m=1;
x_mesh=linspace(xs,xf,nx);t_span=linspace(ts,tf,nt);
u=pdepe(m,@PDE,@IC,@BC,x_mesh,t_span);
u_table=u(:,1:7);
subplot(2,2,1)
plot(x_mesh,u)
xlabel('Radial coordinate, nondim')
ylabel('Velocity, nondim')
title({'Velocity vs coordinate';'at different times'})
text(0.5,u(nt,round(nx/2)),['time=',num2str(tf)])
axis square tight
grid on
subplot(2,2,2)
plot(t_span,u(:,1)), grid on
xlabel('Time, nondim'),ylabel('Velocity, nondim')
title({'Velocity vs time'; 'at pipe center'})
axis square tight
subplot(2,2,[3 4])
surf(x_mesh,t_span,u)
xlabel('Radial coordinate,nondim')
ylabel('Time, nondim'),zlabel('Velocity, nondim')
title('Velocity vs coordinate and time')
axis square tight
function [c,f,s]=PDE(x,t,u,DuDx)                              % PDE
c=1;
f=DuDx;
s=1;
function u0=IC(x)                                % initial condition
u0=0;
function [pa,qa,pb,qb]=BC(xa,ua,xb,ub,t)% boundary
conditions
pa=0;qa=0;
pb=ub;qb=0;
```

After saving this program in the ApExample_8_3 file and entering the running command in the Command Window, the following table and plot (Figure 8.6) appear.

```
>> u_table=ApExample_8_3(0,1,0,1,6,8)
u_table =
      0      0      0      0      0      0      0
 0.1618 0.1593 0.1514 0.1377 0.1170 0.0881 0.0496
 0.2217 0.2174 0.2045 0.1828 0.1519 0.1115 0.0610
 0.2409 0.2361 0.2215 0.1972 0.1631 0.1189 0.0646
 0.2470 0.2420 0.2270 0.2018 0.1666 0.1213 0.0658
 0.2490 0.2440 0.2287 0.2034 0.1678 0.1221 0.0661
```

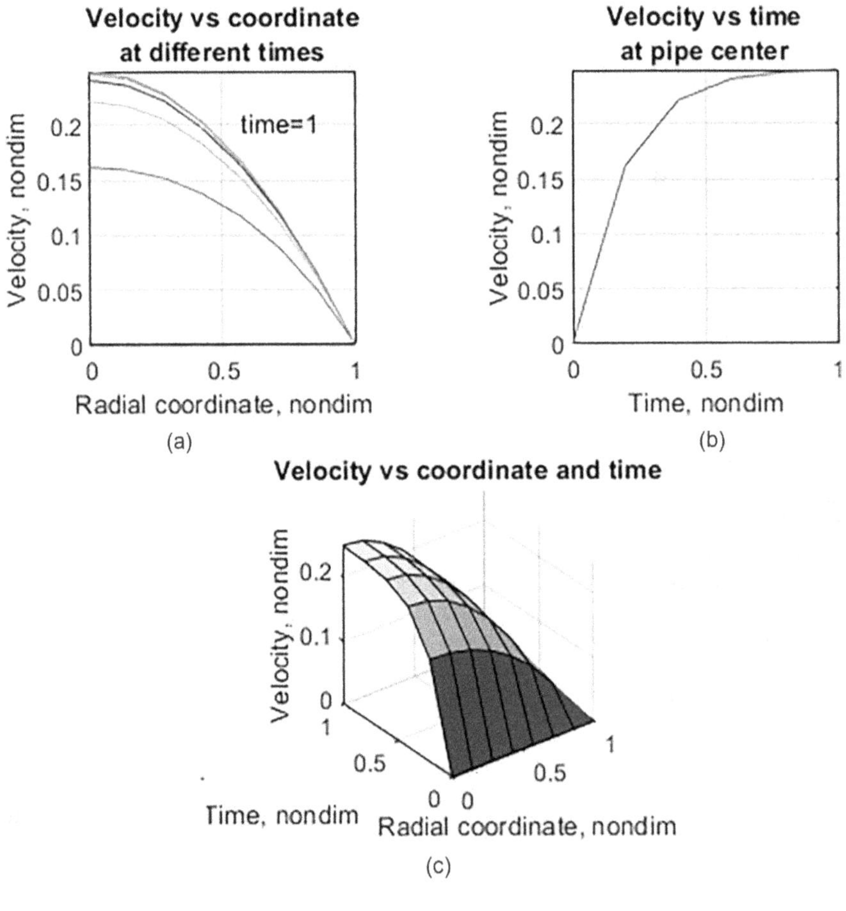

FIGURE 8.6 Flow velocity vs coordinate (a), versus time (b) and as function of both – time and coordinate (c).

The presented function ApExample_8_3 includes:

- The function definition line, in which the input arguments *ts*, *tf*, *xs*, *xf*, *nt*, and *nx* denote the start and end of both times and coordinate, as well as the numbers of point *t*- and *x* intended for the solution; the output argument u_table is for displaying a table of calculated velocities.
- The xmesh and tspan vectors are created by two linspace commands.
- The pdepe function, that is written next, invokes three sub-functions: PDE comprising the terms of the flow equation, IC defining the initial condition, and BC defining the boundary conditions. The calculated values are stored in the u array, which in our case is 2D and is used later to display tables and graphics.
- Then u values at *r*>0 and for each calculated time are assigned to the u_table array.
- The subplot commands are used to plot three graphs in one Figure window; for this, two plot commands generate 2D plots *u*(*r*) and *u*(*t*), and the surf command generates a3D plot *u*(*x*,*t*). The graphs are represented with captions using the regular formatting commands. The axis commands use the square and tight options to better show the results.

8.4.4 Bateman-Burgers PDE

Bateman-Burgers equation is a PDE that describes phenomenon in such areas as fluid mechanics, nonlinear acoustics, gas dynamics, traffic flow, and others. Its 1D form for the case of fluid flow is as follows:

$$\frac{\partial u}{\partial t} = v\frac{\partial^2 u}{\partial x^2} - u\frac{\partial u}{\partial x}, 0 \le x \le 1,\ t \ge 0$$

where *u* is the fluid velocity, *x* – coordinate, *t* – time, and ν – fluid viscosity.

Assume the following initial and boundary conditions

$$u(x,0) = \sin\,(\pi x)$$

$$x(0,t) = 0, \quad x(1,t) = 0$$

Problem: Arrange a user-defined function with name ApExample_8_4 that considers the above equation with given IC and BCs. Take the *x* and *t* point numbers equal to 100 and 19 points respectively. The user-defined function should generate the following three graphs, each in a separate Figure window: *u*(*x*) at *t* point numbers 1, 7, 13, and 19, *u*(*t*) at *x* point numbers 3, 50, 75, and 100, and *u*(*x*,*t*). Assume v = 0.5 and the *x* and *t* ranges from 0 to 1 each. The input parameters of the ApExample_8_4 function should be the start points x_s, t_s and the end points x_f, t_f of the ranges *x* and *t*, the point amount n_x and n_t, and the fluid viscosity ν.

To solve the problem, the Bateman-Burgers equation, initial and boundary conditions must be presented in their standard forms (based on Burstein 2021).

Comparing the equation being solved with the standard form of the PDE equation, we obtain that they are identical when:

$$m = 0 \quad \text{(Cartesian coordinates)}$$

$$c\left(x,t,u,\frac{\partial u}{\partial t}\right) = 1$$

$$f\left(x,t,u,\frac{\partial u}{\partial x}\right) = u\frac{\partial u}{\partial x}$$

$$s\left(x,t,u,\frac{\partial u}{\partial x}\right) = -v\frac{\partial u}{\partial x}$$

The IC and BCs in the standard form are

$$u_0 = 0$$

$$p_a = u_a,\ q_a = 0 \quad \text{at} \quad x_a = 0$$

$$p_b = u_b,\ q_b = 0 \quad \text{at} \quad x_b = 1$$

The program solving this problem is:

```
function ApExample_8_4(xs,xf,ts,tf,nt,nx,nu)
%   solves Bateman-Burgers equation, in cylindrical
coordinates
%                  to run: >> ApExample_8_4(0,1,0,1,19,100,0.3)
close all
m=0;
x_mesh=linspace(xs,xf,nx);t_span=linspace(ts,tf,nt);
myPDE=@(x,t,u,DuDx)PDE(x,t,u,DuDx,nu);
u=pdepe(m,myPDE,@IC,@BC,x_mesh,t_span);      % pdepe solution
figure
i_t=1:6:nt;
plot(x_mesh,u(i_t,:)),grid on
xlabel('Coordinate, nondim'),ylabel('Velocity, nondim')
title({'Velocity vs coordinate';'at different times'})
i_x=round([1/2 1/2 1/2 1/2]*nx+5);
str=num2str(round(t_span(i_t)',2));
str1=[['t=';'t=';'t=';'t='],str(1:4,:)];
text(x_mesh(i_x),u(i_t,round(nx/2)),[str1(1,:);str1(2,:);...
str1(3,:);str1(4,:)])
axis square tight
figure
```

```
j_x=[3 50:25:100];
plot(t_span,u(:,j_x)), grid on
xlabel('Time, nondim'),ylabel('Velocity, nondim')
title({'Velocity vs time';'at different coordinates'})
str=num2str(round(x_mesh(j_x)',2));
str1=[['x=';'x=';'x=';'x='],str(1:4,:)];
text(t_span(round([3 3 3 3])),u(round(3),j_x)+0.02,...
[str1(1,:);str1(2,:);str1(3,:);str1(4,:)])
axis square tight
figure
mesh(x_mesh,t_span,u)
xlabel('Coordinate,nondim'),ylabel('Time, nondim')
zlabel('Velocity, nondim'),axis square tight
function [c,f,s]=PDE(~,~,u,DuDx,nu)                      % PDE
c=1;f=nu*DuDx;s=-u*DuDx;
function u0=IC(x)                          % initial condition
u0=sin(pi*x);
function [pa,qa,pb,qb]=BC(~,ua,~,ub,~) % boundary conditions
pa=ua;qa=0;pb=ub;qb=0;
```

The commands in the user-defined function ApExample_8_4 act as follows:

- The definition line of the user-defined function ApExample_8_4 has input arguments xs, xf, ts, tf, nt, and nx denoting the start and final coordinate and time points, and the numbers of these points;
- The help lines of the function contain the function purpose and the command that must be entered to run the ApExample_8_4 function;
- The first command of the function body closes all previously opened Figure windows (if any);
- In the next lines, the m value is assigned and the x_mesh and t_span vectors are created with two linspace commands;
- The pdepe function invokes three sub-functions (presented at the end of the program):
  1. the myPDE anonymous function, which in turn calls the PDE subfunction and passes it an additional parameter nu; the PDE subfunction contains term of the equation to be solved.
  2. the IC subfunction containing the initial condition;
  3. the BC subfunction containing the boundary conditions;
     the pdepe function output u array contains obtained results, which in the considering case is 2D and is used in the subsequent graphic commands;
- The following commands generate a formatted graph $u(x)$ with four *iso*-time curves taken at the required time points 1, 7, 13, and 19; to display the string 't=' followed by the t-value for each of the four curves, a four-row string matrix str1 is generated with the t-values rounded to two decimal digits;

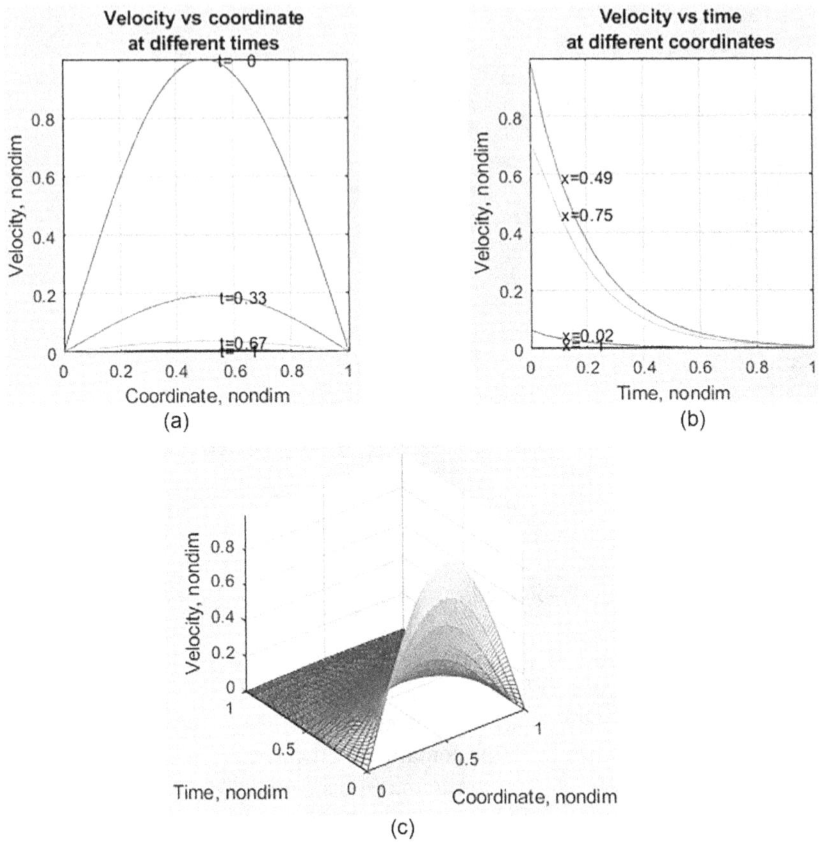

FIGURE 8.7 Velocity distributions as function of coordinates at four time values (a), times at four coordinate values (b), both time and coordinate (c).

- The further commands generate a formatted graph *u*(*t*) with four *iso*-coordinate curves taken for coordinates numbered 3, 50,75, and 100; a four-row string matrix str1 is generated with the x values rounded to two decimal digits to display a string with the 'x=' characters followed by an *x*-value for each of the four curves;
- The mesh and subsequent formatting commands generate a three-dimensional plot *u*(*x*,*t*).
- To run the program, the following command should be entered in the Command Window

```
>> ApExample_8_4(0,1,0,1,19,100,0.5)
```

After this, three resulting plots appear (Figure 8.7a–c), each in a separate Figure window.

9 Coupled 2D PDE Solutions and 3D PDE Solutions

9.1 INTRODUCTION

Solutions of the two-dimensional (2D) and one-dimensional (1D) PDEs have been described in Chapters 6, 7, and 8. Here we introduce two new topics: solving coupled 2D PDEs using the PDE Modeler and solving 3D PDEs using programmatic tools. For the first topic, we present the solution steps and application examples of the PDE Modeler used for two equations that contain two dependent variables in at least one of the PDEs (such PDEs are sometimes called "coupled"). As stated, the PDE Modeler solves the 2D PDEs only and cannot be used for higher dimensional equations. Nevertheless, such a solution is possible with a programmatic tool provided by the PDE toolbox. Therefore, the second direction of the chapter provides basic information and application examples on using the programmatic tool for solving three-dimensional (3D) PDEs. Application examples include:

- Coupled 2D PDEs solution for the deflection of a square plate under an area load;
- Solution for a set of two 2D PDEs describing tri-molecular reaction (Schnakenberg' PDEs);
- Steady-state thermal 3D PDE solution for a hollow cuboid;
- Unsteady thermal 3D PDE solution for disk;
- Solution of 3D PDE vibration problem for a slab with an elliptical hole;
- 3D PDE solution of the electric potential distribution across the plate with variable conductivity.

The commands used for solutions were explained previously or in this chapter; nevertheless, the chapter contains also a table of supplement commands that can be used for 2D and 3D PDE solutions.

9.2 SOLVING A SET OF TWO PDEs USING PDE MODELER

PDE Toolbox programmatic tool allows you to solve a set of two or more 2D PDEs. PDE Modeler provides ability to solve a set of two PDEs only. For this, the same steps should be performed for solving one PDE, but it is needed to select another available option and provide some different answers. Therefore below we

DOI: 10.1201/9781003200352-9

describe the steps for solving a set of two PDEs using the PDE Modeler interface, detailing the differences and referring to Chapter 7 when the step or operation is the same.

9.2.1 Steps to Solve a Set of Two 2D PDEs with PDE Modeler

Let's present the solution steps with PDE Modeler by example. Solve the following system of two 2D Poisson equations that describe, for example, the deflection of a square plate under area load (see Chapra and Canale, 2014):

$$\nabla(\nabla u) = \frac{q}{D} \quad -\nabla(\nabla v) + u = 0$$

The equations are solved with Dirichlet BCs, $u = 0$ and $v = 0$, at the edges of the square plate.

In these equations: $\nabla = \frac{\partial}{\partial x} + \frac{\partial}{\partial y}$; v – deflection; x and y – coordinates; q –load, D – the flexural rigidity.

Problem: Solve the above PDEs assuming q = 33.6 kN/m^2, D = 18.32 kN/m, and a square edge of 2 m. The coordinate center (0,0) is placed at the center of the plate. Plot resulting u and v values in 2D and 3D plots. Save the automatically generated program in a file named ExTwoPDEs_Modeler.

Step 1. We solve the system; therefore, the Generic System line in the Application option should be selected in the popup menu of the Options button located in the main menu (Figure 9.1).

Step 2. At this stage, the 2D model geometry should be drawn. To do this, check the Grid and Snap options (see Figure 7.2). Set the limits for x and y to −1.25 and 1.25 in the Axis Limits dialog box. After that, we can draw the required geometry. To do this, click the "Rectangle/Square (centered)" button in the bar of frequently used buttons, then click the mouse + pointer at point (0,0) and drag the mouse pointer with the pressed button to the grid point (1,1); after releasing the mouse button, the square with its name appears (Figure 9.2).

Step 3. BCs must be specified at this stage. In the case of two equations, PDE Modeler assumes that Dirichlet boundaries are:

$$h_{11}u_1 + h_{12}u_2 = r_1$$

$$h_{21}u_1 + h_{22}u_2 = r_2$$

where in our example, u_1 is the u of our PDEs and u_2 is v; h_{11}- h of the first BC equation; and h_{22}- h of the second BC equation, the coefficients h_{12} and h_{21} are equal to zero. The first digit in the subscript denotes the equation number, and the second denotes the dependent variable number, for example, h_{21} means the coefficient h of the second equation for the first dependent variable u_1.

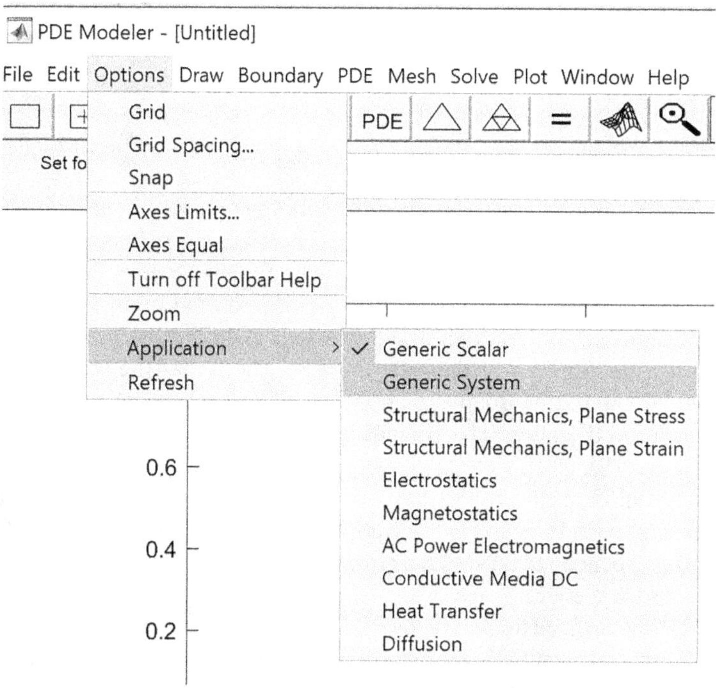

FIGURE 9.1 The first step in two PDEs set solution: selecting the Generic System application.

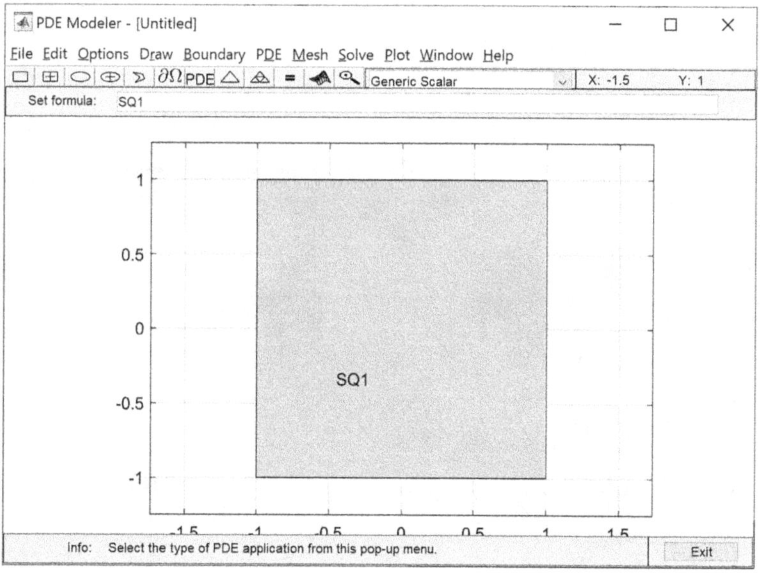

FIGURE 9.2 PDE Modeler window with a drawn square.

Neumann boundaries for a set of two PDEs:

$$\vec{n}(c_{11}\nabla u_1)+\vec{n}(c_{12}\nabla u_1)+q_{11}u_1+q_{12}u_2=g_1 \quad \vec{n}(c_{21}\nabla u_1)+\vec{n}(c_{22}\nabla u_1)+q_{21}u_1+q_{22}u_2=g_2$$

where $\vec{n}$ is the outward unit normal; c, q, and g are coefficients; c is the same as in PDE (see equations in step 4).

When boundary conditions are different for the equations to be solved, mixed boundary conditions can be used. It is not considered case here; for more information, read the PDE Toolbox documentation.

To enter BC, click the Boundary button on the main menu, select the Boundary Mode line, and then check the Show Edge Labels line to see the numbers of the square edges. The BC of the selected edge can be entered by double click the mouse when the mouse pointer is placed on the edge. The Boundary Condition dialog box will appear (Figure 9.3).

In our case, the BC is the Dirichlet condition type with h = 1 and r = 0 for each PDE and on each edge. These BCs are installed by default and do not need to be entered. However, it is advisable to check them for each square edge.

Step 4. Now we need to specify the PDE. To do this, press the PDE menu button, select sequentially the lines PDE Mode and the "PDE Specification …". Alternatively, after checking the PDE Mode, simply click inside the drawn shape (square, in our example). After this, the PDE Specification dialog panel appears (Figure 9.4).

The panel corresponds to the case of a set of the two PDEs, which have the form

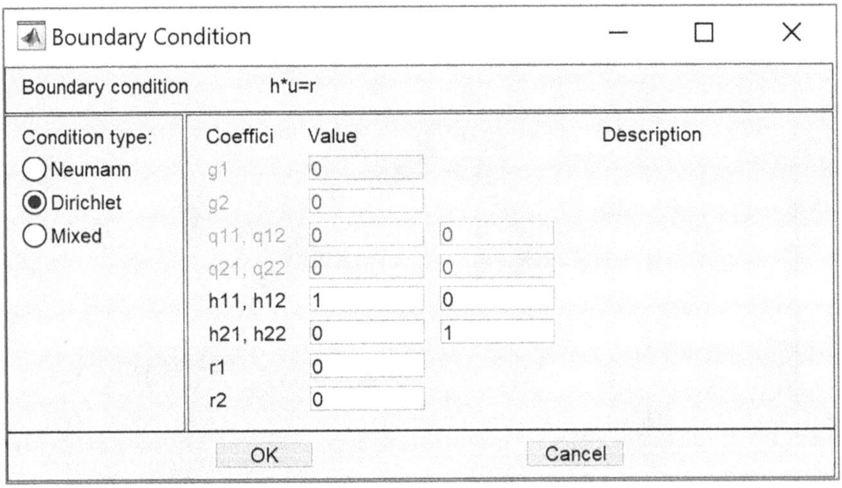

FIGURE 9.3 Boundary Condition dialog box for a set of two PDEs; the box filled for Dirichlet type boundary (default) for one edge of the square.

PDE Specification			— □ ✕
Equation:	-div(c*grad(u))+a*u=f		

Type of PDE:	Coefficie	Value	Value
⦿ Elliptic	c11, c12	1.0	0.0
○ Parabolic	c21, c22	0.0	1.0
○ Hyperbolic	a11, a12	0.0	0.0
○ Eigenmodes	a21, a22	1.0	0.0
	f1, f2	-33.6/18.32	0
	d11, d12	1.0	0.0
	d21, d22	0.0	1.0

OK	Cancel

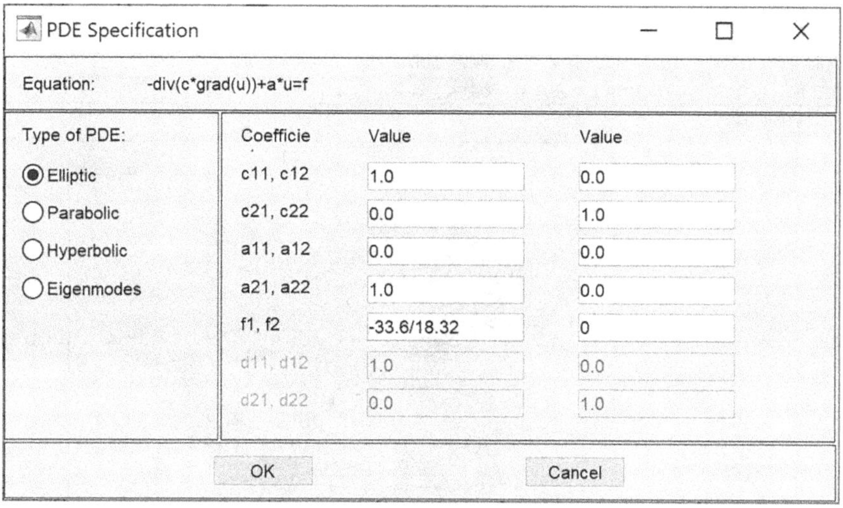

FIGURE 9.4 PDE Specification dialog panel for a set of two PDEs completed for the example problem.

$$-\nabla(c_{11}\nabla u_1) - \nabla(c_{12}\nabla u_2) + a_{11}u_1 + a_{12}u_2$$
$$= f_1 \ -\nabla(c_{21}\nabla u_1) - \nabla(c_{22}\nabla u_2) + a_{21}u_1 + a_{22}u_2 = f_2$$

To mark a PDE type in the PDE Specification dialog panel, it should be determined first. Comparison of the solved equation with the standard equation form (Table 6.1) shows that in our case we have an elliptic equation; the Elliptic type is default, and it appears marked. Further, to enter PDE coefficients in the appropriate fields of this panel, we need to determine them by comparing each of our equations and the equation written at the top of the panel. The equation Equation: -div(c*grad(u))+a*u=f appearing in the panel is written in a general vector form via the divergence, *div*, and gradient, *grad*; for scalar case, it can be rewritten as $-\nabla(c\nabla(u)) + au = f$. Therefore, $c_{11} = c_{22} = 1$, $c_{12} = c_{21} = 0$, $a_{21} = 1$, $a_{11} = a_{12} = a_{22} = 0$, $f_1 = -q/D = -33.6/18.32$, $f_2 = 0$. These coefficients were entered in the appropriate fields of the "Value" column, as shown in Figure 9.3.

All other possible uses of the PDE Specification are identical to the single equation case (Section 7.2.1).

Step 5. Now, a mesh should be created to get a solution at the mesh nodes. The operations at the step are completely identical to the case with one equation (Section 7.2.1), and after they are performed, the square with the refined mesh looks like in Figure 9.5.

Step 6. As with the single equation, the Solve PDE line of the popup menu of the PDE menu button should be selected. For the example problem,

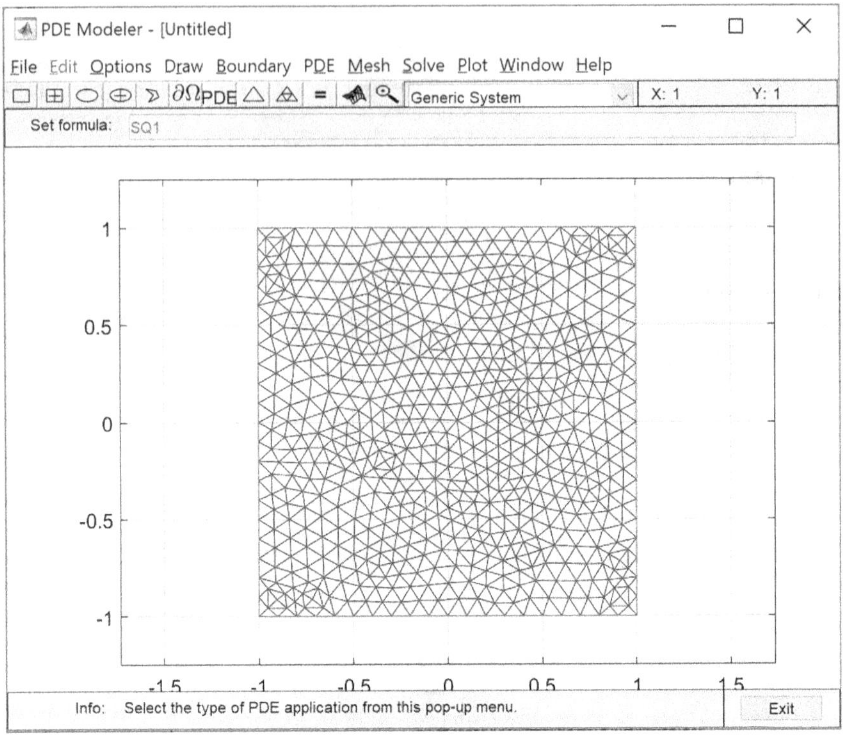

FIGURE 9.5 The Mesh Mode: rectangle with the generated mesh.

it is not necessary to make any changes in the Solve Parameters panel. Therefore, after selecting the Solve PDE mode, the solution is performed and the result for the first equation (u) is presented by default (Figure 9.6).

Since it is desirable to also see the solution of the second PDE (variable v), therefore go to the next step "Plot".

Step 7. To make some changes to the resulting graph or generate 3D graph, select the "Parameters …" line in the popup menu of the Plot button (see Section 7.2.1). To visualize the results for the second equation, the v character should be selected in the first field below the "Property" caption in the Plot Selection window. Also, mark the "Show Mesh" box at the bottom of this window. The result is shown in Figure 9.7.

To generate a 3D graph, the "Height (3D)" box should be checked. The triangular mesh can be shown by marking the "Show mesh" box. To change the color map, select "jet" from the "Colormap:" popup menu. The resulting graph for the u variable is presented in Figure 9.8.

Step 8. To save the automatically generated program in a file, select the "Save As …" line in the popup menu of the main menu File option and type the name `ExTwoPDEs_Modeler` in the "File name:" field of the panel that appears.

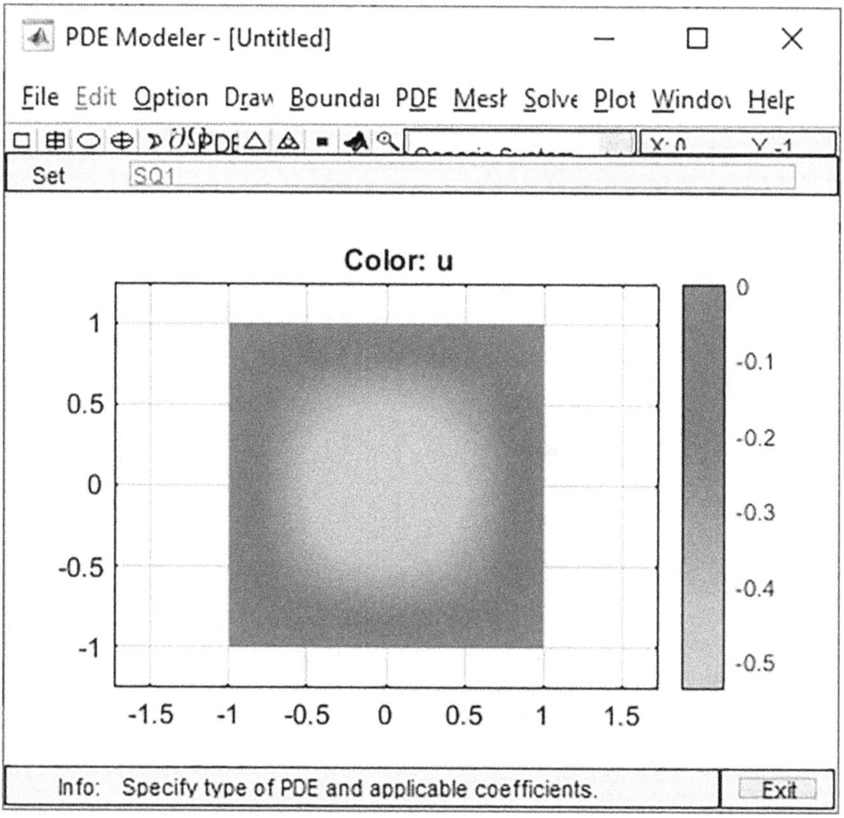

FIGURE 9.6 2D solution for the first PDE of the two PDEs example set.

9.2.2 About Exporting a Solution and Mesh for a Set of Two PDEs

Analogously to the case of a single PDE, to export a solution, select the "Export Solution …" line in the popup menu of the Solve main menu button. In the appeared small box, enter a new name, or leave the default variable name and click the OK. The variable with the selecting name appears in the Workspace window of the MATLAB® desktop. For the non-stationary PDEs, u is the matrix with columns corresponding to the current time each, but in stationary cases, u is a column vector (in our example, it contains the 1378 u-values), the first half of which is the solution for the first dependent variable and the second half – for the second variable.

The solution is obtained at certain x,y mesh nodes; the export of these data and their designations does not differ from that described earlier for the case with one PDE (Section 7.2.2).

After the solution and node coordinates have been exported, their values can be displayed. For the example problem, to display in the `shortg` format (best of fixed or floating point, five-digit format) the coordinates y and x together with

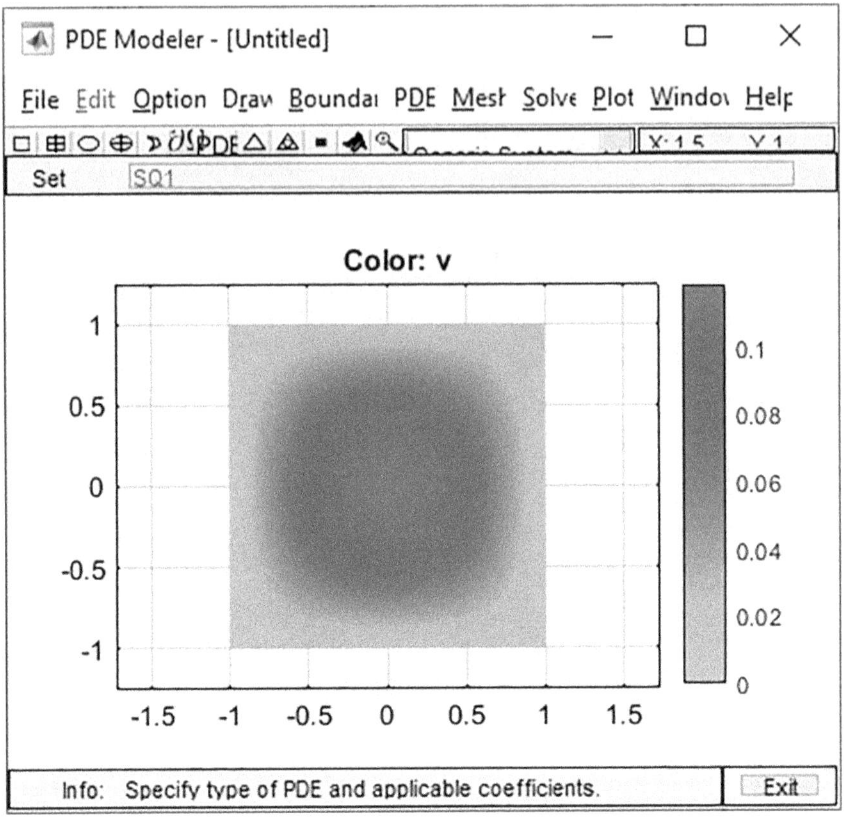

FIGURE 9.7 2D solution for the second PDE of the set of two PDEs of the example.

u- and v-values for, e.g., points 1:100:601 and for the final point, the following commands can be entered in the Command Window:

```
>>node = [1:100:601 length(p)];format shortg
>>disp('x y u v'),disp([p(:,node)' u(node)
u(length(p)+node)])
        x           y           u           v
       -1          -1           0           0
  0.56242    -0.63646    -0.25286    0.043699
        1         0.5           0           0
 -0.89188    -0.20425    -0.11995    0.020196
-0.064398      0.3255    -0.48919     0.10427
  0.63103     -0.1101    -0.34151    0.066427
 -0.20393     0.48576    -0.41333    0.083737
 -0.37152     0.34832    -0.42694    0.086907
```

Each obtained *u*- or *v*-value corresponds to the specific node coordinates of the triangular mesh. The tri2grid command can be used to convert the triangle mesh

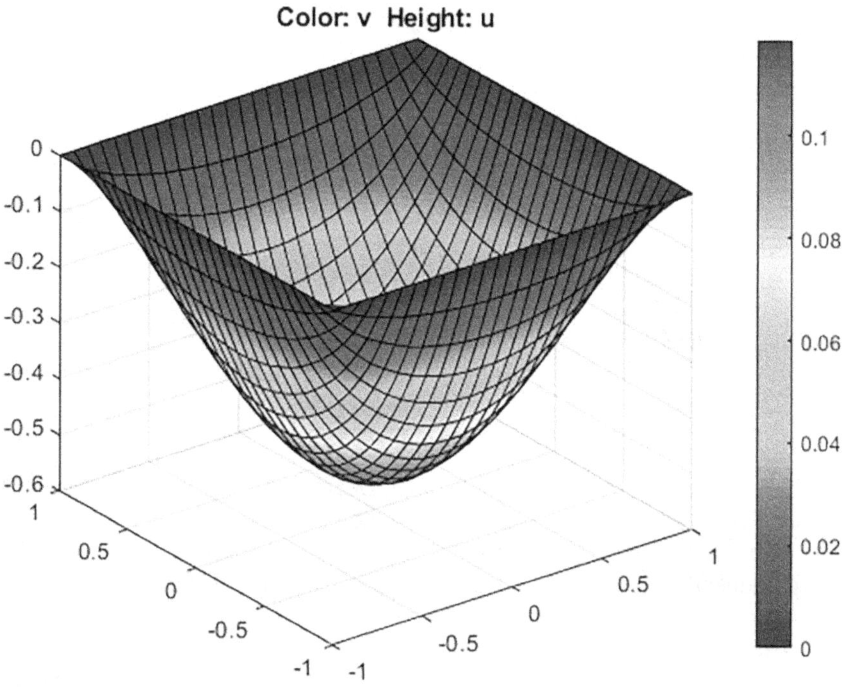

FIGURE 9.8 3D plot with the solution for the first PDE generated using the Plot Selection window.

to a rectangular grid in the same way as described in Section 7.2.3. However, note that you should not use the entire vector u, but only the first or second half of it, depending of whether you want u or v.

9.3 3D PDEs AND STEPS TO ITS SOLUTION

In principle, to solve 3D PDE, we need to perform the same steps that were described early for the 2D PDEs in Chapter 6. Nevertheless, at certain stages, it is required to add some input/output arguments in the studied commands or use the new specified commands. Below, the steps to 3D PDE solving with the necessary commands are described.

9.3.1 Steps 1 and 2: Create a PDE Model Object and 3D Geometry

1. Creation of a model object

   To create a 3D model as an object, you must use the same createpde command as in case of a 2D object. Thus the possible command view is

```
model=createpde;
```

2. Creation and plotting of 3D model geometry

   There are three options for creation of a 3D geometry object of the technical part: import an STL file (the file is created in any CAD system, outside the MATLAB software), restore geometry from a mesh (a special case when the mesh to some part was created and saved in an STL file), and use the commands of the PDE Toolbox that can generate some simple 3D objects. The latter are discussed in this primer. There are three commands that generate the geometry of a 3D model – multicuboid, multicylinder, and multisphere.

The form of the multicuboid command is as follows:

```
gm=multicuboid(Width,Depth,Height,'Name_1',Value_1,
'Name_2',Value_2,…)
```

where Width is the x-axis size of the cuboid, Depth is the y-axis size, and Height is the z-axis size; 'Name_1',Value_1, 'Name_2',Value_2,... - property pairs (optional), where Name_1, Name_2, ... must be written in single quotes. The most frequently used pair is 'Void', true (or false), and it designates that the cuboid is empty (or full, default).

To include geometry in the previously created model object, the model.Geometry=gm command should be used. In our case, the command includes cuboid /s into the Geometry property of the created PDE model. To draw the 3D geometry, the pdegplot command should be used, and the simplest form is

```
pdegplot(model, 'Name_1',Value_1, 'Name_2',Value_2,…)
```

where model is a model created above in Step 1, 'Name_1',Value_1, 'Name_2',Value_2,... are property pairs in which Name_1, Name_2, ... must be written in single quotes. The frequently used pairs are 'CellLabels','on' and 'FaceLabels','on' (the 'off' value is default in these properties), and also 'FaceAlpha', 1 (default) or a number between 0 and 1. These pairs are used to show numbers assigned to the object (cell) geometry (e.g., cuboid), to its faces, as well as to introduce transparency to the object.

Let's create and plot, for example, a $5 \times 3 \times 8$ filled cuboid. Write the following commands for this:

```
>>model=createpde;             % creates PDE model object
>>gm=multicuboid(5,3,8);       % cube with width 5, depth
                               3 and height 8
>>model.Geometry=gm            % sets cuboid to the
                               Geometry property
model =
  PDEModel with properties:
           PDESystemSize: 1
         IsTimeDependent: 0
                Geometry: [1×1 DiscreteGeometry]
```

```
    EquationCoefficients: []
      BoundaryConditions: []
       InitialConditions: []
                    Mesh: []
          SSolverOptions: [1×1 pde.PDESolverOptions]
>>pdegplot(model,'CellLabels','on','FaceLabels','on','FaceAl
pha',0.5)
```

The cuboid generated by these commands is shown in Figure 9.9.

If we need to produce the hollow cuboid (cuboid with centered cuboid-shaped hole), we can use the same multicuboid command with arguments specified as two-element row vectors. So, the command

```
>> gm=multicuboid([5 3],[3 1.5],8,'Void',[false,true]);
```

written into the above script instead of previous one results in the 3D geometry shown in Figure 9.10.

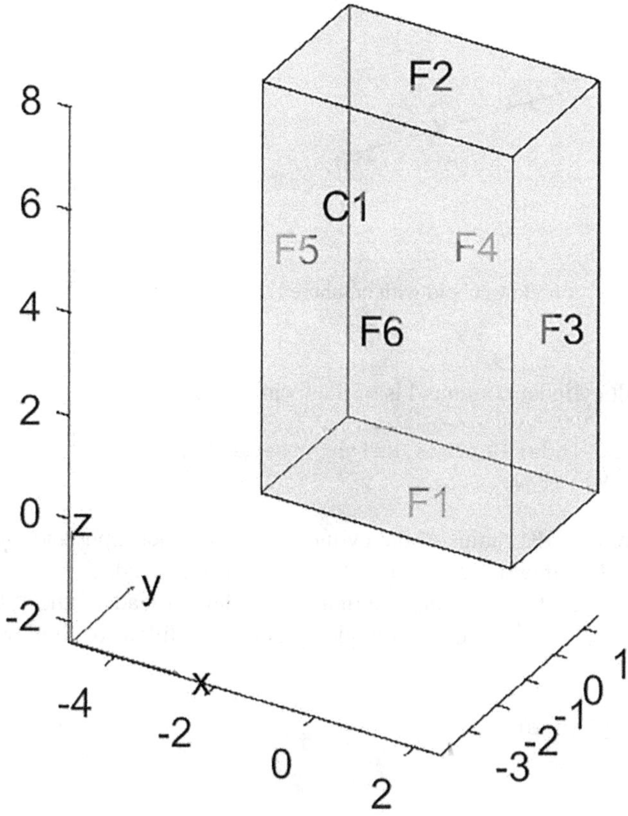

FIGURE 9.9 A filled cuboid with numbered cell and faces.

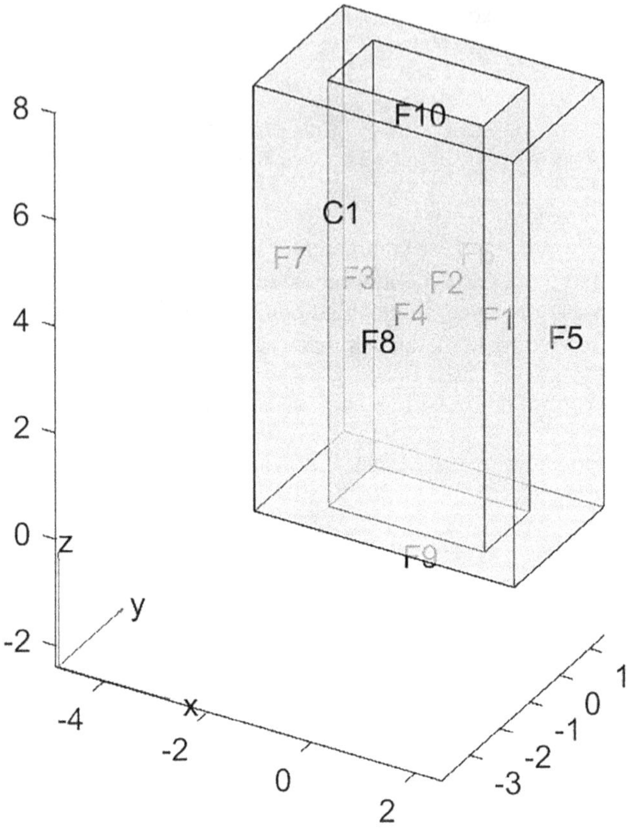

FIGURE 9.10 A hollow cuboid with numbered cell and faces.

The multicylinder command is of the form:

```
gm=multicylinder(Radius,Height,'Name_1',Value_1,
'Name_2',Value_2,…)
```

where Radius is the radius of the cylinder, and all other input and output arguments have the same sense as in the multicuboid command.

For example, the following commands written instead of the multicuboid commands in the previous example generate a filled or hollow cylinder (Figure 9.11a,b).

```
gm=multicylinder(3,10);
```

or

```
gm=multicylinder([3 1.5],10,'Void',[false,true]);
```

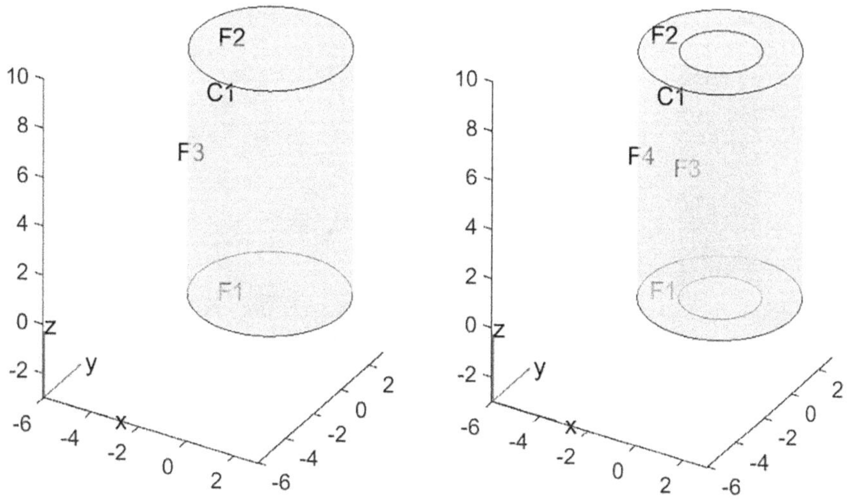

FIGURE 9.11 Filled (a) and hollow (b) cylinders with labeled cells and faces.

The multisphere command creates the geometry of one or more spherical cells. The command looks like this:

```
gm=multisphere(Radius,'Void',Void_value)
```

where Radius is the radius of the sphere and 'Void',Void_value is the property pair in which the Void_value can be true (hollow sphere) or false (filled sphere). All input arguments can be row vectors when we need to create multiple spheres. In the case of a single sphere, this command property pair can be omitted.

For example, the following commands produce the filled and hollow spheres in the same Figure window (see Figure 9.12a,b):

```
model=createpde;
gm=multisphere(10);
```

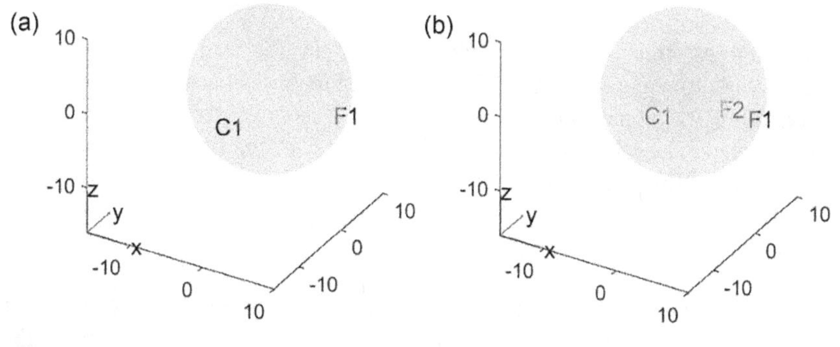

FIGURE 9.12 Filled (a) and hollow (b) spheres with labeled cells and faces.

```
model.Geometry=gm;
subplot(1,2,1)
pdegplot(model,'CellLabels','on','FaceLabels','on','FaceAl
pha',0.5)
gm=multisphere([10,6],'Void',[false,true]);
model.Geometry=gm;
subplot(1,2,2)
pdegplot(model,'CellLabels','on','FaceLabels','on','FaceAl
pha',0.5)
```

Additional tool to create 3D geometry are the extrude command that allows you to vertically extrude previously created 2D geometry or any of its faces along the z-axis. The simplest form of this command is

```
extrude(g,Height)
```

where g is the 2D model geometry and Height is a number indicating the size of the vertical extension of the 2D model.

As an example, we create a 2D model of the square with an off-center elliptic hole and convert it to 3D model using the extrude command. The following commands solve this problem

```
model=createpde;
R1=[3,4,-1,-1,1,1,-1,1,1,-1]';            % square
e2=[4,0,0.5,0.2,0.4,0,0,0,0,0]';          % elliptic slot
gd = [R1,e2];                             % geometry matrix
sf = 'R1-e2';                             % formula set
ns = char('R1','e2')';                    % geometric shape names
g=decsg(gd,sf,ns);                        % decomposed geometry
gm=geometryFromEdges(model,g);            % geometry to model
subplot(1,2,1)
pdegplot(model,'EdgeLabels','on','FaceLabels','on') % 2D plot
axis equal
g3=extrude(gm,1);        % extrude vertically to a height of 1
subplot(1,2,2)
pdegplot(g3,'FaceLabels','on','FaceAlpha',0.5)      % 3D plot
```

After running this script, two plots appear in the same Figure window (Figure 9.13a,b) – one shows the 2D geometry of the model and the second shows the resulting 3D geometry of the model obtained using the extrude command.

Generated 3D geometry object can be rotated with the rotate command; one of the forms of this command is

```
rotate(g, theta, point_1,point_2)
```

where g – is the model object geometry that should be rotated, theta – rotating angle in degrees, point_1 and point_2 are three-element point vectors that specify axis about which we want to rotate the model.

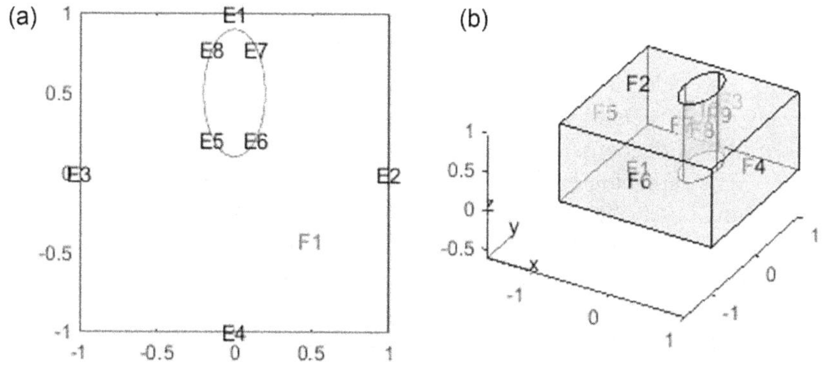

FIGURE 9.13 2D model (a) converted to 3D (b) using the extrude command.

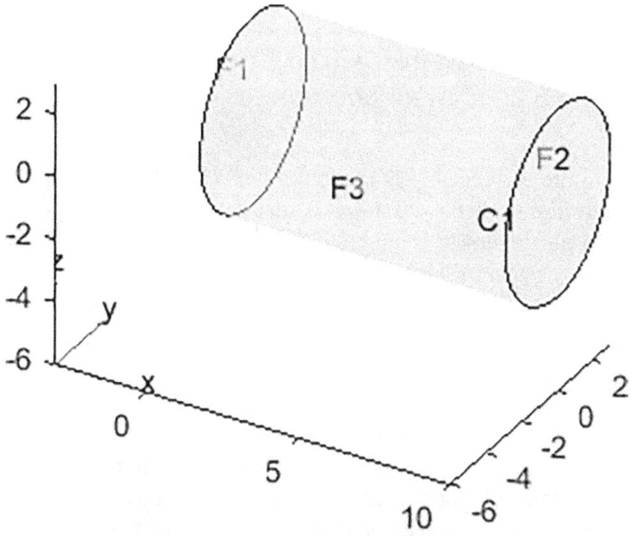

FIGURE 9.14 The cylinder position after rotating 90 degrees around the *y*-axis.

For example, rotate(gm, 90,[0,0,0],[0,1,0]) rotates gm geometry by 90° around the y-axis. If the gm here is the vertically oriented cylinder (Figure 9.10a), thus we receive this cylinder in horizontal position (Figure 9.14).

Note: If we need to rotate around the *z*-axis, the simplified version of this command can be used: rotate(gm, theta).

9.3.2 Steps 3 and 4: BC, IC, and PDE Coefficients

3. Specification of BC and IC

   To set the boundary and initial conditions for a 3D model, the same commands are used for the 2D model (Section 6.4.1). But BCs for 3D objects are specified for faces (not for edges), for example applyBoundaryCondition(model,'dirichlet','Face',1,'u',0); this command sets BC Dirichlet, u = 0, for face 1 of the model. If the boundary condition for some face/s was/were not specified, the boundary condition is 'neumann' with coefficients g and q are equal to zero.

The IC should be specified for unsteady PDEs using the same setInitialConditions command as in 2D modeling. For example, the command setInitialConditions(model, −5) sets an initial value of −5 for the entire body of the model.

4. Setting the coefficients of the PDE terms

   The specifyCoefficients command for 3D geometry is identical to the 2D case. For example, the command specifyCoefficients(model,'m',0,'d',0,'c',1,'a',0,'f',10) sets the m, d, c, a, and f PDE coefficients for the current model, which can be 2D or 3D.

9.3.3 Steps 5 and 6: Creating a Tetrahedral Mesh and Perform the Solution

5. Creation of a mesh

   The finite element in 3D modeling has a tetrahedral shape (triangular pyramid, in other words) that is different from a triangle. However, the same generateMesh command and the same pdemesh command are used to generate and show a tetrahedral mesh as for a triangular mesh. Possible maximum size of a mesh element is 7.3485 (default). For example, generateMesh(model,'Hmax',4) generates a finer mesh than the default.

6. Execution of the PDE solution.

   To solve a 3D problem, the same commands are used for the steady, time-dependent, and eigenvalue 2D PDEs (Subsection 6.4.1). For example, res=solvepde(model,0:20) solves the unsteady PDE for the model at 21 time values and assigns all the results to the res object, in which the res.NodalSolution matrix contains the obtained u values and the mesh node coordinates at the specified times.

9.3.4 Step 7: Presentation of the Solution

The plotpde3D command is used for graphical representation of the solution obtained for a 3D model. The command is as follows:

```
plotpde3D(model,'ColorMapData',res.NodalSolution)
```

where 'ColorMapData' is the property name that designates obtained data needed for plotting (contained in the property value res.NodalSolution).

For example plotpde3D(model,'ColorMapData',res.NodalSolution(:,21)) generates a 3D solution surface.

The interpolateSolution command (see Section 6.4.1) is used to retrieve numeric values at the desired cross-sections, lines, or points. For example, the command uintrp=interpolateSolution(res,x,y,z,3,1:length(tlist)), written for some fixed x and z coordinates and for 21time values, interpolates for y, and assigns uintrp the results for component number 3; the solution is 20 × 1 × 21 matrix, and the squeeze command – uintrp=squeeze(uintrp) – is used to easily remove the matrix singleton dimension.

9.3.5 Steady-State 3D Heat Equation

Problem: For hollow cuboid (Figure 9.2), solve the steady-state PDE $-\nabla k \nabla T = 0$ (heat equation, $\nabla = \frac{\partial}{\partial x} + \frac{\partial}{\partial y} + \frac{\partial}{\partial z}$) when the temperatures of the top and bottom faces of the cylinder are 100 and 300C, respectively. All other sides are insulated (default). Compose a live script named Elliptic3D. Assume the dimensions of the cuboid are given in cm, and thermal conductivity k is constant and equal to 80 cm^3/s. Present the results as two 3D hollow cuboid plots, one showing the model geometry and the other showing the resulting temperatures. Display a table of temperatures in vertical central cross-section of the cuboid for seven equidistant points z and y.

To solve the problem follow the above steps.

1. The first command is model=createpde, which creates the PDE model object.
2. Now we need to create and plot geometry of the object, as described in step 2 above. This can be done using the following commands:

```
gm=multicuboid([5 3],[3 1.5],8,'Void',[false,true]);
% hollow cuboid
model.Geometry=gm;   % sets cuboid to the Geometry
                       property
subplot(2,1,1)
pdegplot(model,'CellLabels','on','FaceLabels','on',
'FaceAlpha',0.5)
axis equal
```

   The resulting geometry is presented in Figure 9.7a.
3. According to the conditions of the problem, the boundaries are the Dirichlet-type on the bottom and top faces of the model. As shown in Figure 9.6a, these boundaries are numbered 9 and 10. BC on each of the other faces is set by default. With this in mind, the commands that specify BC in our model are as follows:

```
applyBoundaryCondition(model,"dirichlet","Face",10,
'u',100);
applyBoundaryCondition(model,"dirichlet",'Face',9,
'u',300);
```

Here u denotes the T of the equation to be solved.

4. To specify the PDE terms we need to compare our and the standardized PDEs. Comparison shows that our equation is elliptical in which T denotes u, and the coefficients are $m = 0$, $d = 0$, $c = 80$, $a = 0$, and $f = 0$, so the command that sets the PDE coefficients reads:

```
specifyCoefficients(model,'m',0,'d',0,'a',0,'c',80,'f',
0,"Cell",1);
```

5. Generate now a mesh of tetrahedral elements using the command

```
generateMesh(model);
```

6. Now we can solve the PDE with commands

```
res=solvepde(model);
u=res.NodalSolution;
```

The res.NodalSolution array contains defined u values and mesh nodes that are passed to u variable.

7. The resulted graph and the required table can be generated using the following commands:

```
subplot(2,1,2)
pdeplot3D(model,"ColorMapData",u)
yq=linspace(-1.5,1.5,7);zq=linspace(0,8,7);
[Y,Z]=meshgrid(yq,zq);
x=zeros(size(Y));
uintrp_x_0=interpolateSolution(res,x,Y,Z); % cross-
section at x=0
uintrp_x_0=reshape(uintrp_x_0,size(Y))
```

The resulting plot and table are presented in Figures 9.15 and 9.16.

The complete live script Elliptic3D with the resulting graphs and table is presented in Figure 9.17.

9.3.6 Unsteady 3D Heat Equation

Problem: For a disk with $R = 25$ and $h = 10$, solve the unsteady dimensionless PDE $\frac{\partial T}{\partial t} - \nabla k \nabla T = 0$ (heat equation, $\nabla = \frac{\partial}{\partial x} + \frac{\partial}{\partial y} + \frac{\partial}{\partial z}$, x, y, and z – Cartesian coordinates) at 51 time values in the range 0 … 50. We assume that IC is $u0 = 0.5 \cdot 10^{-3}(xyz)^2$, and BC on the bottom faces of the disk is of the Dirichlet type with $T = 0$. BCs on the top and side faces are of the Neumann type with zero values for "g" and "q" (default). Compose a live script named Parabolic3D. Assign k = 1. Generate the following graphs: disk geometry, gridded disk, and three 3D temperature distribution plots, at times equal to 0, 20, and 50. Display a table of final temperatures in vertical cross-section for the six y- and five z- equidistantly spaced points.

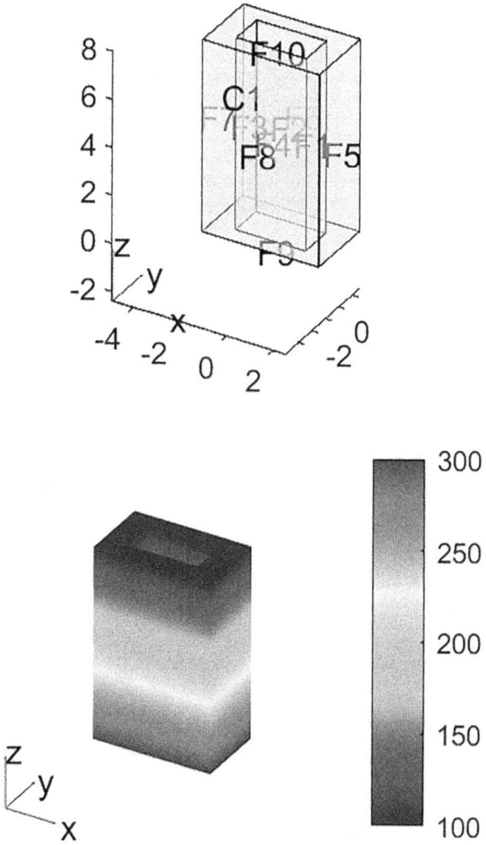

FIGURE 9.15 The hollow cuboid geometry and resulting temperature distribution.

Elliptic3D.mlx | uintrp_x_0 = 7×7

	1	2	3	4	5	6	7
1	300.0000	300.0000	NaN	NaN	NaN	300.0000	300.0000
2	266.6667	266.6667	NaN	NaN	NaN	266.6667	266.6667
3	233.3333	233.3333	NaN	NaN	NaN	233.3333	233.3333
4	200.0000	200.0000	NaN	NaN	NaN	200.0000	200.0000
5	166.6667	166.6667	NaN	NaN	NaN	166.6667	166.6667
6	133.3333	133.3333	NaN	NaN	NaN	133.3333	133.3333
7	100.0000	100.0000	NaN	NaN	NaN	100.0000	100.0000

FIGURE 9.16 A live-script-generated table representing the resulting temperatures in the central longitudinal section of a hollow cuboid at seven equidistant points x (columns) and z (rows). NaN means empty space inside the cuboid.

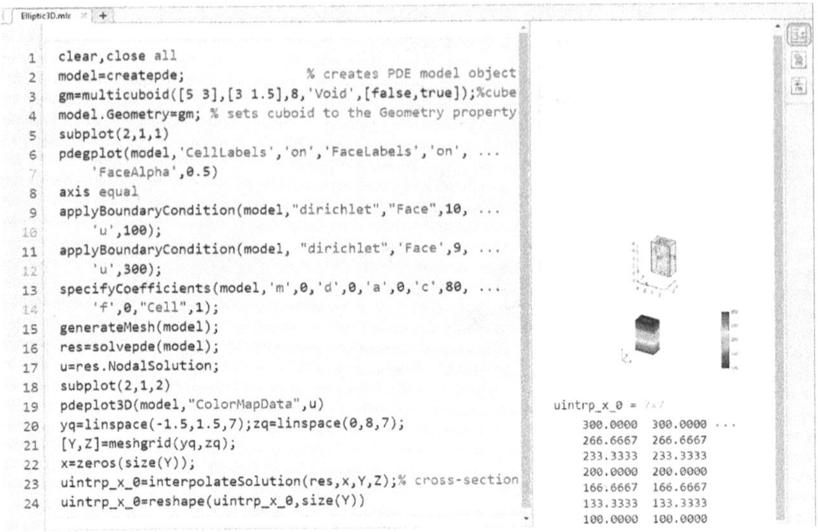
Elliptic3D.mlx
clear,close all
model=createpde;   % creates PDE model object
gm=multicuboid([5 3],[3 1.5],8,'Void',[false,true]);%cube
model.Geometry=gm; % sets cuboid to the Geometry property
subplot(2,1,1)
pdegplot(model,'CellLabels','on','FaceLabels','on', ...
'FaceAlpha',0.5)
axis equal
applyBoundaryCondition(model,"dirichlet","Face",10, ...
'u',100);
applyBoundaryCondition(model, "dirichlet",'Face',9, ...
'u',300);
specifyCoefficients(model,'m',0,'d',0,'a',0,'c',80, ...
'f',0,"Cell",1);
generateMesh(model);
res=solvepde(model);
u=res.NodalSolution;
subplot(2,1,2)
pdeplot3D(model,"ColorMapData",u)
yq=linspace(-1.5,1.5,7);zq=linspace(0,8,7);
[Y,Z]=meshgrid(yq,zq);
x=zeros(size(Y));
uintrp_x_0=interpolateSolution(res,x,Y,Z);% cross-section
uintrp_x_0=reshape(uintrp_x_0,size(Y))
uintrp_x_0 =
300.0000 300.0000 ...
266.6667 266.6667
233.3333 233.3333
200.0000 200.0000
166.6667 166.6667
133.3333 133.3333
100.0000 100.0000

FIGURE 9.17 Live script Elliptic3D showing the commands and results of solving the steady-state heat equation.

To solve this problem, we repeat the steps described above in the chapter.

1. Enter the `model=createpde` command, which creates the model as a PDE object.
2. Now we create the geometry of our object and draw it with the following commands:

```
R=25;h=10;tf=50;
gm=multicylinder(R,h);  % disk
model.Geometry=gm;      % sets disk to the Geometry
property
pdegplot(model,'CellLabels','on','FaceLabels','on',
'FaceAlpha',0.5)
```

   The resulting geometry is shown in Figure 9.18.
3. According to the problem conditions, we have the Dirichlet boundaries, u=0, on the bottom face – numbers 1 (Figure 9.10); and the Neumann boundary (default) at the other faces. Bearing this in mind, we should specify BC only for one face. The appropriate command is as follows:

```
applyBoundaryCondition(model,"dirichlet","Face",1,
'u',0);
```

   Here u denotes the T of the equation to be solved.

   IC in our problem is given by expression, therefore we use the anonymous function (Section 6.4.1) that should be called by the setInitialCondition command. The commands are as follows:

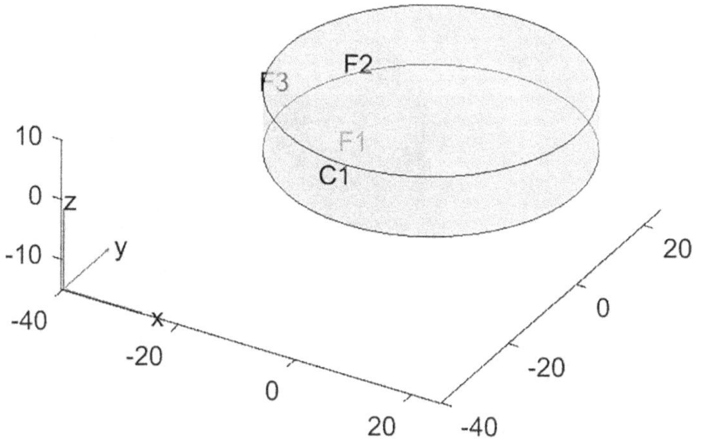

FIGURE 9.18 A graph of the 3D geometry of a disk with faces and cell labels; origin at the center of the lower face of the disk.

```
u0=@(location,state)5e-4*(location.x.*location.y.
*location.z).^2;
setInitialConditions(model,u0);
```

4. To specify the PDE terms, we need to compare our and standard PDEs (see Table 6.1). The comparison shows that our equation is parabolic, and its coefficients are $m = 0$, $d = 1$, $c = 1$, $a = 0$, and $f = 0$, so the command that specifies the PDE coefficients is

```
specifyCoefficients(model,'m',0,'d',1,"a",0,'c',1,'f',0
,"Cell",1);
```

5. Now generate a mesh of tetrahedral elements and generate the mesh plot. This can be done with the following commands:

```
generateMesh(model);
pdemesh(model)
```

   The generated mesh plot is shown in Figure 9.19.

6. We can solve now the unsteady PDE for the given times that should be specified in the tlist vector. The commands are

```
tlist=linspace(0,tf,51);
res=solvepde(model,tlist);
u=res.NodalSolution;
```

   The last command passes the defined temperatures and nodal coordinates from the res.NodalSolution matrix to the u variable.

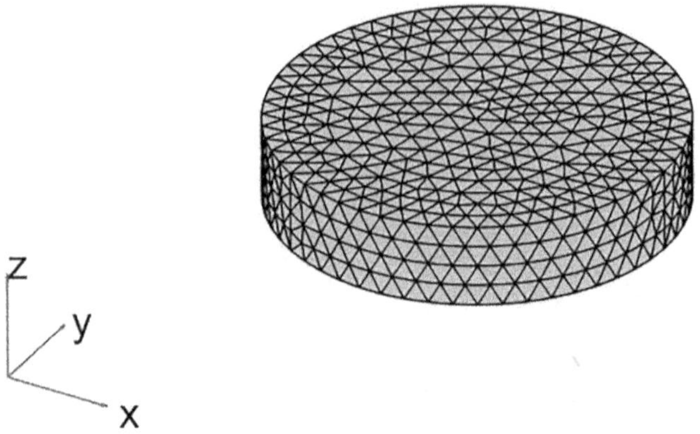

FIGURE 9.19 Generated tetrahedral mesh.

7. The required figure with three resulting graphs showing temperature distribution at three times, 0, 25, and 50, can be produced as follows:

```
subplot(2,2,1)
pdeplot3D(model,"ColorMapData",u(:,1))
title('t=0')
colorbar off
subplot(2,2,2)
pdeplot3D(model,"ColorMapData",u(:,26))
title('t=25')
colorbar off
subplot(2,2,[3 4])
pdeplot3D(model,"ColorMapData",u(:,41))
title('t=40')
colorbar off
```

The commands allowing obtain a table with final temperatures above the vertical cross-section for six y- and five z- equally spaced points are

```
yq=linspace(-0.99*R,0.99*R,6);zq=linspace(0,h,5);
[Y,Z]=meshgrid(yq,zq);
x=zeros(size(Y));
uintrp_x_0=interpolateSolution(res,x,Y,Z,length(tlist));
% vertical cross-section
T_x_0=reshape(uintrp_x_0,size(Y))
```

Since R is a real number, we take $-0.99R$ and $0.99R$ in the yq to avoid NaN at $-R$ at $+R$.

The resulted graphs and a table are presented in Figure 9.20.

The complete Parabolic3D live script together with the generated graphs and table is shown in Figure 9.21.

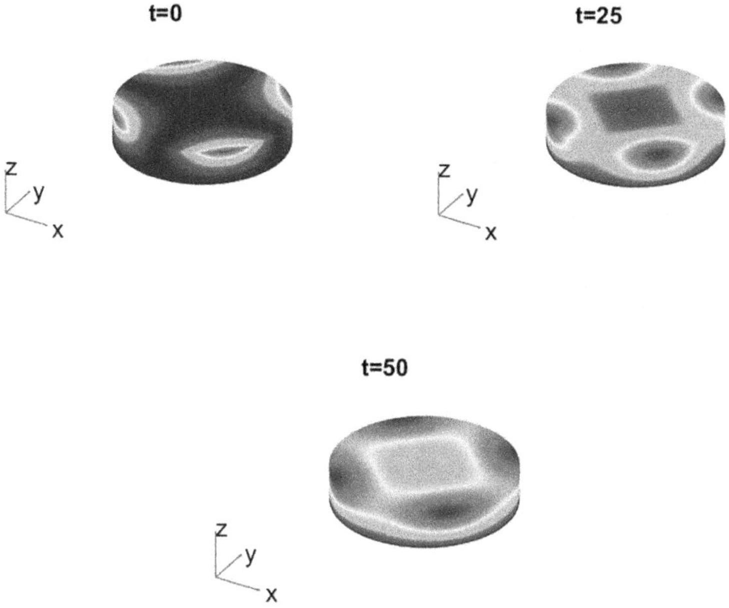


```
T_x_0 = 5×6

         0    -0.0000     0.0000          0          0          0
   54.4979    46.3696    27.7969    27.7891    46.3819    54.7061
  101.2338    85.7530    51.3928    51.3632    85.7345   101.4076
  132.5998   111.9769    67.1168    67.1287   112.0351   132.5260
  143.3956   121.2060    72.6531    72.6662   121.2230   143.5449
```

FIGURE 9.20 Solution of the unsteady 3D PDE for disk. Three plots for t = 0, 25, and 50, and final temperature distribution in the vertical central cross-section for six y-points (columns) and five z-points (rows).

9.4 SUPPLEMENTARY COMMANDS OF THE 2D/3D PDE PROGRAMMATIC TOOL

In addition to the PDE Toolbox commands that were described in previous chapters, this toolbox provides many other commands for 2D and 3D PDE solutions. By entering the command >>help pde in the Command Window, you can get a complete list of the available commands for defining, solving, plotting, and other PDE purposes. Table 9.1 presents some supplementary commands that can be useful in 2D and 3D PDE programming. The table shows the simplest possible command forms, with brief explanations, examples, and the resulting graphs.

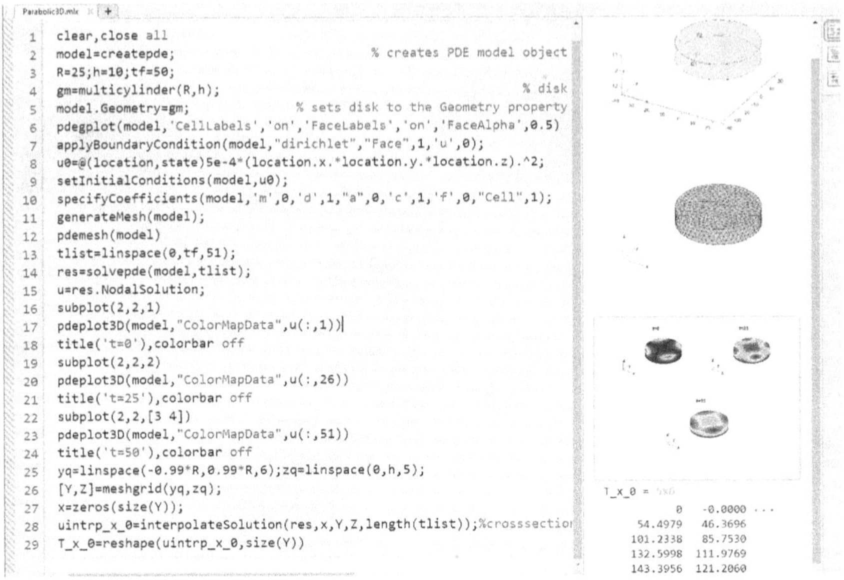

Parabolic3D.mlx
clear,close all
model=createpde; % creates PDE model object
R=25;h=10;tf=50;
gm=multicylinder(R,h); % disk
model.Geometry=gm; % sets disk to the Geometry property
pdegplot(model,'CellLabels','on','FaceLabels','on','FaceAlpha',0.5)
applyBoundaryCondition(model,"dirichlet","Face",1,'u',0);
u0=@(location,state)5e-4*(location.x.*location.y.*location.z).^2;
setInitialConditions(model,u0);
specifyCoefficients(model,'m',0,'d',1,"a",0,'c',1,'f',0,"Cell",1);
generateMesh(model);
pdemesh(model)
tlist=linspace(0,tf,51);
res=solvepde(model,tlist);
u=res.NodalSolution;
subplot(2,2,1)
pdeplot3D(model,"ColorMapData",u(:,1))
title('t=0'),colorbar off
subplot(2,2,2)
pdeplot3D(model,"ColorMapData",u(:,26))
title('t=25'),colorbar off
subplot(2,2,[3 4])
pdeplot3D(model,"ColorMapData",u(:,51))
title('t=50'),colorbar off
yq=linspace(-0.99*R,0.99*R,6);zq=linspace(0,h,5);
[Y,Z]=meshgrid(yq,zq);
x=zeros(size(Y));
uintrp_x_0=interpolateSolution(res,x,Y,Z,length(tlist));%crosssectio
T_x_0=reshape(uintrp_x_0,size(Y))
T_x_0 =
0 -0.0000 ...
54.4979 46.3696
101.2338 85.7530
132.5998 111.9769
143.3956 121.2060

FIGURE 9.21 Live script Parabolic3D with the commands and some results of its running.

Note that the PlateHolePlanar.stl, trimesh, and l'shaped items used in this table represent 2D or 3D decomposed geometry contained in a file, matrix, and data function, respectively, that come with the MATLAB® PDE Toolbox software.

9.5 APPLICATION EXAMPLES

9.5.1 Tri-Molecular Reaction: The Schnakenberg Coupled 2D PDEs

Schnakenberg tri-molecular reaction model contains two transient PDEs, the dimensionless form of which (based on Beentjes 2010) is as follows:

$$\frac{\partial u}{\partial t}=\Delta u+\gamma\left(a-u+u^2v\right)\ \frac{\partial v}{\partial t}=D\Delta v+\gamma\left(b-u^2v\right)$$

where u is concentration of the auto-catalyst, v – concentration of the reactant; $\nabla=\frac{\partial}{\partial x}+\frac{\partial}{\partial y}$; a, b, D, and γ are positive coefficients.

Problem: Use PDE Modeler to solve the above set of 2D PDEs for rectangular reactor cross-section 1×0.4 and the a, b, D, and γ coefficients are 0.2, 1.3, 820, and 40, respectively. BCs along the left, top and bottom edges are Neumann with $g=q=0$ for u and v, and BC on the right edge is Dirichlet with $u=v=0$. Assume IC for u is $2\sin(\pi xy)$, and for v it is $0.4\cos(\pi xy)$. Take the times t in the range 0 … 1 with step 0.25. Present the obtained concentration u and v distributions in the form of 2D and 3D graphs at the final value of t. Export the solution to the

TABLE 9.1
Additional Commands That Can Be Used for Solving 2D and 3D PDEs

Command in Simplest Form	Parameters	Description	Example
`importGeometry (model,STL_file)`	`model` – model object, `STL_file` – path to STL file	Imports 2D or 3D geometry to the model from STL file previously created using the CAD or other software	`model=createpde;` `importGeometry(model,'PlateHolePlanar.stl');` `pdegplot(model)` 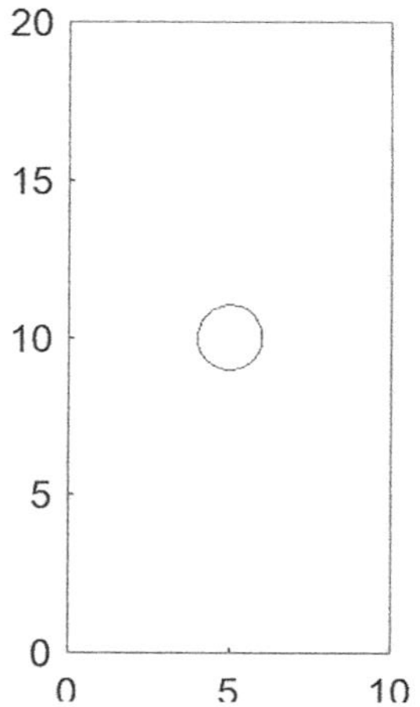

(*Continued*)

TABLE 9.1 (*Continued*)
Additional Commands That Can Be Used for Solving 2D and 3D PDEs

Command in Simplest Form	Parameters	Description	Example

`geometryFromMesh (model,nds,elmnts)`	`model` – as above, `nds` – 2xN node matrix for 2D case, and 3xN matrix for 3D case (N – node number); `elmnts` – an integer matrix (for details see Mesh Data in PDE Toolbox documentaion)	Generates 2D or 3D geometry from mesh previously created	`load tetmesh`
			`geometryFromMesh(model,X',tet');`
			`pdegplot(model,'FaceAlpha',0.5)`
			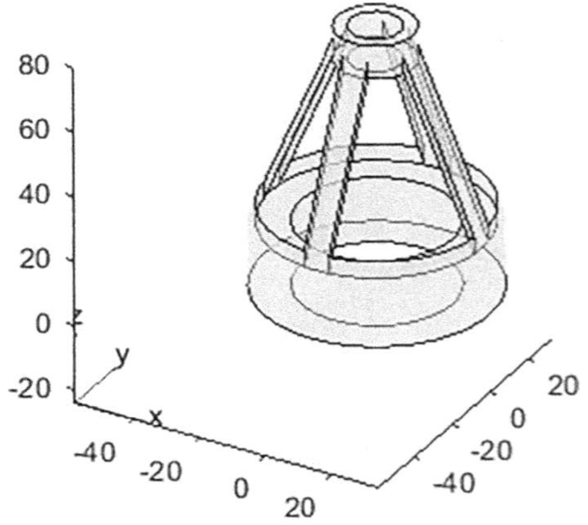

(*Continued*)

TABLE 9.1 (*Continued*)
Additional Commands That Can Be Used for Solving 2D and 3D PDEs

Command in Simplest Form	Parameters	Description	Example

`[p,e,t]=initmesh(g)`	`g` – matrix that specifies the model geometry;
	`p,e,t` – are the mesh data, see Section 7.2.3
		Generates and returns in p,e,t the data of the initial triangular 2D mesh	`model=createpde;`
			`R1=[3,4,-1,-1,1,1,-1,1,1,-1]';`
			`g=decsg(R1,'R1',char('R1')');`
			`gm=geometryFromEdges(model,g);`
			`model.Geometry=gm;`
			`[p,e,t]=initmesh(model.Geometry);`
			`pdemesh(p,e,t)`
			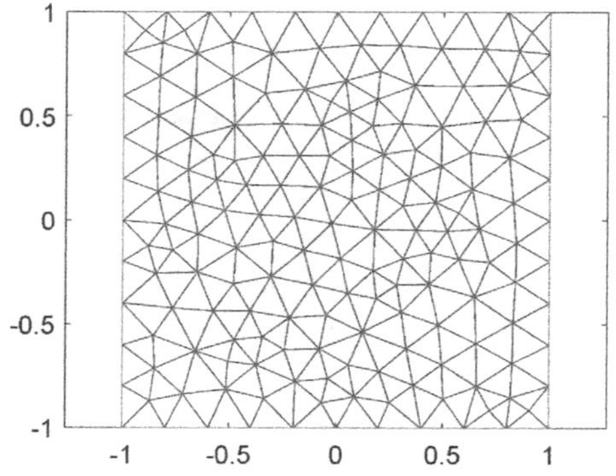

(*Continued*)

TABLE 9.1 (*Continued*)
Additional Commands That Can Be Used for Solving 2D and 3D PDEs

Command in Simplest Form	Parameters	Description	Example
`[p,e,t]=refinemesh (g,p,e,t)`	The same notations as for `initmesh`	Refines the triangular 2D mesh	`[p,e,t] = initmesh('lshapeg');` `[p,e,t]=refinemesh('lshapeg',p,e,t);` `pdemesh(p,e,t)` 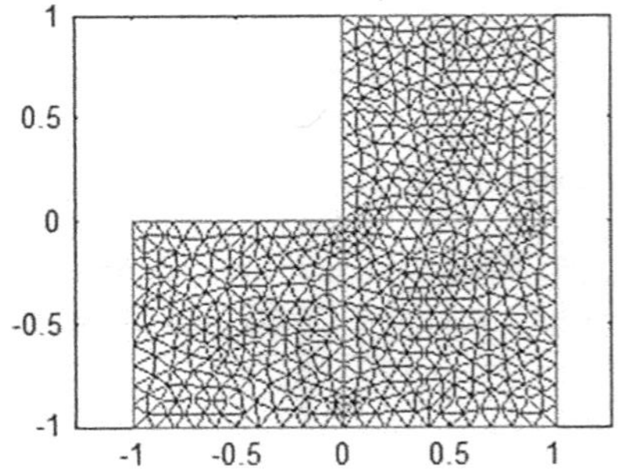

(*Continued*)

TABLE 9.1 (*Continued*)
Additional Commands That Can Be Used for Solving 2D and 3D PDEs

Command in Simplest Form	Parameters	Description	Example

`findBoundaryConditions(model.BoundaryConditions,'Face',FaceNumber)`	`model.BoundaryConditions` – property of the created PDE model;
	`'Face',FaceNumber` – region and its number for which the boundary conditions are searched
		Finds BC assignment for some geometric region	`model = createpde;`
			`gm=multicuboid(5,3,8); model.Geometry=gm`
			`pdegplot(model,'FaceLabels','on','FaceAlpha',0.5)`
			`applyBoundaryCondition(model,'neumann','Face',1,"g",0,"q",0);`
			`findBoundaryConditions(model.BoundaryConditions,'Face',1)`
			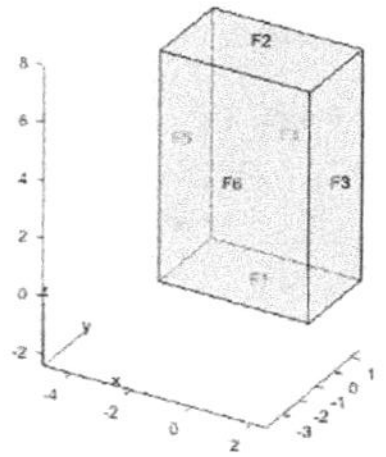
			`ans =`
			`BoundaryCondition with properties:`
			`BCType: 'neumann'`
			`RegionType: 'Face'`
			`RegionID: 1`
			`r: []`
			`h: []`
			`g: 0`
			`q: 0`
			`u: []`
			`EquationIndex: []`
			`Vectorized: 'off'`

(*Continued*)

TABLE 9.1 (*Continued*)

Additional Commands That Can Be Used for Solving 2D and 3D PDEs

Command in Simplest Form	Parameters	Description	Example

`u=assempde (model,c,a,f)` or `u=assempde (model,p,e,t,c,a,f)`	`Model` – pre-created model object; `c, a, and f` – coefficients of elliptic PDE (see Section 6.3) `p,e,t` – mesh data(see section 7.2.3)	Solves elliptic PDE (legacy, not recommended currently)	`model=createpde`
			`R1=[3,4,-1,-1,1,1,-1,1,1,-1]';`
			`g=decsg(R1,'R1',char('R1')');`
			`geometryFromEdges(model,g);`
			`applyBoundaryCondition(model,'Dirichlet','Edge',1,"u",60);`
			`generateMesh(model,'GeometricOrder','linear');`
			`c = 1;a = 0;f = 5;`
			`u = assempde(model,c,a,f);`
			`pdeplot(model,'XYData',u,"ColorMap",'jet')`
			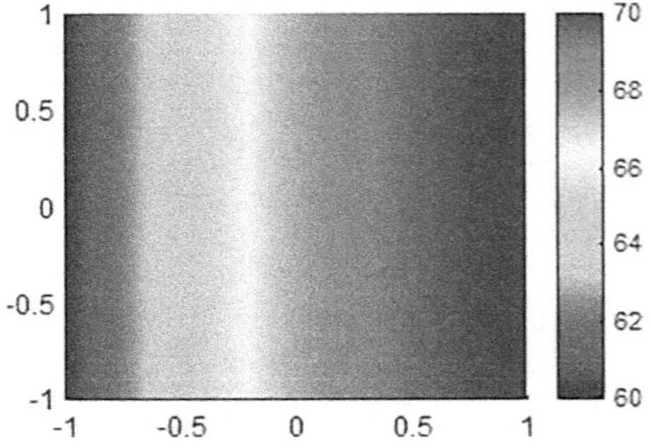

(Continued)

TABLE 9.1 (*Continued*)

Additional Commands That Can Be Used for Solving 2D and 3D PDEs

Command in Simplest Form	Parameters	Description	Example

`findCoefficients (model. EquationCoefficients, 'Cell', CellNumber)`	`model.EquationCoefficients` –property of the created PDE model; 'Cell',`CellNumber` – region for wich the coefficient are searched	Defines active coefficients	`model = createpde;`
			`gm=multicuboid(5,3,8);`
			`model.Geometry=gm`
			`pdegplot(model,'CellLabels','on','FaceAlpha',0.5)`
			`specifyCoefficients(model,'m',0,'d',0,'c',12,'a',0,'f',1,'Cell',1);`
			`findCoefficients(model.EquationCoefficients,'Cell',1)`

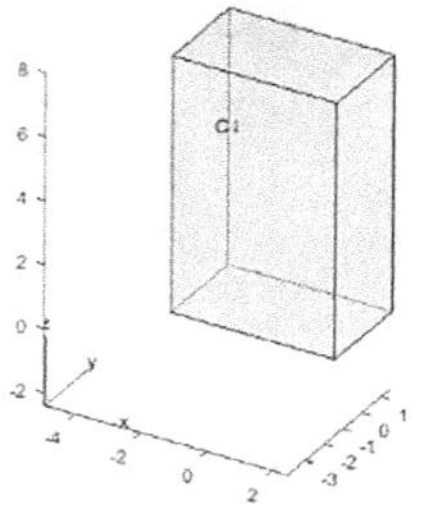


```
ans =
  CoefficientAssignment with properties:

    RegionType: 'cell'
      RegionID: 1
             m: 0
             d: 0
             c: 12
             a: 0
             f: 1
```

(*Continued*)

TABLE 9.1 (*Continued*)
Additional Commands That Can Be Used for Solving 2D and 3D PDEs

Command in Simplest Form	Parameters	Description	Example

`pdesurf(p,t,u)`	`p` and `t` – are the mesh data, see Section 7.2.3;
	`u` – obtained solution
		Generates surface plot rapidly	`model=createpde;`
			`R1=[3,4,-1, -1,1,1, -1,1,1, -1]';`
			`G=decsg(R1,'R1',char('R1')');`
			`geometryFromEdges(model,g);`
			`applyBoundaryCondition(model,'Dirichlet','Edge',1,"u",60);`
			`[p,e,t]=initmesh(model.Geometry);`
			`c=1;a=0;f=5;`
			`u=assempde(model,p,e,t,c,a,f);`
			`pdesurf(p,t,u)`
			`colormap('jet')`
			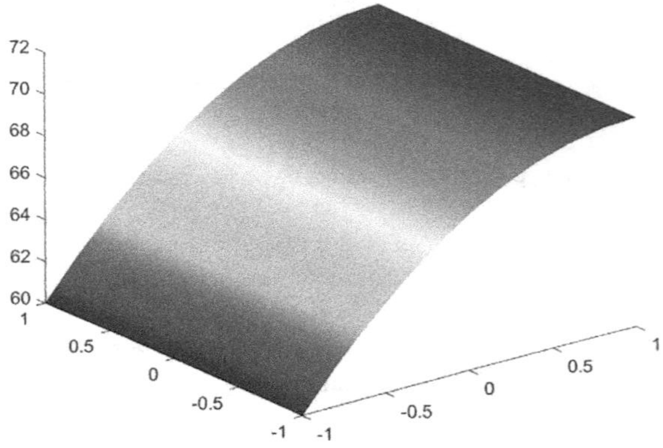

workspace and present the concentrations in two tables: one for u and second for v; take concentrations at five equidistant points along x and y. Save the auto-generated program in a file named ApExample_9_1.

Each of the Schnakenberg equations is of parabolic type (Table 6.1), therefore for each of them $m = 0$ and $d = 1$. To compare these equations with the required form (Section 9.2.1, step 4), present it as

$$\frac{\partial u}{\partial t} - \Delta u + \gamma u = \gamma\left(a + u^2 v\right) \quad \frac{\partial v}{\partial t} - D\Delta v = \gamma\left(b - u^2 v\right)$$

These equations are identical to the standard form when $c_{11} = 1$, $c_{22} = \mathrm{D} = 40$, $a_{11} = 820$, $a_{22} = 0$, $f_1 = \gamma\left(a + u^2 v\right) = 820\left(0.2 + u^2 v\right)$, $f_2 = \gamma\left(b - u^2 v\right) = 820\left(1.3 - u^2 v\right)$, and $c_{12} = c_{21} = a_{12} = a_{21} = 0$.

To solve the problem, activate PDE Modeler by typing in the Command Window

```
>> pdeModeler
```

Mark the Grid and Snap lines in the popup menu of the Options menu button, and enter the limits [–0.5 1.5] for the x-axis and [–0.25 0.75] for the y-axis (after selecting the Axis Limits line in the same popup menu). Check Axes Equal and select the Generic System line in the appearing popup list, after selecting the Application option in the same popup menu of the Option button.

Activate Draw Mode and draw a rectangle using the 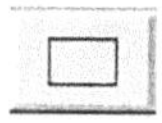 Rectangle/Square button. Place the mouse pointer at point (0, 0) and drag the mouse to the point (1, 0.4). Click inside the rectangle to check its geometric parameters with the Object Dialog panel.

Now switch to boundary mode by selecting the Boundary Mode line from the popup menu of the Boundary button in the main menu. Click the bottom, left, and top edges of the rectangle in sequece and mark the Neumann box in the Boundary Conditions panel that appears, check that each coefficient value is zero (default). Make sure the boundaries on the right rectangle meet the required Dirichlet conditions (default).

Now select PDE Mode of the popup menu of the PDE main menu button. Mark the Parabolic type of PDE in the opened PDE Specifications dialog panel. For a system of two equations u is u(1,:) and v is u(2,:); we need to use these variables in the expressions that should be entered in the f1 and f2 fields. The entered expressions and also the coefficients are shown in Figure 9.22, which represents the PDE Specification panel completed for the PDE system.

Initialize the triangle mesh in the Mesh Mode (line in the popup menu of the main menu Mesh button) and click on the Refine line.

Select the "Parameters …" line in the popup menu of the Solve main menu button and on the Solve Parameters panel that appears, enter 0:0.25:1 in the "Time:" field and the initial conditions [2*sin(pi*x.*y);0.5*cos(pi*x.*y)] in the "u(t0):" field (see Figure 9.23).

PDE Specification			— □ ×
Equation:	d*u'-div(c*grad(u))+a*u=f		

Type of PDE:	Coefficient	Value	Value
○ Elliptic	c11, c12	1.0	0.0
● Parabolic	c21, c22	0.0	40.0
○ Hyperbolic	a11, a12	820.0	0.0
○ Eigenmodes	a21, a22	0.0	0.0
	f1, f2	820*(0.2+u(1,:).^2.*u(2,:))	820*(1.3-u(1,:).^2.*u(2,:))
	d11, d12	1.0	0.0
	d21, d22	0.0	1.0

OK Cancel

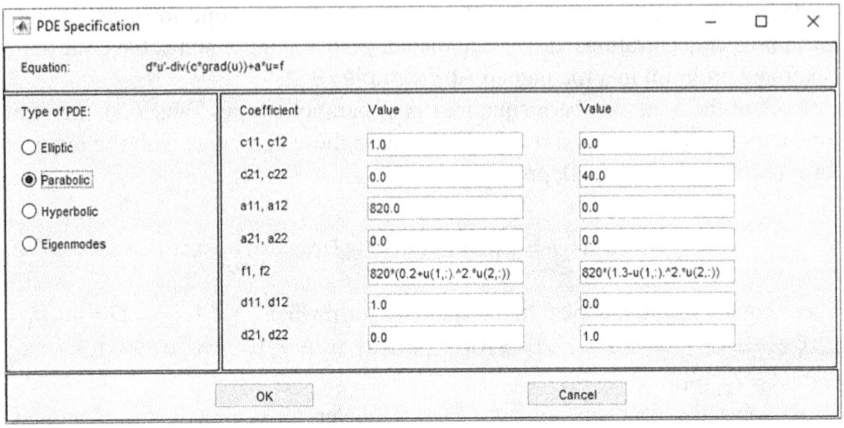

FIGURE 9.22 PDE Specification window completed for Schnakenberg system of two PDEs.

Solve Parameters — □ ×

Time:

`0:.25:1`

u(t0):

`[2*sin(pi*x.*y);0.5*cos(pi*x.*y)]`

Relative tolerance:

`0.01`

Absolute tolerance:

`0.001`

OK Cancel

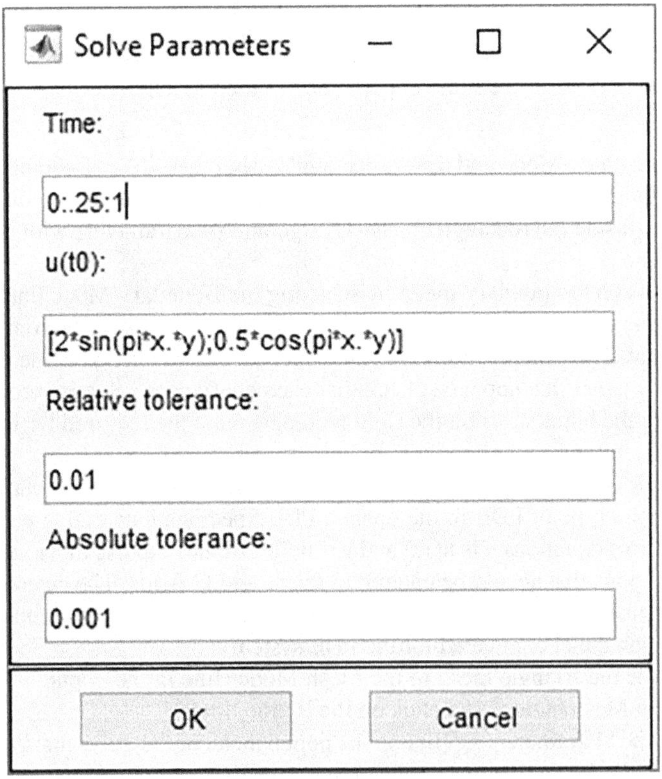

FIGURE 9.23 Solve Parameters dialog box completed for our two-PDE system.

Then click the Solve PDE line of the popup menu of the main menu Solve button. A 2D solution appears with a colored bar. Figure 9.24 represents a final solution for *u* variable after selecting the "jet" in the "Colormap" field and marking the "Show mesh" box (in the "Plot Selection" window that appears after checking the "Parameters …" option of the main menu Plot button).

To visualize obtained concentration distribution for *v*, the v option must be selected in the first and in the last fields below the "Property:" caption of the Plot Selection window. Resulted 2D plot is presented in Figure 9.25.

For each component of the PDE system, we check the "Height (3D plot)" checkbox in the Plot Selection window to generate a 3D graph with solutions for *u* and *v* (Figure 9.26a and b).

Pass the solution and mesh parameters to the MATLAB® workspace: select the Export Mesh and Export Solution lines within the popup menus of the corresponding Solve and Mesh menu buttons. If you have not changed the variable names, the solution for variables *u* and *v* is in the *u* matrix (each matrix column corresponds to certain times, i.e., *u*(:,1) corresponds to t=0 and *u*(:,5) – to the final t = 1), while mesh parameters are within the matrices *p*, *e*, and *t*. After this, in order to get and display *u*-values at the required orthogonal grid points, the following commands should be entered in the Command Window.

```
>> x=0:0.25:1;y=0:0.0999:0.4;
>> T_table=tri2grid(p,t,u(1:length(p),end),x,y)
```

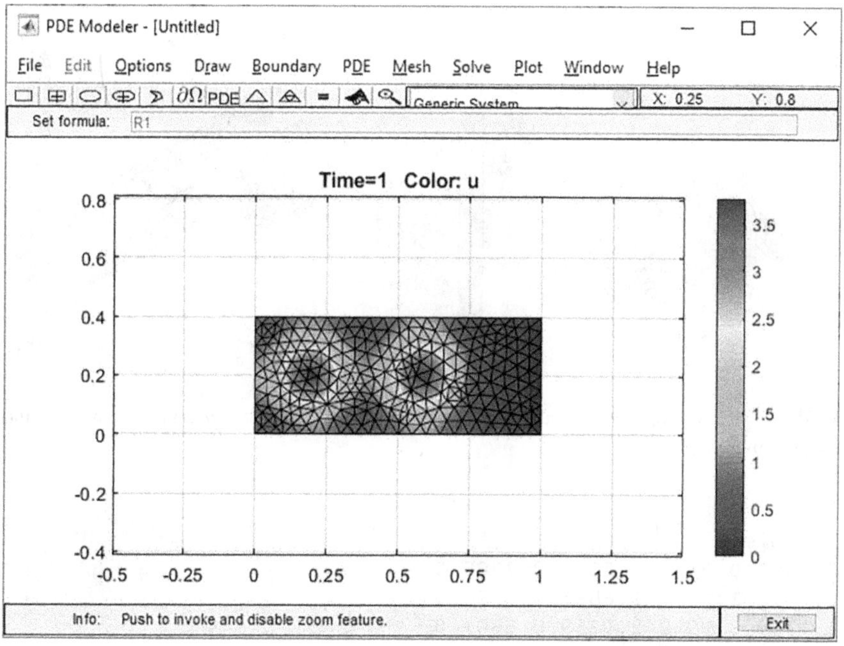

FIGURE 9.24 2D plot showing the *u*-concentration distribution at the final time.

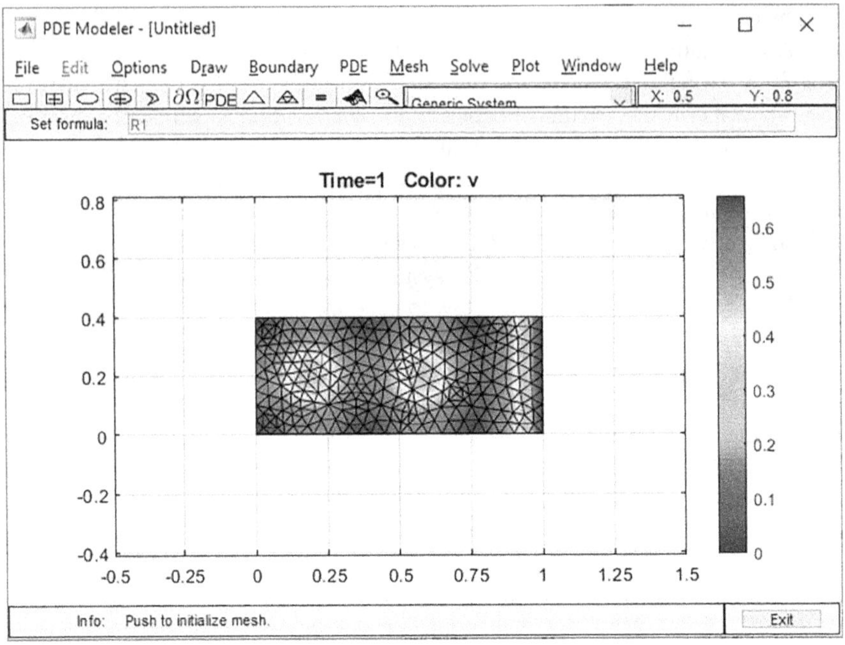

FIGURE 9.25 2D plot showing the final distribution of the *v*-concentration.

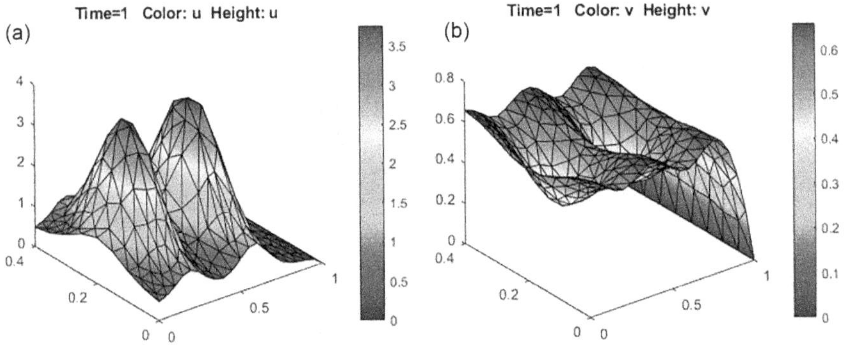

FIGURE 9.26 3D plots with final concentrations of the *u*-component (a) and *v*-component (b).

```
T_table =
0.4809 0.8504 0.8108 0.3985 0
0.7546 1.8586 1.6581 0.5921 0
1.1319 3.0962 2.5359 0.8603 0
0.7564 1.8879 1.6277 0.6030 0
0.4809 0.8526 0.8095 0.4050 0
```

Now, to obtain v-values at the same grid points as u, the following command should be entered:

```
>> T_table=tri2grid(p,t,u(length(p)+1:end,end),x,y)
T_table =
0.6579 0.5915 0.5968 0.6582 0
0.6042 0.4892 0.5073 0.6234 0
0.5435 0.3806 0.4187 0.5825 0
0.6028 0.4878 0.5063 0.6185 0
0.6570 0.5909 0.5954 0.6542 0
```

The PDE Modeler automatically generates a file containing commands that follow the steps in the solution that you complete. To save the program, select the "Save As …" line in the popup File menu and type the name ApExample_9_1 in the "File name:" field of the "Save As" panel that appears. The file contains a user-defined function, and its name must match the function name. To check this, open the saved file in the MATLAB® Editor window and write ApExample_9_1 in the function definition line instead of the default name assigned automatically by PDE Modeler.

9.5.2 Vibrations of a Slab with an Elliptical Hole

Dimensionless 3D wave equation (see also Section 6.5.2)

$$\frac{\partial^2 u}{\partial t^2} - \Delta u = 0$$

describes oscillatory displacements u of a body about of equilibrium plain; here $\Delta = \frac{\partial^2}{\partial x^2} + \frac{\partial^2}{\partial y^2} + \frac{\partial^2}{\partial z^2}$, x, y, and z – Cartesian coordinates, t – time.

Problem: Compose live script named ApExample_9_2 that solves above PDE for a $2 \times 4 \times 0.3$ rectangular slab having an elliptical hole with 0.3 and 0.4 semi-axes centered at $x = 1$ and $y = 2.6$. First, build the body geometry as 2D object and then convert it to 3D using the extrude command; show 2D and 3D object geometry and the generated tetrahedral mesh. Consider BC as Dirichlet with $u = 0$ on the left vertical face and as Neumann with g = 0 and q = 0 on each other face of the body. Suppose IC as $u0 = 0$ and $du0/dt = x/2$. Take 21 time points from the range 0 … 5. Show the generated mesh and present the results on the four resulting plots in the same Figure window at times 1, 2.5, 4, and 5.

The ApExample_9_2 live script solving this problem is as follows:

```
clear,close all
model=createpde;
a_r=2;b_r=4;c_r=0.3;                           % slab parameters
R1=[3,4,0,0,a_r,a_r,0,b_r,b_r,0]';             % slab
e2=[4,a_r/2,b_r/2+0.6,0.3,0.4,0,0,0,0,0]'; % elliptic hole
```

```
gd = [R1,e2];                               % geometry matrix
sf = 'R1-e2';                               % formula set
ns = char('R1','e2')';                      % geometric shape names
g=decsg(gd,sf,ns);                          % decomposed geometry
gm=geometryFromEdges(model,g);              % geometry to model
pdegplot(model,'EdgeLabels','on','FaceLabels','on')
                                            % 2D plot
axis equal
gm=extrude(gm,c_r);                         % extrudes 2D object
                                            vertically to a
                                            height=c_r
model.Geometry=gm;                          % passes created 3D
                                            discrete geometry to
                                            model
pdegplot(gm,'FaceLabels','on','FaceAlpha',0.5)
                                            % 3D geometry plot
applyBoundaryCondition(model,"dirichlet",'Face',5,'u',0);
u0=0;du0=@(location,state)location.x./2;
setInitialConditions(model,u0,du0);
specifyCoefficients(model,'m',1,'d',0,"a",0,"c",1,"f",0,'
Cell',1);
generateMesh(model);
figure
pdemesh(model)
tlist=linspace(0,5,21);
res=solvepde(model,tlist);
u=res.NodalSolution;
figure
subplot(2,2,1)
pdeplot3D(model,'ColorMapData',u(:,5))
title(['Time =',num2str(tlist(5))])
subplot(2,2,2)
pdeplot3D(model,'ColorMapData',u(:,11))
title(['Time =',num2str(tlist(11))])
subplot(2,2,3)
pdeplot3D(model,'ColorMapData',u(:,17))
title(['Time =',num2str(tlist(17))])
subplot(2,2,4)
pdeplot3D(model,'ColorMapData',u(:,end))
title(['Time =',num2str(tlist(end))])
```

The commands perform the following operations:

- The commands at first two lines clear the workspace, close all Figure windows, and generate the model object;
- The next thirteen lines contain commands that draw a rectangular slab with an elliptic hole, extrude 2D object geometry vertically, and show the first 2D geometry created and the resulting 3D geometry with faces and cell labels;

- BCs and ICs were specified by the commands in lines 16 … 18. Slab face 5 specified as the Dirichlet type, and all other bounds are Neumann type by default and do not need to be specified. The starting displacement rate *du0/dt* applied to the slab was specified using the anonymous function du0 with location and state arguments; the function uses the location.x variable as *x* in the IC expression.
- The coefficients of the wave equation are specified in the standardized PDE form with the specifyCoefficients command on line 19 as follows: $m = 1$, $d = 0$, $c = 1$, $a = 0$, and $f = 0$;
- The tetrahedral mesh was generated and plotted with commands on lines 20 and 22;
- The solution was obtained for the required time values, and the results were assigned to the variable *u* by the commands on lines 23 … 25. The solvepde command is used in a non-stationary form with the tlist vector containing the required time values;
- The graphical representation of the results for times 1, 2.5, 4, and 5 is plotted in four 3D graphs at the same figure using commands in the thirteen final lines.

The live script generated plots showing the 2D and 3D model geometries with the faces and cell labels, mesh, and four 3D solution views for times 1, 2.5, 3, and 5, are presented in Figure 9.27.

9.5.3 Electric Potential, Plate with Varying Conductivity

Distribution of the electric potential *u* in the 3D plate with coordinate-varying conductivity *c* can be defined solving the following PDE

$$\nabla\left[c(x,y,z)\nabla u\right]=0$$

where $\nabla = \frac{\partial}{\partial x}+\frac{\partial}{\partial y}+\frac{\partial}{\partial z}$; *x*, *y*, and *z* – Cartesian coordinates; the equation given in the dimensionless form.

Problem: Create script named ApExample_9_3 that solves the above PDE for a 1x1x0.1 conducting plate with irregular conductivity that varies with coordinates as $c(x,y,z) = 3.1 + \sin(2\pi x) + \cos(3\pi y) + \sin(\pi z)$. The plate is isolated by perimeter and has two circular holes, one at point x = 0.625, y = 0.75, and second at point x = 0.375, y = 0.25; radius of holes is 0.125. Assume the potential at face of the first hole is 1 while at the second 0, this boundary conditions here are the Dirichlet-type. First, build the plate geometry as 2D object and then convert it to 3D using the extrude command; show 2D and 3D object geometry and the generated tetrahedral mesh in the same Figure window. Present resulted potential distribution in 3D plot.

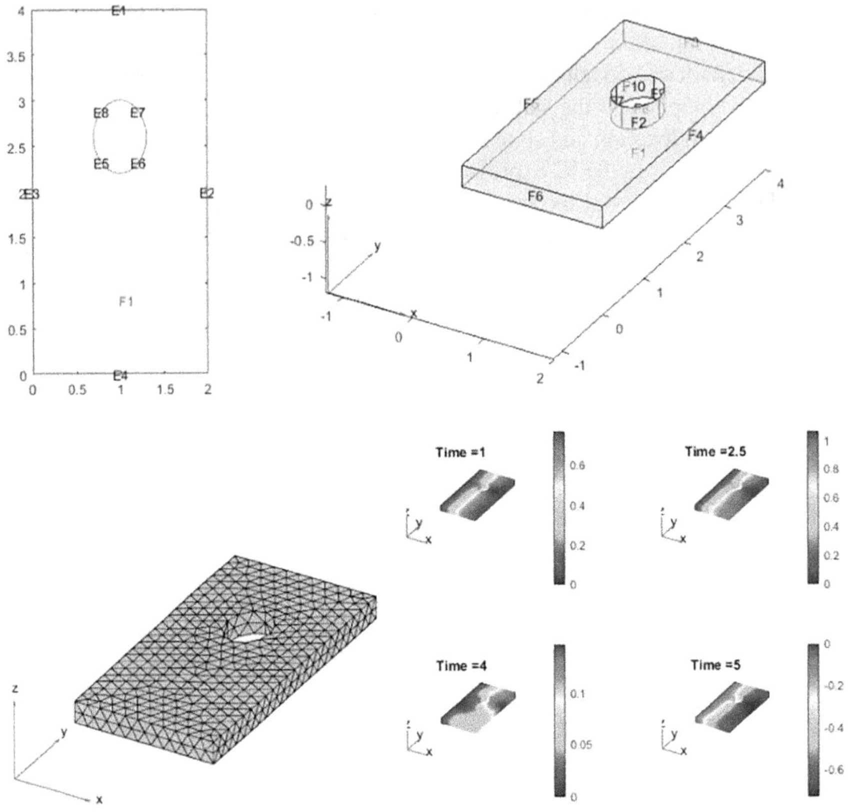

FIGURE 9.27 2D and 3D slab geometries with edge, face, and cell labels (top plots), as well as mesh and solution views at t = 1, 2.5, 4, and 5 (bottom plots).

The ApExample_9_3 script solving this problem is as follows:

```
clear,close all
model=createpde;
a_r=1;b_r=1;c_r=0.1;                          % plate parameters
R1=[3,4,0,0,a_r,a_r,0,b_r,b_r,0]';            % square plate
c1=[1,0.625,0.75,0.125,0,0,0,0,0,0]';         % circle hole 1
c2=[1,0.375,0.25,0.125,0,0,0,0,0,0]';         % circle hole 2
gd = [R1,c1,c2];                              % geometry matrix
sf = 'R1-c1-c2';                              % formula set
ns = char('R1','c1','c2')';              % geometric shape names
g=decsg(gd,sf,ns);                       % decomposed geometry
gm=geometryFromEdges(model,g);           % geometry to model
subplot(2,2,1)
pdegplot(model,'EdgeLabels','on','FaceLabels','on')
                                         % 2D plot
axis equal
```

```
gm=extrude(gm,c_r); % extrudes 2D vertically to a height=c_r
model.Geometry=gm;
subplot(2,2,2)
pdegplot(gm,'FaceLabels','on','FaceAlpha',0.5)      % 3D plot
applyBoundaryCondition(model,'dirichlet','Face',[11:14],'u',1);
applyBoundaryCondition(model,'dirichlet','Face',[7:10],'u',0);
c=@(location,state)3.1+sin(2*pi*location.x)+cos(3*pi*locatio
n.y)+sin(pi*location.z);
specifyCoefficients(model,'m',0,'d',0,'a',0,'c',c,'f',0,
'Cell',1);
generateMesh(model);
subplot(2,2,[3 4])
pdemesh(model)
res=solvepde(model);
u=res.NodalSolution;
figure
pdeplot3D(model,'ColorMapData',u)
title('Electric potential distribution')
```

The commands perform the following operations:

- The commands in the first two lines clear the workspace, close all the Figure windows, and generate the model object;
- The next fourteen lines contain commands that draw a square plate with two round holes, extrude 2D object geometry vertically, pass the 3D geometry to the model object, and transfer the first 2D geometry created and the resulting 3D geometry, containing the faces and cell labels, to the same Figure window;
- BCs were specified by the commands in the next two lines. The hole faces with numbers 11 …14 are specified as 'dirichlet' with $u = 1$, and faces 7 … 10 of the second hole are specified as 'dirichlet' with $u = 0$. All other boundaries are isolated and by default refer as Neumann type with g = 0 and q = 0 and do not need to be specified.
- Potential equation coefficients are specified in the standardized PDE form using the specifyCoefficients command on line 22 as follows: $m = 1$, $d = 0$, $c = c(x,y,z)$, $a = 0$, and $f = 0$. Material conductivity $c(x,y,z)$ was specified using the anonymous function c with location and state arguments; the function uses the variables location.x, location.y, and location.z as the x, y, and z variables of the c expression.
- The commands in lines 23 … 25 generate a tetrahedral mesh and pass their view to the Figure window containing the two previous graphs, after which the Figure window is complete.
- The solution was performed by the solvepde command, and its results were assigned to the u variable by the commands on lines 26 and 27.
- The 3D graphical presentation of the results is carried out with commands in three final lines.

- The 2D and 3D model geometries, containing faces and cell labels, and mesh of the model, are presented in Figure 9.28, and 3D solution is shown in Figure 9.29.

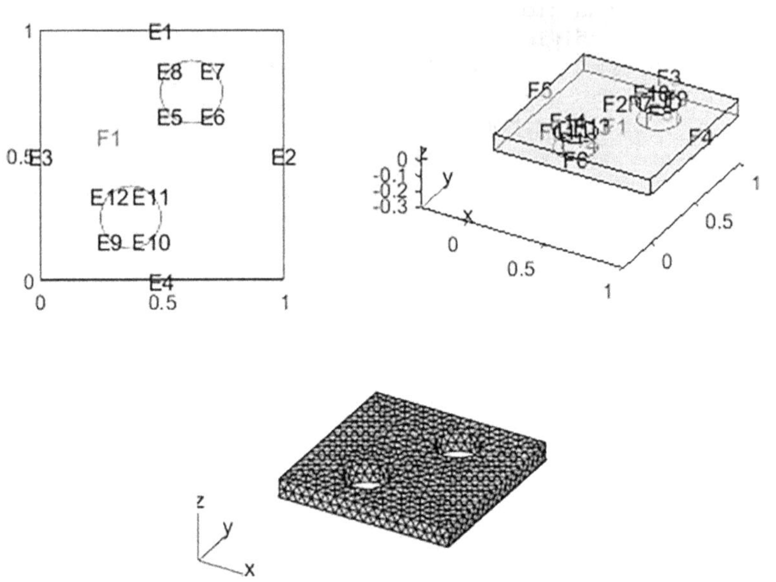

FIGURE 9.28 2D and 3D model geometry with edge, face, and cell labels (top plots), and generated mesh (bottom plots).

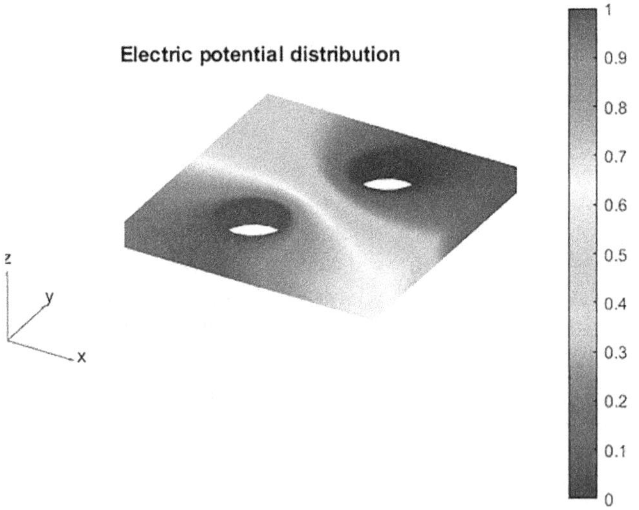

FIGURE 9.29 Resulting 3D distribution of the electric potential across the plate with two holes.

10 Toward Solving ODE and PDE Problems in the Life Sciences

10.1 INTRODUCTION

As in other scientific and engineering fields, differential equations (Des) play an important role in life science and bioengineering. Many processes and phenomena such as technological fields as chemical and biokinetics, biomedicine, epidemiology, pharmacology, genetics, and other biosciences can be described by their means. These equations are often unsolvable analytically, in which cases a numerical approach is called for, but universal numerical method does not exist. The tools for numerical solutions, described in the previous chapters, are applicable to solve ordinate (ODE) and partial (PDE) differential equations arising in the bio-area. This chapter provides example solutions of various bio-oriented problems for both categories of differential equations using the ode-, bvp-, and pdepe-solvers, as well as the programmatic and Modeler tool of the PDE Toolbox. Among the DE problems solved in the chapter are biomolecular reaction rate, diffusion-reaction problem, enzyme kinetics model, tumor growth, steady-state concentration distribution in the reactor, diffusion-Brusselator coupled problem, and others. Since the bio-audience is perhaps less prepared for programming, there is more space in the explanation of the programs than in the other chapters. Herewith, both basic familiarity with mentioned categories of DEs and knowledge of the previous material in this book are assumed.

10.2 SOLVING ODEs WITH ODE- AND BVP- SOLVERS

10.2.1 Rate Equation for Biomolecular Reaction

The rate equation for a bimolecular reaction A + B → C has form

$$\frac{dx}{dt} = k(a - x)(b - x)$$

where x is the amount of the product C at time t; k is the reaction rate constant; a and b are the initial amounts of the A and B reactants.

Problem: Write live script named ApExample_10_1 that solves the above equation and represents the result in an $x(t)$ plot and in a table. Consider $k = 0.007$ sec^{-1}, $a = 2$ and $b = 3$ mol, the reaction time range is 0 … 10 min, and the initial

DOI: 10.1201/9781003200352-10

amount of the product C at the beginning of the reaction is 0 mol. Take 101 points of time for ODE solution and six points for displaying numerical results. Plot the obtained values on the concentration-time graph and display them in a time-concentration table.

In the variables adopted to the ode-solvers, we have to rename x to y in our equation.

To solve the problem, a function files containing the user-defined function with the solving equation should be created. It is rational to compose such a function as an anonymous function, since we have only one short ODE, and, moreover, such a function does not have local variables/coefficients that should be passed to it in a special way. The input argument of this function must include variables t and y. The function should have an output parameter named, for example, dy (stands for $\frac{dy}{dt}$). The function body must contain ODEs. In our example, such a function is

```
dy = @(t,y)k*(a-y).*(b-y);
```

Now the solver should be selected. For our example, when there are no specific recommendations for a suitable method, the ode45 solver should be chosen. And the appropriate command should look like

```
[t,y] = ode45(dy,t_span,y0)
```

To fulfillment the program, we firstly need to assign the coefficients k, a, and b, the time range t_span, and, at the end, add the commands for plotting and displaying the results. A live script that performs all of this solves ODE, generates graph, and displays the resulted table, shown in Figure 10.1.

The commands in the ApExample_10_1 live script act as follows:

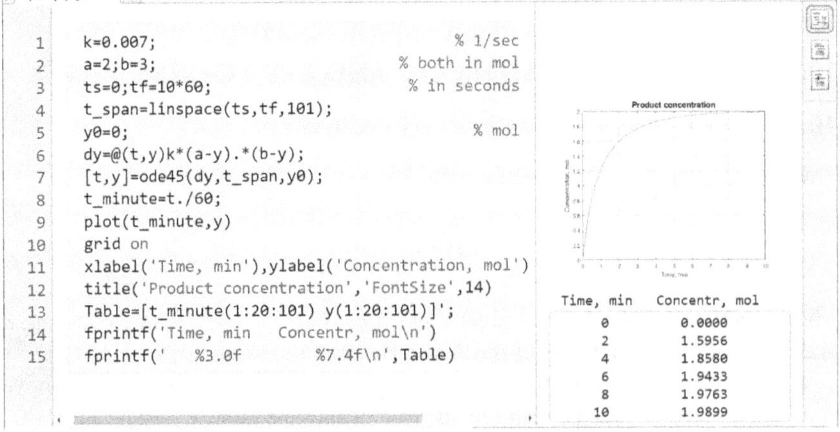

FIGURE 10.1 Live script with results of the biomolecular reaction solution.

- ODE coefficients k, a, and b, as well as the start and end times ts and tf are assigned first; the unit of time is converted from minutes to seconds.
- The 101 time values are specified in t_span vector by the linspace command in the ts … tf range.
- The initial concentration value y0 is assigned.
- The anonymous function dy containing ODE is introduced in the form described above.
- The ode45 command of the ode-solver is involved to solve our equations;
- To generate the resulting graph *y*(*t*), the plot command is used with several graph formatting commands, between which the title command with the SizeFont property introduces a font size 14 for better visibility of the title;
- Two fprintf commands are used to display the table header and two columns with time values (in minutes and no decimal places) and concentration values (with four decimal places). Because the fprintf command converts rows to columns, the t_minute and y vectors have been combined into a two-row matrix named Table; in both vectors, every 20th value is selected for printing.

10.2.2 Steady-State Concentration Distribution in a Short Tube

The diffusion with reaction equation for steady-state concentration distribution is

$$D\frac{d^2A}{dx^2} - kA = 0$$

where A is the amount of a compound (in M), that diffuses through tube and reacts in this process, at the end of compounds there is an adsorbent that absorbs A; k is the reactor rate constant (in s^{-1}); and D is the diffusion coefficient (in cm^2/s).

Problem: Compose a live script named ApExample_10_2 that solves the above ODE for a short 4 cm tube when $D = 1.47\cdot 10^{-6} cm^2/s$, $k = 4.9\cdot10^{-6} s^{-1}$, and for the in- and out-concentrations equal 0.2 and 0 M, respectively. Use the bvp4c command and take the initial concentration and rate of the concentration change equal to 0.2 M and −0.1 M/s at six equidistant points each. Plot the obtained values in the concentration-coordinate graph and display them in the table of two-column: coordinate-concentration.

In notations of the ode solvers, the A and x in the original ODE are respectively y and t in the solver. For convenience, rename t of the solver as x and the A of the original equation as y_1. To solve the second-order equation, we convert it into a set of two first-order equations:

$$\frac{dy_1}{dx} = y_2 \quad \frac{dy_2}{dx} = \frac{k}{D}x$$

To solve these equations, function files containing the user-defined function with the solving equations should be created. Compose this function file as an anonymous function, since we have only two short ODEs. Such a function is

```
myODE=@(x,y) [y(2);k/D*y(1)];
```

The bvp4c command that solves this equation includes three input arguments (Section 5.3.1). In our case this is a call to the myODE function, a call to a function with boundary conditions, and initial values for y_1 and y_2 at six x-points. The commands that enter the initial conditions and the bvp4c command are

```
xs = 0;xf = 4;
x_mesh = linspace(xs,xf,6);
solinit = bvpinit(x_mesh,[0.2-0.1]);
xy_sol = bvp4c(my ODE,@BC,solinit)
```

The function containing boundary conditions should contain the y data for the two end points a and b in the form $g(y(a),y(b)) = 0$ (Section 5.3.1.); the function should be placed at the bottom of program. The BC function is

```
function res = BC(ya,yb)
res = [ya(1)-0.2;yb(1)];
```

To complete the program, we need to set the coefficients k, and D at the beginning, and the commands for plotting and displaying the results at the end of the program.

The live script that performs all of the above: solves the ODEs, generates a graph, and displays the resulting table, shown in Figure 10.2.

The commands in the ApExample_10_2 live script operate as follows:

- First, values are assigned to the ODE coefficients k and D and to the start and final coordinates xs and xf are assigned.

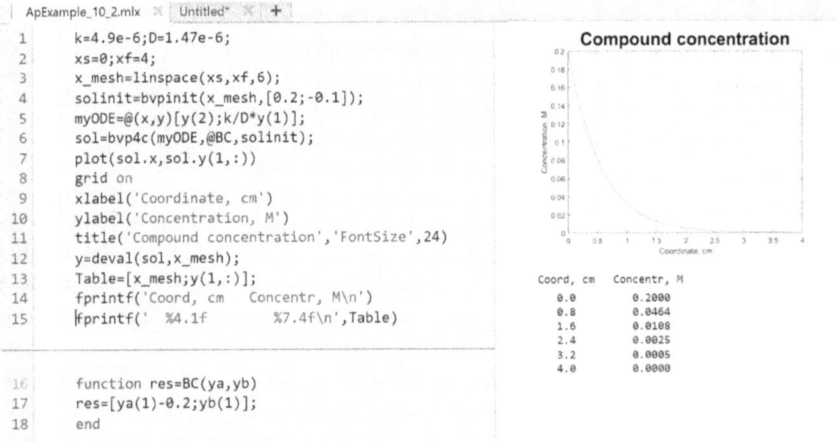

FIGURE 10.2 The ApExample_10_2 live script with results of the diffusion-reaction equation solution.

- Six coordinate values are specified in x_mesh by the linspace command in the range *xs* … *xf*.
- The initial values of the concentration and its rate of change are assigned to the solinit vector using the bvpinit command.
- The anonymous function myODE containing the ODEs is introduced in the form described above.
- The bvp4c command of the bvp-solver solves two first-order ODEs;
- To generate the resulting graph *y*(*x*), the plot command is used together with several graph formatting commands, between which the title command with the FontSize property introduces a font size of 24 for better visibility of the title;
- Two fprintf commands are used to display the table header and two columns with coordinate values (in cm and with one decimal digit) and concentration values (in M and with four decimal digits). Because the fprintf command converts rows to columns, the 1×6 x_mesh vector and the first column of the y matrix containing the solution are concatenated into a two-row matrix named Table.

10.2.3 A Model of Enzyme Kinetics

In biochemistry, a model of the enzyme-catalyzed reactions E+S ⇌ ES→E+P was developed by L. Michaelis and M. Menten. The model is represented by the following set of four first-order ODEs:

$$\frac{d[S]}{dt} = k_{1r}[ES] - k_1[S][E]$$

$$\frac{d[E]}{dt} = (k_{1r} + k_2)[ES] - k_1[S][E]$$

$$\frac{d[ES]}{dt} = k_1[S][E] - (k_{1r} + k_2)[ES]$$

$$\frac{d[P]}{dt} = k_2[ES]$$

where square brackets [ ] denote an amount of the following substances: substrate *S*, free enzyme *E*, substrate-bound enzyme *ES*, and product *P*; k_1 and k_{1r} are the forward and reverse reaction rate constants in the E+S ⇄ ES reaction, k_2 is the rate constant of the ES→P reaction.

Problem: Compose live script named ApExample_10_3 that solves the above set of ODEs in the time range 0…6 sec with initial $[S]=8$ mol, $[E]=4$ mol, $[ES]=0$, and $[P]=0$. Use 101 time points for solution. Reaction rate constants: $k_1=2\,\text{mol}^{-1}\text{s}^{-1}$, $k_{1r}=1\,\text{s}^{-1}$, $k_2=1.5\,\text{s}^{-1}$. Plot the amount of substances as a function of time. Display the obtained data as table with the time values in the first column and corresponding amounts in the four other columns, show in table each 10th value only.

Rename [*S*], [*E*], [*ES*], and [*P*] as y_1, y_2, y_3, and y_4 respectively, and rewrite the set of original equations in a form suitable for the ode-solver:

$$\frac{dy_1}{dt} = k_{1r}y_3 - k_1y_1y_2$$

$$\frac{dy_2}{dt} = (k_{1r} + k_2)y_3 - k_1y_1y_2$$

$$\frac{dy_3}{dt} = k_1y_1y_2 - (k_{1r} + k_2)y_3$$

$$\frac{dy_4}{dt} = k_2y_3$$

According to the conditions of the problem, this set of equations is solving in t range 0 … 6, and with the following initial substances values $y_1 = 8$, $y_2 = 8$, $y_3 = 0$, and $y_4 = 0$.

To solve these equations, a user-defined function with the solving equations should be created. Compose this function in such a way that constants of the ODEs should be passed to the function. Therefore, the input argument of this function, in addition to the arguments *t* and *y*, should contain also k1, k1r, and k2. The function should have an output parameter named, for example, *dy* (stands for $\frac{dy}{dt}$). The function body must contain ODEs. In our example, such a function is

```
function dy = ODEs(~,y,k1,k1r,k2)
dy=[k1r*y(3)-k1*y(1)*y(2);
(k1r+k2)*y(3)-k1*y(1)*y(2);
k1*y(1)*y(2)-(k1r+k2)*y(3);
k2*y(3)];
end
```

Now the solver should be selected. Since there is no prior reason, the ode45 solver should be chosen. As we need to pass additional constants to the function with solving equations, thus use an anonymous function. The appropriate commands should look like

```
myODE=@(t,y)ODEs(t,y,k1,k1r,k2);
t,y]=ode45(myODE,t_span,y0);
```

To complete the program, we need to set the variables k1, k1r, and k2, the time range t_span, and the vector y0 containing each substance initial concentration. At the end, the commands for plotting and displaying the resulting table should be added. The live script that performs all of this, solves ODEs, generates a graph, and displays the resulted table, shown in Figures 10.3 and 10.4.

```
% calculates substance amounts in the reactions E+S<=>ES->E+P
k1=2;k1r=1;k2=1.5;
S0=8;E0=4;ES0=0;P0=0;
ts=0;tf=6;
t_span=linspace(ts,tf,101);
y0=[S0,E0,ES0,P0];
myODE=@(t,y)ODEs(t,y,k1,k1r,k2);
[t,y]=ode45(myODE,t_span,y0);
plot(t,y),grid on
title('Enzyme - catalyzed reactions S<=>E->SE->E+P')
xlabel('Time, sec'),ylabel('Amounts of substances, mol')
legend('S','E','SE','P')
Table=[t(1:10:101,:) y(1:10:101,:)]';
fprintf('    t        S         E         ES        P\n')
fprintf('  %5.1f  %7.4f  %7.4f  %7.4f  %7.4f\n',Table)

function dy=ODEs(~,y,k1,k1r,k2)
dy=[k1r*y(3)-k1*y(1)*y(2);
        (k1r+k2)*y(3)-k1*y(1)*y(2);
        k1*y(1)*y(2)-(k1r+k2)*y(3);
        k2*y(3)];
end
```

FIGURE 10.3 The ApExample_10_3 live script that solve the ODEs describing the reagent concentration changes for reactions E + S ⇌ ES→E + P.

The commands in the ApExample_10_3 live script act as follows:

- First, the reaction rates k1, k1r, and k2, the initial substance amounts S0, E0, ES0, P, as well as the start and end calculation time ts and tf are specified.
- 101 values of time are specified in the t_span vector by the linspace command in the *ts* … *tf* range.
- The initial concentration values are collected in the vector y0.
- The anonymous function myODE, which calls the ODE-containing sub-function, is introduced in the form described above.
- The ode45 command of the ode-solver solves our equations;
- The plot command, along with several graph formatting commands, is used to generate the resulting graph of $y(x)$ curves.
- Two fprintf commands are used to display the table header and five columns with time values (in seconds and with one decimal digit) and amount values (in moles and with four decimal digits). Because the fprintf command converts rows to columns, the vector t and matrix y were concatenated into a five-row matrix named Table; in both vectors, every 10th value was selected for printing.

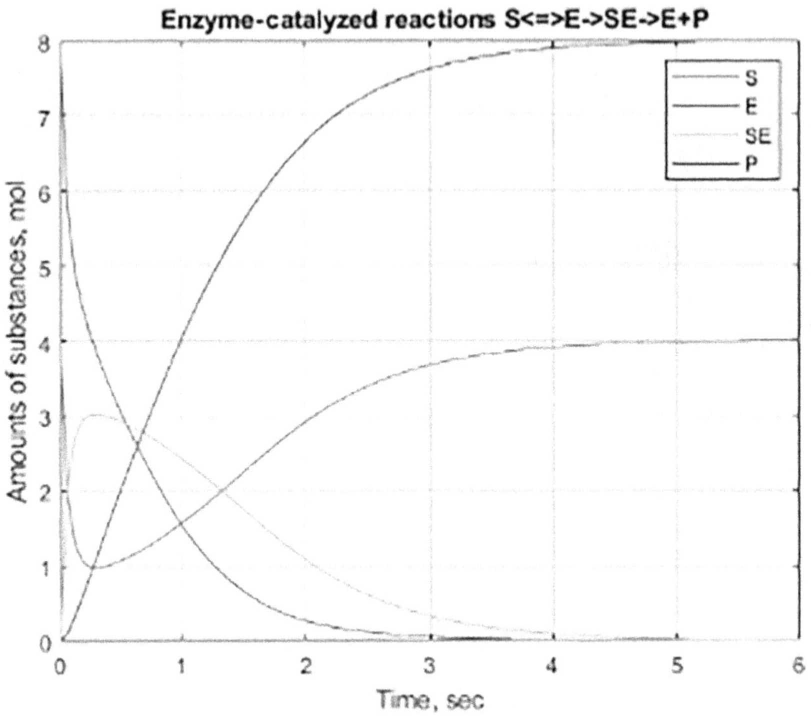

t	S	E	ES	P
0.0	8.0000	4.0000	0.0000	0.0000
0.6	2.6996	1.1564	2.8436	2.4568
1.2	1.1034	1.8362	2.1638	4.7328
1.8	0.3740	2.6747	1.3253	6.3007
2.4	0.1352	3.3167	0.6833	7.1815
3.0	0.0552	3.6732	0.3268	7.6180
3.6	0.0240	3.8479	0.1521	7.8239
4.2	0.0107	3.9299	0.0701	7.9192
4.8	0.0049	3.9679	0.0321	7.9630
5.4	0.0022	3.9853	0.0147	7.9831
6.0	0.0010	3.9933	0.0067	7.9923

FIGURE 10.4 Generated plot and resulting table displayed by the live script ApExample_10_3. Reagent concentration changes for reactions E + S ⇌ ES→E + P.

10.2.4 Two Reactors in Series

Two reactors connected in series with reaction $A \rightarrow B$ in each are in an unsteady state. The mass balance for the stirred tank of each reactor is as follows (based on Chapra and Canale 2015):

$$\frac{d[A_1]}{dt} = \frac{1}{\tau}([A_0]-[A_1]) - k[A_1]$$

$$\frac{d[B_1]}{dt} = -\frac{1}{\tau}[B_1] + k[A_1]$$

$$\frac{d[A_2]}{dt} = \frac{1}{\tau}([A_1]-[A_2]) - k[A_2]$$

$$\frac{d[B_2]}{dt} = \frac{1}{\tau}([B_1]-[B_2]) + k[A_2]$$

where square brackets [ ] denote the concentration of the reagent; A_0 is the amount of A at the inlet of the first reactor; A_1 – amount of A at the outlet of the first reactor and in inlet of the second; A_2 – amount of A at the outlet of the second reactor; B_1 – amount of B at the outlet of the first reactor and in inlet of the second; B_2 – amount of B in the second reactor; τ is the residence time for each reactor; k is the rate constant for the $A \rightarrow B$ reaction.

Problem: write a live script named ApExample_10_4 that solves the above set of ODEs in the time range 0 …20 min with initial $[A_0] = 20$ gmol/L, and initial values of all other reactants are zero. Consider $k = 0.12$ 1/min, $\tau = 5$ min, and number of time points equal to 101. Plot the concentrations as function of time, display the obtained data in the form of a table with the time values in the first column and the concentrations in the other four columns, and show only every 10th value in the table.

Rename $[A_1]$, $[B_1]$, $[A_2]$, and $[B_2]$ as y_1, y_2, y_3, and y_4, respectively, and rewrite the set of original equations in a form suitable for the ode-solver:

$$\frac{dy_1}{dt} = \frac{1}{\tau}([A_0] - y_1) - ky_1$$

$$\frac{dy_2}{dt} = -\frac{1}{\tau}y_2 + ky_1$$

$$\frac{dy_3}{dt} = \frac{1}{\tau}(y_1 - y_3) - ky_3$$

$$\frac{dy_4}{dt} = \frac{1}{\tau}(y_2 - y_4) + ky_3$$

where $[A_0]$, τ, and k are the specified constants. This set of equations is solving in t range 0 … 20, and with the following initial substances values $y_1 = 0$, $y_2 = 0$, $y_3 = 0$, and $y_4 = 0$.

To solve these equations, a user-defined function with the solving equations should be created. Compose this function in such a way that constants of the solving equations should be passed to the function. Therefore, the input argument of this function, in addition to the arguments t and y, should contain also A0, tau, and k. The function should have an output parameter named, for example, dy (stands for $\frac{dy}{dt}$). The function body must contain ODEs. In our example, such a function is

```
function dy = ODEs(~,y,A0,tau,k)
dy = [1/tau*(A0-y(1))-k*y(1);
-1/tau*y(2)+k*y(1);
1/tau*(y(1)-y(3))-k*y(3);
1/tau*(y(2)-y(4))+k*y(3)];
end
```

Since there is no prior reason, the ode45 solver can be chosen. As we need to pass additional constants to the function with ODEs, thus we use an anonymous function. The appropriate commands should look like

```
myODE = @(t,y)ODEs(t,y,A0,tau,k);
t,y] = ode45(myODE,t_span,y0);
```

To complete the program, we need to specify the coefficients A0, tau, and k, the time range t_span, and the vector y0 containing initial concentrations for each reagent. At the end, the commands for plotting and displaying the resulting table should be added. The live script that performs all of this, solves ODEs, generates a graph, and displays the resulted table, shown in Figures 10.5 and 10.6.

The commands in the live script ApExample_10_5 act as follows:

- First, the appropriate values are assigned to the variables k1, k1r, k2, S0, E0, ES0, P, ts and tf.
- 101 values of time are specified in the t_span vector by the linspace command in the ts … tf range.
- The initial concentration values are collected in the vector y0.
- The anonymous function myODE, which calls the ODE-containing sub-function, is introduced in the form described above.
- The ode45 command of the ode-solver solves our equations;
- The plot command together with several graph formatting commands is used to generate the resulting graph of $y(x)$ curves; the 'LineWidth',3 property pair is used for better visibility of the generated curves.
- Two fprintf commands are used to display the table header and five columns with time values (in minutes and with one decimal digit) and amount values (in gmol/L and with four decimal digits). Because the fprintf command converts rows to columns, so the vector t and matrix y were concatenated into a five row matrix named Table; in both vectors every 10th value was selected.

```
ApExample_10_4.mlx

% calculates reagent concentrations for two reactors in series
k=0.12;
tau=5;
A0=20;
A1_0=0;B1_0=0;A2_0=0;B2_0=0;
ts=0;tf=20;
t_span=linspace(ts,tf,101);
y0=[A1_0,B1_0,A2_0,B2_0];
myODE=@(t,y)ODEs(t,y,A0,tau,k);
[t,y]=ode45(myODE,t_span,y0);
plot(t,y,'LineWidth',3),grid on
title({'Two reactors in series';'Reagent concentrations'})
xlabel('Time, min'),ylabel('Reagent concentrations, gmol/L')
legend('A1','B1','A2','B2')
Table=[t(1:10:101,:) y(1:10:101,:)]';
fprintf('      t       A_1        B_1      A_2       B_2\n')
fprintf('  %5.1f  %7.4f  %7.4f  %7.4f  %7.4f\n',Table)

function dy=ODEs(~,y,A0,tau,k)
dy=[1/tau*(A0-y(1))-k*y(1);
        -1/tau*y(2)+k*y(1);
        1/tau*(y(1)-y(3))-k*y(3);
        1/tau*(y(2)-y(4))+k*y(3)];
end
```

FIGURE 10.5 The ApExample_10_4 live script, which solves the ODEs describing the concentration changes for two reactors connected in series.

10.3 SOLVING 1D PDEs USING THE PDEPE COMMAND

10.3.1 A 1D Reactor Model

The mass balance based model of an elongated reactor with one entry and one exit can be described by the adhesion-diffusion equation with a first-order reaction term:

$$\frac{\partial c}{\partial t} = D\frac{\partial^2 c}{\partial x^2} - U\frac{\partial c}{\partial x} - kc$$

where c is concentration, mol/m^3; D is a dispersion coefficient, m^2/h; k is the first-order decay coefficient, h^{-1}; x is the coordinate, m; and t –time, h.

The boundary conditions, BC, and initial conditions, IC, are:

$$c(0,t) = c_0 - D\frac{\partial u(0,t)}{\partial x},\ D\frac{\partial u(L,t)}{\partial x} = 0$$

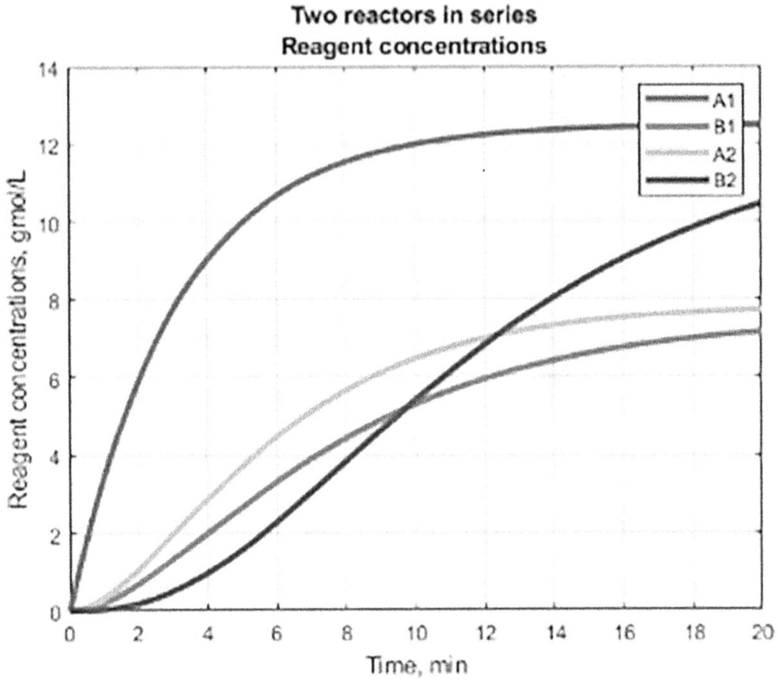

t	A_1	B_1	A_2	B_2
0.0	0.0000	0.0000	0.0000	0.0000
2.0	5.9089	0.6847	1.0565	0.1745
4.0	9.0245	1.9889	2.8600	0.9642
6.0	10.6675	3.3087	4.4680	2.2794
8.0	11.5337	4.4284	5.6626	3.8388
10.0	11.9904	5.3029	6.4751	5.4048
12.0	12.2313	5.9543	6.9998	6.8314
14.0	12.3583	6.4255	7.3273	8.0511
16.0	12.4253	6.7595	7.5268	9.0492
18.0	12.4606	6.9929	7.6461	9.8401
20.0	12.4792	7.1545	7.7164	10.4520

FIGURE 10.6 Generated plot and resulting table displayed by the live script ApExample_10_4. Reagent concentration vs time for two reactors connected in series.

$$c(x,0) = 0 \ at < 0x \le L$$

where L is the reactor length in m and c_0 is the inlet concentration in mol/m^3.

Problem: Compose a script with name ApExample_10_5 that displays a resulting table and generates two graphs in one Figure window: one 3D plot of the concentrations as function of coordinate *x* and time *t*, and the second – 2D plot of *u*(*x*) with four *iso*-time curves. Consider $D = 2$, $U = 1$, $k = 0.2$, and $c_{0=}$ 100. Take the 21 points of *x* and 20 points of *t*. To reduce the table volume, set the four iso-time rows (No 2, 6, 13, and 20) for six equally spaced coordinate values.

To solve the problem, the PDE, IC, and BC must be represented in standard form.

Rewriting the above equation

$$\frac{\partial u}{\partial t} = \frac{\partial}{\partial x}\left(D\frac{\partial u}{\partial x}\right) - U\frac{\partial c}{\partial x} - kc$$

and comparing with the form required by the pdepe solver (Section 8.2), we obtain that they are identical when:

$$m = 0$$

$$c\left(x,t,u,\frac{\partial u}{\partial t}\right) = 1$$

$$f\left(x,t,u,\frac{\partial u}{\partial x}\right) = D\frac{\partial u}{\partial x}$$

$$s\left(x,t,u,\frac{\partial u}{\partial x}\right) = -U\frac{\partial c}{\partial x} - kc$$

The IC in standard form is

$$u_0 = 0$$

The BCs in standard form ($p + gf = 0$) are

$$u - c_0 + 1 \cdot D\frac{\partial u}{\partial x} = 0 \quad \text{at} \quad x = x_a = 0, \quad \text{and} \quad 0 + 1 \cdot D\frac{\partial u}{\partial x} = 0 \quad \text{at} \quad x = x_b = \mathrm{L}$$

and in standard notations:

$$p(0,t,u) = u - c_0, \; p(1,t,u) = 0$$

$$q(0,t) = q_a = 1, \; q(1,t) = q_b = 1,$$

$$f\left(x,t,u,\frac{\partial u}{\partial x}\right)=D\frac{\partial u}{\partial x}$$

The commands for solving this problem are:

```
%             1D elongated reactor with first-order kinetics
n_x=21;n_t=20;
D=2;U=1;k=0.2;c0=100;L=10;
m=0;xs=0;xf=L;ts=0;tf=3.2;
x_mesh=linspace(xs,xf,n_x);t_span= linspace(ts,tf,n_t);
myPDE=@(x,t,u,DuDx)PDE(x,t,u,DuDx,D,U,k);
myBC=@(xa,ua,xb,ub,t)BC(xa,ua,xb,ub,t,c0);
u=pdepe(m,myPDE,@IC,myBC,x_mesh,t_span);
i_t=[2 6:7:n_t];j_t=1:5:n_x;
u_tab=[t_span(i_t)' u(i_t,j_t)];
fprintf(' t,h x=0,m x=2.5,m x=5,m x=7.5,m x=10,m\n')
fprintf('% 5.2f %7.4f %7.4f %7.4f %7.4f %7.4f\n',u_tab')
[X,T]=meshgrid(x_mesh,t_span);
surf(X,T,u)
xlabel('Coordinate'),ylabel('Time'), zlabel('Concentration')
u=pdepe(m,myPDE,@IC,myBC,x_mesh,t_span);
title('1D reactor with first order kinetics')
plot(x_mesh,u(i_t,:)','LineWidth',2),grid on
str=num2str(round(t_span(i_t)',2));
str1=[['t=';'t=';'t=';'t='],str(1:4,:)];
text(([L/3 L/3 L/3 L/3]-0.3),u(i_t,round(n_x/3))+2,...
[str1(1,:);str1(2,:);str1(3,:);str1(4,:)])
xlabel('Coordinate, m'),ylabel('Concentration, mol/m^3')
function [c,f,s]=PDE(x,t,u,DuDx,D,U,k)                    % PDE
c=1;
f=D*DuDx;
s=-U*DuDx-k*u;
end
function u0=IC(x)                                  % initial condition
u0=0;
end
function [pa,qa,pb,qb]=BC(xa,ua,xb,ub,t,c0)        % boundaries
pa=c0-ua;pb=0;
qa=1;qb=1;
end
```

The commands perform the following:

- Assign values to variables n_x, n_t, D, U, k, c0, L, m, xs, xf, ts, and tf;
- Generate vectors x_mesh and t_span (using two linspace commands) in range *xs* … *xf* and *ts* … *tf*, respectively;

- Refer to functions PDE and BC using two anonymous functions myPDE and myPC;
- Calculate the resulting vector of concentrations u using the pdepe command, the command calls to three sub-functions, among which the BC sub-function containing boundary conditions organized correspondingly to Table 8.2.
- Constitute the resulting table (by the two fprintf commands) in such a way that the first column contains the required time values, and other columns contain the resulting concentrations for five equidistant coordinates. The table has also a header.
- Generate and format two plots in the one Figure window in accordance with the problem conditions. Between these commands, the str, str1, and text commands add texts with the time values at four *x*,*y* points selected so that they are close to the corresponding curve.

After saving this program in the ApExample_10_5 file and entering the running command in the Command Window, the following table and graph (Figure 10.7) appear.

```
>> ApExample_10_5
t,h      x=0,m    x=2.5,m    x=5,m   x=7.5,m   x=10,m
0.17   27.5191    0.0885    0.0000    0.0000    0.0000
0.84   52.1392    9.3819    0.4838    0.0073    0.0001
2.02   65.6568   28.2444    7.6106    1.1775    0.1736
3.20   70.9653   38.8028   16.6416    5.2348    1.8205
```

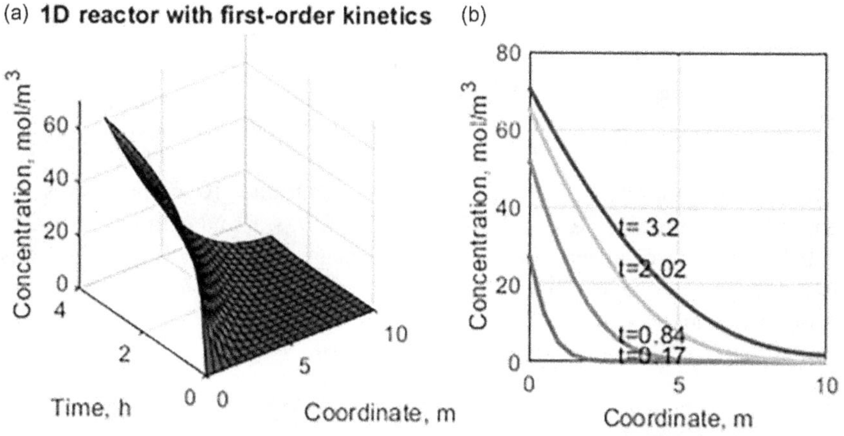

FIGURE 10.7 Resulting concentrations as a function of the coordinate and time (a), and as a function of the coordinate for four time values (b).

10.3.2 Initial Stage of Tumor Growth

In oncology researches, tumor-related angiogenesis is an important object of study. The initial stage of this phenomenon is described by the following set of PDEs (based on Orme and Chaplain 1996):

$$\frac{\partial n}{\partial t}=\frac{\partial}{\partial x}\left(d\frac{\partial n}{\partial x}-a\cdot n\frac{\partial c}{\partial x}\right)+S\cdot r\cdot n\cdot(N-n)\quad \frac{\partial c}{\partial t}=\frac{\partial}{\partial x}\left(\frac{\partial c}{\partial x}\right)+S\left(\frac{n}{n+1}-c\right)$$

where N is the total cell population, n is the concentration of the capillary endothelial cells (EC), c is that of fibronectin, t is the time, and x-coordinate; d, a, s, and r are experimentally determined constants.

The ICs have light alterations along x and represented by the piecewise functions as follows:

$$n(x,0)=\begin{cases}1.05, & at\ 0.3\le x\le 0.6\\ 1, & otherwise\end{cases}$$

$$c(x,0)=\begin{cases}0.5025, & at\ 0.3\le x\le 0.6\\ 0.5, & otherwise\end{cases}$$

The BCs for studied case are

$$\frac{\partial n(0,t)}{\partial x}=\frac{\partial n(1,t)}{\partial x}=0$$

$$\frac{\partial c(0,t)}{\partial x}=\frac{\partial c(1,t)}{\partial x}=0$$

Problem: Compose the live script named ApExample_10_6 that solves the above PDEs with given ICs and BCs. Consider d, a, s, r, and N to be 0.001, 3.8, 3, 0.88, and 1, respectively. Use 41 x points in range $x_s=0\ldots\ x_f=1$, and 21 t points in the range $t_s=0\ldots\ t_f=100$. Plot the solution on one page in four graphs: $c(x,t)$, $n(x,t)$, $c(x,t_f)$, and $n(x,t_f)$.

Rewrite the PDEs in the required standard form:

$$\begin{bmatrix}1\\1\end{bmatrix}.*\frac{\partial}{\partial t}\begin{bmatrix}u_1\\u_2\end{bmatrix}=\frac{\partial}{\partial x}\begin{bmatrix}d\dfrac{\partial u_1}{\partial x}-a\cdot u_1\dfrac{\partial u_2}{\partial x}\\ \dfrac{\partial u_2}{\partial x}\end{bmatrix}+S\begin{bmatrix}r\cdot u_1\cdot(N-u_1)\\ \dfrac{u_1}{u_1+1}-u_2\end{bmatrix}$$

where u_1 and u_2 are n and c, respectively; the element-wise multiplication is used here to represent the vector $\begin{bmatrix} \frac{\partial u_1}{\partial t} \\ \frac{\partial u_2}{\partial t} \end{bmatrix}$. From this equation follows that c, f, and s terms of the standard form are c = $\begin{bmatrix} 1 \\ 1 \end{bmatrix}$, f = $\begin{bmatrix} d\frac{\partial u_1}{\partial x} - a \cdot u_1 \frac{\partial u_2}{\partial x} \\ \frac{\partial u_2}{\partial x} \end{bmatrix}$, and s = $S\begin{bmatrix} r \cdot u_1 \cdot (N - u_1) \\ \frac{u_1}{u_1 + 1} - u_2 \end{bmatrix}$.

Now, the ICs in standard form are:

$$\begin{bmatrix} u_1(x,0) \\ u_2(x,0) \end{bmatrix} = \begin{bmatrix} \begin{cases} 1.05, & at\ 0.3 \le x \le 0.6 \\ 1, & otherwise \end{cases} \\ \begin{cases} 0.5025, & at\ 0.3 \le x \le 0.6 \\ 0.5, & otherwise \end{cases} \end{bmatrix}$$

Comparing the BC at x = 0 and at x = 1 with the required form (see Table 8.2), we obtain pa = pb = $\begin{bmatrix} 0 \\ 0 \end{bmatrix}$ and qa = qb = $\begin{bmatrix} 1 \\ 1 \end{bmatrix}$.

The live script solving this problem[1]:

```
%       calculates the fibronectin and endoterial cell
        concentrations
m = 0;xpoints=41;tpoints=21;
d = 1e-3; a = 3.8;S = 3; r = 0.88; N = 1;
x_mesh = linspace(0,1,xpoints); t_span =
linspace(0,100,tpoints);
myPDE=@(x,t,u,DuDx)PDE(x,t,u,DuDx,d,a,S,r,N);
sol = pdepe(m,myPDE,@IC,@BC,x_mesh,t_span);
n = sol(:,:,1); c = sol(:,:,2);
subplot(2,2,1)
surf(x_mesh,t_span,c);
title('Fibronectin, c(x,t)');
xlabel('Distance x'); ylabel('Time t');
zlabel('Concentration c')
axis square
```

[1] The live script ApExample_10_6 follows the pdex5 example given in MATLAB demo; to view it - type and enter >>edit pdex5.

```
subplot(2,2,2)
surf(x_mesh,t_span,n);
title('Endothelial cells, n(x,t)');
xlabel('Distance x'); ylabel('Time t');
zlabel('Concentration n')
axis square
subplot(2,2,3)
plot(x_mesh,c(end,:)),grid on
title('Fibronectin, final concentration');
xlabel('Distance x'); ylabel('Concentration c');
subplot(2,2,4)
plot(x_mesh,n(end,:)),grid on
title('EC, final concentration');
xlabel('Distance x'); ylabel('Concentration n');
function [c,f,s] = PDE(x,t,u,DuDx,d,a,S,r,N)      % PDEs for
solution
c = [1; 1];
f = [ d*DuDx(1) - a*u(1)*DuDx(2);DuDx(2)];
s = [S*r*u(1)*(N - u(1)); S*(u(1)/(u(1) + 1) - u(2))];
end
function u0 = IC(x)                              % Initial conditions
u0 = [1; 0.5];
if x >= 0.3 & x <= 0.6
u0(1) = 1.05 * u0(1);
u0(2) = 1.0005 * u0(2);
end
end
function [pa,qa,pb,qb] = BC(xl,ul,xr,ur,t)      % Boundary
conditions
pa = [0; 0]; qa = [1; 1];
pb = [0; 0]; qb = [1; 1];
end
```

The commands act as follows:

- Assign values to variables m, xpoints, tpoints, d, a, S, r, and N.
- Generate vectors x_mesh and t_span (using two linspace commands) in the required ranges each.
- Calls the PDE function using the anonymous function myPDE that passes additional parameters to it.
- Calculate the resulting 3D matrix of concentrations sol (using the pdepe command, the command calls to three sub-functions containing terms of PDEs, ICs, and BCs), the first page of this matrix contains solution for the concentrations of c-component, and the second for n-component. Note that the arguments in the pdepe sub-functions are assigned as column vectors, with the first lines containing the terms of the first reagent and the second containing the terms of the second reagent.

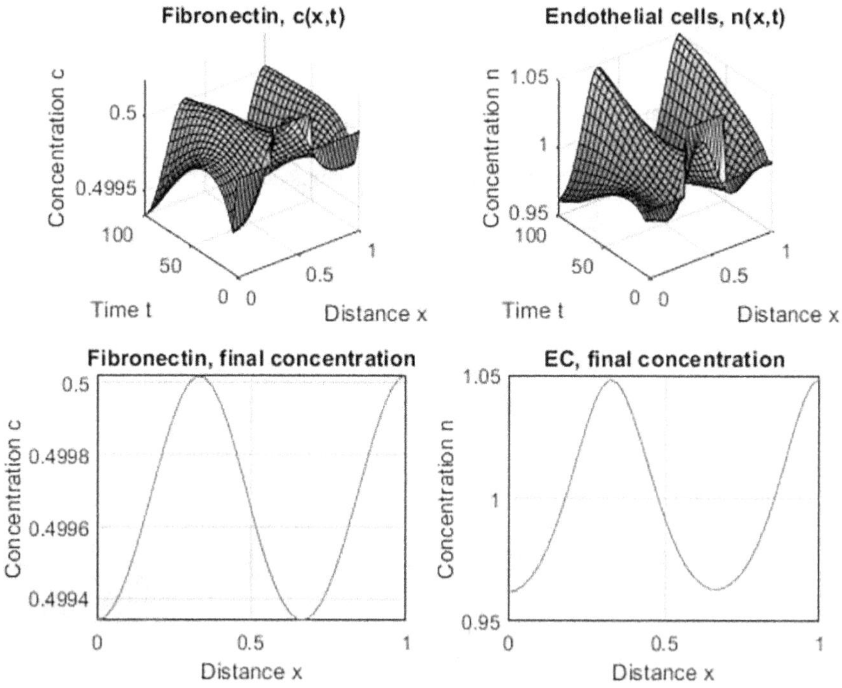

FIGURE 10.8 Resulting plots, generated by the ApExample_10_6 live script, with concentrations of the fibronectin and endothelial cells in a tumor.

- Generate and format four plots in the one Figure window (using the subplot, surf, and plot commands) according to the conditions of the problem.

The plots generated by this live script are presented in Figure 10.8.

10.3.3 Bacterial Culture, Diffusing and Reproducing

Consider the concentration n of bacterial culture which is diffusing (due to the random motion) and reproducing, depending on the nutrient local concentration c. The process can be described by a couple of following PDEs:

$$\frac{\partial n}{\partial t} = D_n \frac{\partial^2 n}{\partial x^2} + \left(\frac{k_{max} c}{k_n + c} - k \right) n$$

$$\frac{\partial c}{\partial t} = D_c \frac{\partial^2 c}{\partial x^2} - \alpha \frac{k_{max} c}{k_n + c} n$$

where D_n and D_c are diffusion coefficients for the bacterial and nutrient substances, respectively, $\frac{k_{max}c}{k_n+c}$ is the reproduction rate (of the Michaelis-Menten type), k_{max} and k_n are constants, k is the bacteria death rate coefficient; we treat all parameters as dimensionless.

Suppose those initial concentrations of reagents are distributed irregularly along x as follows:

$$n_0 = \begin{cases} 1, \ at \ 0.4 < x < 0.6 \\ 0, \ otherwise \end{cases}$$

$$c_0 = 1 + \sin(2\pi x)$$

The BC is

$$\frac{\partial n(0,t)}{\partial x} = \frac{\partial n(1,t)}{\partial x} = 0$$

$$\frac{\partial c(0,t)}{\partial x} = \frac{\partial c(1,t)}{\partial x} = 0$$

Problem: Create a live script named ApExample_10_7 that solves the above PDEs with their ICs and BCs. It is assumed that $D_n = 2$, $D_c = 1.2$, $k_{max} = 2$, $k_n = 1.9$, $k = 0.8$, and $\alpha = 0.3$. Employ 41 time values in the 0 … 0.1 range and 41 coordinate values in the 0 … 1 range. Plot the surfaces of the obtained concentrations $n(t,x)$ and $c(t,x)$ on the same page. For the $n(t,x)$ plot, take the azimuth and elevation angles to be 130 and 45 degrees, respectively.

Rewrite the PDEs in the required standard form:

$$\begin{bmatrix} 1 \\ 1 \end{bmatrix} .* \frac{\partial}{\partial t}\begin{bmatrix} u_1 \\ u_2 \end{bmatrix} = \frac{\partial}{\partial x}\begin{bmatrix} D_n \frac{\partial u_1}{\partial x} \\ D_c \frac{\partial u_2}{\partial x} \end{bmatrix} + \begin{bmatrix} \left(\frac{k_{max}u_2}{k_n + u_2} - k\right)u_1 \\ \alpha \frac{k_{max}u_2}{k_n + u_2} u_1 \end{bmatrix}$$

where u_1 and u_2 are n and c, respectively; the element-wise multiplication represents the vector $\begin{bmatrix} \frac{\partial u_1}{\partial t} \\ \frac{\partial u_2}{\partial t} \end{bmatrix}$. From this equation follows that c, f, and s terms of the standard PDE form are $c = \begin{bmatrix} 1 \\ 1 \end{bmatrix}$, $f = \begin{bmatrix} D_n \dfrac{\partial u_1}{\partial x} \\ D_c \dfrac{\partial u_2}{\partial x} \end{bmatrix}$, and s = $\begin{bmatrix} \left(\dfrac{k_{max} u_2}{k_n + u_2} - k \right) u_1 \\ \alpha \dfrac{k_{max} u_2}{k_n + u_2} u_1 \end{bmatrix}$.

Now, the ICs in standard form are:

$$\begin{bmatrix} u_1(x,0) \\ u_2(x,0) \end{bmatrix} = \begin{bmatrix} \begin{cases} 1.05, & at\ 0.3 \le x \le 0.6 \\ 1, & otherwise \end{cases} \\ \begin{cases} 0.5025, & at\ 0.3 \le x \le 0.6 \\ 0.5, & otherwise \end{cases} \end{bmatrix}$$

Comparing the BC equations at x = 0 and at x = 1 with the required form (see Table 8.2), we obtain pa = pb = $\begin{bmatrix} 0 \\ 0 \end{bmatrix}$ and qa = qb = $\begin{bmatrix} 1 \\ 1 \end{bmatrix}$.

The live script commands solving the problem are as follows:

```
% calculates the concentrations for bacterial culture and
nutrient
alfa=0.3;Dn=2;Dc=1.2;
kmax=1.2;kn=0.9;k=1.8;
m=0;
x_mesh=linspace(0,1,41);
t_mesh=linspace(0,0.1,41);
myPDE=@(x,t,u,DuDx)PDEs(x,t,u,DuDx,Dn,Dc,alfa,kmax,kn,k);
sol=pdepe(m,myPDE,@IC,@BC,x_mesh,t_mesh);
n=sol(:,:,1);
c=sol(:,:,2);
subplot(1,2,1)
mesh(x_mesh,t_mesh,n)
view(-130,45);
title('n(t,x)')
xlabel('Coordinate'),ylabel('Time'),zlabel('u')
axis square
subplot(1,2,2)
mesh(x_mesh,t_mesh,c)
title('c(t,x)')
xlabel('Coordinate'),ylabel('Time'),zlabel('v')
axis square
```

```
function [c,f,s]=PDEs(x,t,u,DuDx,Dn,Dc,alfa,kmax,kn,k) % PDEs
c=[1;1];
f=[Dn;Dc].*DuDx;
Kc=kmax*u(2)./(kn+u(2));
s=[(Kc-k)*u(1);-alfa*Kc*u(1)];
end
function u0=IC(x)                                  % Initial conditions
u0=[x>0.4&&x<0.6;1+sin(2*pi*x)];
end
function [pa,qa,pb,qb]=BC(xa,ua,xb,ub,t) % Boundary conditions
pa=[0;0];qa=[1;1];
pb=[0;0];qb=[1;1];
end
```

These commands do the following:

- Assign values to variables alpha, Dn, Dc, kmax, kn, k, and m.
- Generate vectors x_mesh and t_span (using two linspace commands) in the required ranges each.
- Refer to the myPDE function using an anonymous function that passes additional parameters to it.
- Calculate the resulting 3D matrix of concentrations sol (using the pdepe command that calls to three sub-functions containing the PDEs terms, the ICs, and BCs), the first page of this matrix contains the solution for the concentrations of *c*-component, and the second for *n*-component. In the IC sub-function, u0 is a column vector whose first element is the conditional statement that gives required (by our ICs) value equal to one when it is true and zero when it is false. Note that the arguments in the pdepe sub-functions are assigned as column vectors, with the first lines containing the terms of the first reagent and the second containing the terms of the second reagent.
- Generate and format two plots in the one Figure window (using the subplot, and mesh commands) according to the conditions of the problem.

The plots generated by this live script are presented in Figure 10.9.

10.3.4 A Model of the Reaction-Diffusion System

A bio-mathematical model of a non-linear system of two reaction-diffusion equations is presented as follows:

$$\frac{\partial u}{\partial t} = \frac{1}{2}\frac{\partial^2 u}{\partial x^2} + \frac{1}{1+v^2}$$

$$\frac{\partial v}{\partial t} = \frac{1}{2}\frac{\partial^2 v}{\partial x^2} + \frac{1}{1+u^2}$$

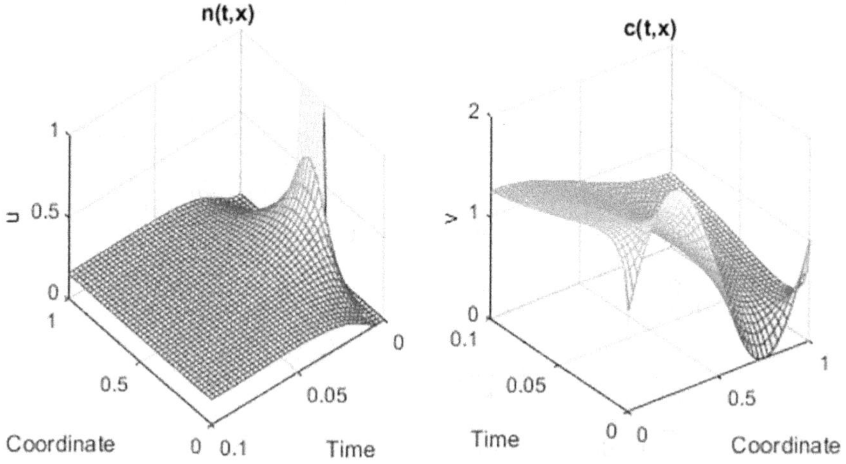

FIGURE 10.9 Resulting plots, generated by the ApExample_10_7 live script, with bacterial culture and nutrient concentrations versus coordinate and time.

where u and v are the reagent concentrations; all parameters are dimensionless.

The IC equations are assigned as

$$u(x,0) = 1 + \frac{1}{2}\cos(2\pi x)$$

$$u(x,0) = 3$$

and the BCs are

$$\frac{\partial u(0,t)}{\partial x} = \frac{\partial u(1,t)}{\partial x} = 0$$

$$\frac{\partial v(0,t)}{\partial x} = \frac{\partial v(1,t)}{\partial x} = 0$$

<u>Problem</u>: Compose program as user-defined function ApExample_10_8 without parameters that solves the above PDEs with its ICs and BCs. Employ 21 x values in the range 0 … 1 and 51 t values of in the range 0 … 0.4. Present the concentration changes for each reagent as a function of time t and coordinate x in two plots on the same page.

Rewrite first the set in the standard form

$$\begin{bmatrix} 1 \\ 1 \end{bmatrix} .* \frac{\partial}{\partial t}\begin{bmatrix} u_1 \\ u_2 \end{bmatrix} = \frac{\partial}{\partial x}\begin{bmatrix} \frac{1}{2}\frac{\partial u_1}{\partial x} \\ \frac{1}{2}\frac{\partial u_2}{\partial x} \end{bmatrix} + \begin{bmatrix} \frac{1}{1+u_2^2} \\ \frac{1}{1+u_1^2} \end{bmatrix}$$

where u_1 is u and u_2 is v; the argument m is 0 (in Cartesian coordinates), and terms with x^0 are 1 and are therefore not present in the above expression; element-wise multiplication is used here to present the left hand as a $\begin{bmatrix} \frac{\partial u_1}{\partial t} \\ \frac{\partial u_2}{\partial t} \end{bmatrix}$ vector. From this equation follows that c, f, and s terms of the standard PDE form are $c = \begin{bmatrix} 1 \\ 1 \end{bmatrix}$, $f = \begin{bmatrix} \frac{1}{2}\frac{\partial u_1}{\partial x} \\ \frac{1}{2}\frac{\partial u_2}{\partial x} \end{bmatrix}$, and $s = \begin{bmatrix} \frac{1}{1+u_2^2} \\ \frac{1}{1+u_1^2} \end{bmatrix}$.

The ICs in the standard form are

$$\begin{bmatrix} u_1(x,0) \\ u_2(x,0) \end{bmatrix} = \begin{bmatrix} 1+\frac{1}{2}\cos(2\pi x) \\ 3 \end{bmatrix}$$

And the BCs at x = 0 and at x = 1 after comparing with the required form (see Table 8.2) can be written as pa = pb = $\begin{bmatrix} 0 \\ 0 \end{bmatrix}$ and qa = qb = $\begin{bmatrix} 1 \\ 1 \end{bmatrix}$.

The commands solving the problem

```
function ApExample_10_8
%                              solves Reaction-diffusion equation
%                                         To run >>ApExample_10_8
m=0;n_x=21;n_t=51;
x_mesh=linspace(0,1,n_x);t_span=linspace(0,0.4,n_t);
sol=pdepe(m,@ODEs,@ICs,@BCs,x_mesh,t_span);
u=sol(:,:,1);
v=sol(:,:,2);
subplot(1,2,1)
surf(x_mesh,t_span,u)
title('First reagent, u(x,t)')
xlabel('Coordinate'),ylabel('Time'),zlabel('Concentration')
axis square
subplot(1,2,2)
surf(x_mesh,t_span,v)
title('Second reagent, v(t,x)')
xlabel('Coordinate'),ylabel('Time'),zlabel('Concentration')
axis square
function [c,f,s]=ODEs(x,t,u,DuDx)          % PDEs for solution
```

```
c=[1;1];
f=[1/2;1/2].*DuDx;
s=[1./(1+u(2).^2);1./(1+u(1).^2)];
function u0=ICs(x)                              % initial conditions
u0=[1+1/2*cos(2*pi*x);3];
function [pa,qa,pb,qb]=BCs(xa,ua,xb,ub,t) % boundary conditions
pa=[0;0];qa=[1;1];
pb=[0;0];qb=[1;1];
```

These commands do the following:

- Assign values to variables n_x, n_t, and m.
- Generate vectors x_mesh and t_span (using two linspace commands) in the required ranges each.
- Calculate the resulting three-dimensional matrix of concentrations sol (using the pdepe command that calls to three sub-functions containing the PDEs terms, the ICs equations, and BCs parameters), the first page of this matrix contains solution for the concentrations of u-component, and the second for v-component. Note that the arguments in the pdepe sub-functions are assigned as column vectors, with the first lines containing the terms of the first reagent and the second containing the terms of the second reagent.
- Generate and format two plots in the one Figure window (using the subplot, and commands) according to the conditions of the problem.

The plots generated by this script are presented in Figure 10.10.

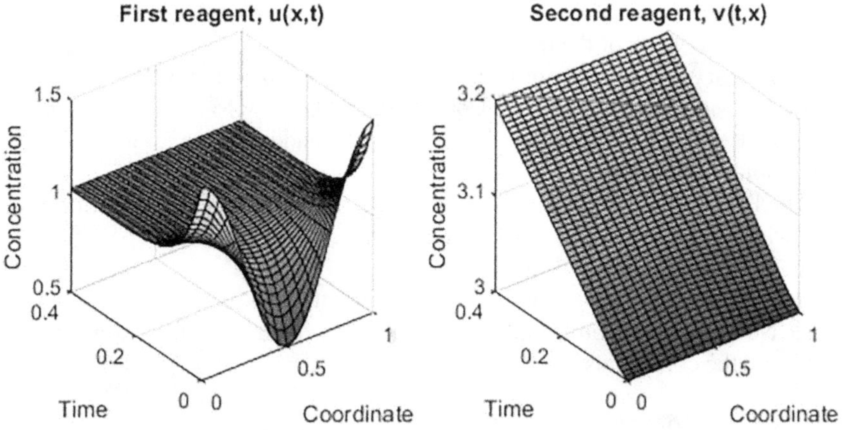

FIGURE 10.10 Resulting concentrations of reagents versus coordinate and time, generated by the live script ApExample_10_8.

After running this function in the Command Window, the following plots are generated:

```
>> ApExample_10_8
```

10.4 SOLVING 2D PDEs USING THE PDE TOOLBOX

10.4.1 Steady-State Concentration Distribution in a Reactor

The dimensionless PDE describing the steady-state concentration C in a tank of the 20×5 rectangular cross-section is

$$D\Delta C - kC = 0$$

where D and k are dispersion and reaction rate coefficients, respectively; Δ is the Laplacian $\frac{\partial^2}{\partial x^2} + \frac{\partial^2}{\partial y^2}$.

The BCs for the design features of an actual tank design are:

$$C(0,y)=C_1$$

$$C(x,0)=\begin{cases} C_2, & L-10\le x\le L \\ 0, & otherwise \end{cases}$$

$$C(x,width)=c(L,y)=0$$

where C_1 and C_2 are the concentrations that should be specified; L and *width* are the length and width of the rectangle.

Problem: Write live script named ApExample_10_9 that solves the above equations with the specified BCs. Consider $L = 20$, $D = 0.5$, $k = 0.1$, $C_1 = 40$, and $C_2 = 100$. Represent the result in 3D and 2D plots, and in a table for a rectangular grid with 8 equidistant x-points and 5 equidistant y-points.

The PDE to be solved is an elliptic type equation, and we should perform the following steps (Section 6.4.2) to solve it.

1. First, create a model object with the command model = createpde.
2. Now we need to build and plot the geometry of our object, for which we use the commands

```
L=20;width=5;
R1=[3,4,0,0,L,L,0,width,width,0]';                  % rectangle
sf = 'R1';                                         % formula set
ns = char('R1')';                   % basic geometric form names
g=decsg(R1,sf,ns);                                 % decomposition
```

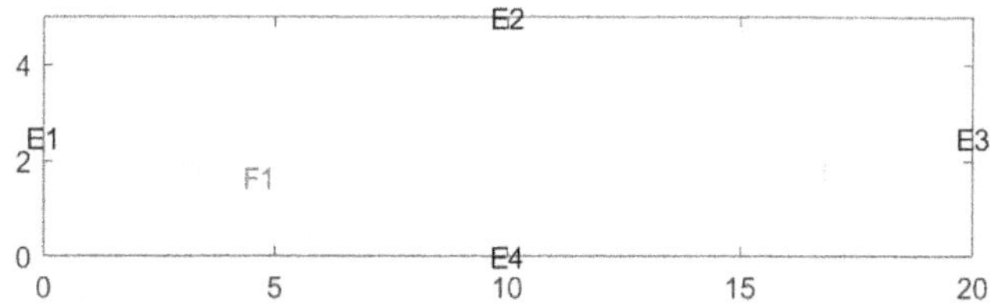

FIGURE 10.11 Tank cross-section geometry, the origin is located in the bottom left corner of the rectangle. The boundaries are designated by labels E1 … E4, and face is labeled as F1.

```
geometryFromEdges(model,g);                    % model geometry
pdegplot(model,'EdgeLabels','on','FaceLabels','on') %
plotting
axis equal
```

The resulting geometry of the tank cross-section with the boundary and face labels is shown in Figure 10.11.

3. The BCs of this problem are the Dirichlet type on the left E1 and bottom E4 edges of the rectangle; *C* notation is *u* of the standard PDE and Dirichlet boundaries (see Table 6.2). Considering this, the commands that specify the boundary conditions in our model are

```
applyBoundaryCondition(model,'dirichlet','Edge',1,'u',40);
applyBoundaryCondition(model,'dirichlet','Edge',[2,3],
'u',0);
BC=@(location,state)100*(location.x>=L-10&&location.x<=L);
applyBoundaryCondition(model,'dirichlet','Edge',4,'h',1,
'r',BC);
```

Note that the piecewise equation at the E4 boundary is represented in an anonymous function where the value 100 is generated when the conditional statement is true and the value 0 is generated otherwise.

4. To specify the PDE coefficients, we need to require the identity of our PDE ($\nabla D \nabla C - kC = 0$) and the standard form PDE(Chapter 6). From their comparison, it follows that the PDE coefficients are $m = 0$, $d = 0$, $c = 0.5$, $a = 0$, and $f = 1$, so the command that sets the coefficient looks like this:

```
specifyCoefficients(model,'m',0,'d',0,'c',-0.5,'a',-
0.1,'f',0, 'Face',1);
```

5. Generate a mesh with triangular elements using the command generateMesh(model).
6. After this, we can solve the PDE with the command res = solvepde(model); the defined *u* values and mesh nodes are contained in the res.NodalSolution matrix.

7. The resulting graphs and the required table can be obtained using the following commands:

```
pdeplot(model,'XYData',res.NodalSolution,...
    'ZData',res.NodalSolution,'Mesh','on','ColorMap','
    jet')
xlabel('x'),ylabel('y'),zlabel('Concentration')
axis tight
pdeplot(model,'XYData',res.
NodalSolution,'Mesh','on',...
'ColorMap','jet')
[x,y]=meshgrid(linspace(0,L,8),linspace(0,width,5));
xlabel('x'),ylabel('y')
axis equal tight
uintrp=interpolateSolution(res,x,y);
uintrp = reshape(uintrp,size(x))
```

All of the above commands were assembled into the ApExample_10_9 live script that is shown in Figure 10.12.

The graphs created by the above program are shown in Figure 10.13.

The results table is presented in Figure 10.14.

ApExample_10_9.mlx

```
%  calculates the steady-state distribution of the concentration in the tank
clear, close all
model = createpde;
L=20;width=5;
R1=[3,4,0,0,L,L,0,width,width,0]';                       % rectangle
sf = 'R1';                                                % formula set
ns = char('R1')';                                         % basic geometric form names
g=decsg(R1,sf,ns);                                        % decomposition
geometryFromEdges(model,g);                               % model geometry
pdegplot(model,'EdgeLabels','on','FaceLabels','on')       % plotting
axis equal
applyBoundaryCondition(model,'dirichlet','Edge',1,'u',40);
applyBoundaryCondition(model,'dirichlet','Edge',[2,3],'u',0);
BC=@(location,state)100*(location.x>=L-10&&location.x<=L);
applyBoundaryCondition(model,'dirichlet','Edge',4,'h',1,'r',BC);
specifyCoefficients(model,'m',0,'d',0, ...
    'c',-0.5,'a',-0.1,'f',0,'Face',1);
generateMesh(model);
res = solvepde(model);                                    % pde solution
pdeplot(model,'XYData',res.NodalSolution, ...
    'ZData',res.NodalSolution,'Mesh','on','ColorMap','jet')
xlabel('x'),ylabel('y'),zlabel('Concentration')
axis tight
pdeplot(model,'XYData',res.NodalSolution,'Mesh','on','ColorMap','jet')
[x,y]=meshgrid(linspace(0,L,8),linspace(0,width,5));
xlabel('x'),ylabel('y')
axis equal tight
uintrp=interpolateSolution(res,x,y);
uintrp = reshape(uintrp,size(x))                          % 2D table
```

FIGURE 10.12 The real-time program ApExample_10_9 solves the PDE in steady state for the concentration distribution in the tank.

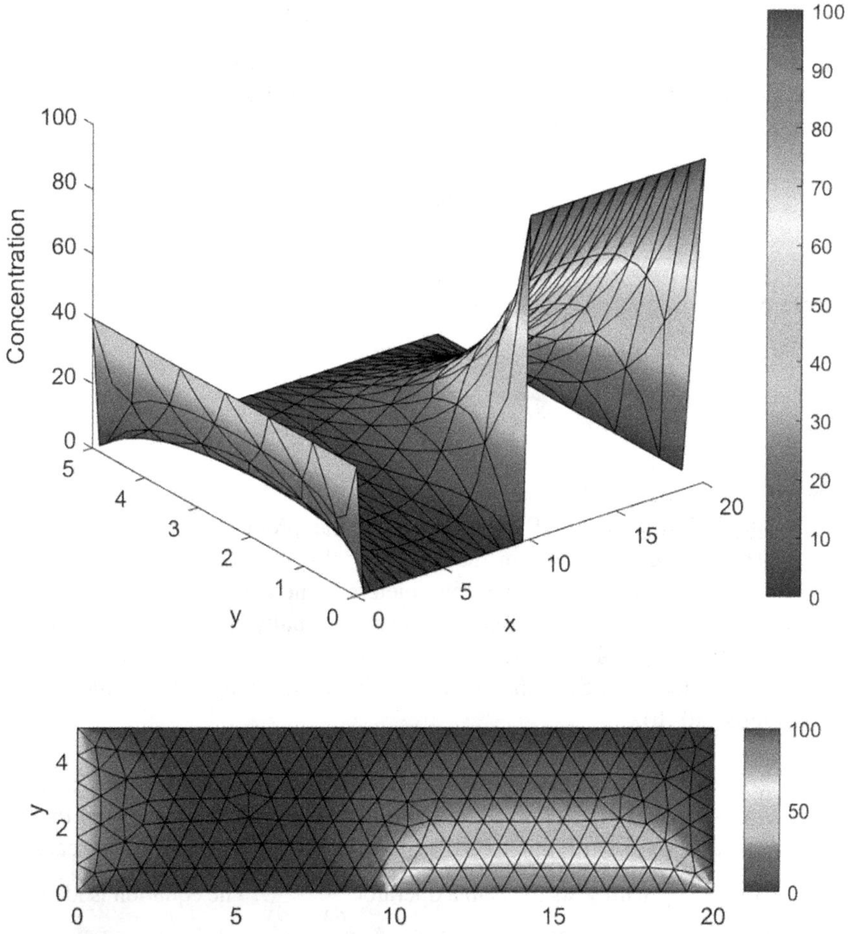

FIGURE 10.13 3D and 2D graphs with solution of the steady-state PDE for the concentration distribution in the tank.

ApExample_10_9.mlx | uintrp = 5×8

	1	2	3	4	5	6	7	8
1	40.0000	0.0000	0	0.0000	100.0000	100.0000	100.0000	100.0000
2	40.0000	4.0793	1.1028	8.5345	49.2987	54.8959	51.7774	0.0000
3	40.0000	5.6482	1.4755	7.0951	23.6653	28.3308	24.8840	-0.0000
4	40.0000	4.0804	0.9931	3.5696	9.6743	11.9294	9.9563	0.0000
5	40.0000	0.0000	0.0000	0.0000	0.0000	0.0000	0.0000	-0.0000

FIGURE 10.14 Resulting table with concentrations represented for eight equidistant coordinate values (columns) and for five equidistant times (rows).

10.4.2 Displacement of Homogenous Membrane

A uniform pressure *P* on a homogenous membrane causes it to tense *T* and displace *u*. The displacement of square membrane can be described by the following expression:

$$\Delta u = \frac{P}{T}$$

where Δ is the Laplacian $\frac{\partial^2}{\partial x^2} + \frac{\partial^2}{\partial y^2}$.

The membrane is clamped by the perimeter so that BCs are

$$u(0,y) = u(x = L, y) = u(x,0) = u(x,L) = 0$$

where L is the side length of the square.

Problem: Using the PDE Modeler solve the problem for $L = 1\,\text{cm}$ and P/T = 0.6/cm. Present results in the 2D view plot with the five contour lines. Transfer defined *P*-values and mesh parameters to the workspace and display the *P*(X,Z) table for the orthogonal grid with seven equally spaced values of *X* and *Y*, each in the range –L/2 … L/2. Find the maximum displacement value and its coordinate location. Save the automatically created program in a file named ApExample_10_10.

- First, determine the type of the solving equation by matching this equation with the required standard form (Table 6.1); it is easy to see that our equation is an elliptic PDE type since in the standard form reads $\nabla(\nabla P) = \frac{P}{T}$ with ∇ as the Nabla operator $\frac{\partial}{\partial x} + \frac{\partial}{\partial y}$. The equation is identical to the standard PDE form when $m = 0$, $d = 0$, $c = 1$, $a = 0$, and $f = \frac{P}{T}$.
- Now activate the PDE Modeler with the command

```
>>pdeModeler
```

In the PDE Modeler window that opens, mark the Grid and Snap lines in the popup menu of the Options button of the PDE Modeler menu, type the *x* and *y* limits as [–1 1] and [–1 1] in the Axis Limits dialog box, and mark the Axis Equal line. Check the general Application option – Generic Scalar (default).

- Then, activate the Draw Mode and draw a rectangle using the 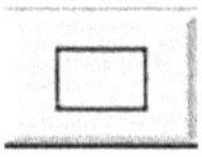 rectangle/square button. Now place the mouse arrow on point (–0.5,–0.5) and drag it to point (0.5, 0.5) in the Modeler window. Check the rectangle parameters in the Object Dialog panel that appears after clicking within the drawn rectangle.

- Activate the Boundary button and in the appeared "Boundary Conditions" panel mark the "Boundary Mode" and "Show edge labels" options. Click the first square boundary line and select the Dirichlet option in the appeared "Boundary Condition" dialog box. Verify the one in the h field and zero in the r field; do the same for each boundary of the rectangle (these actions are optional, since Dirichlet boundaries with the corresponding *h* and *r* values are set by default).
- Now select PDE Mode from the popup menu of the PDE Modeler main menu button. Studying body has one subdomain, thus there is no need to mark the "Show domain labels" option in the popup menu. Click the mouth button within the square, and in the PDE Specification panel that opens, check/mark the "Elliptic" type of PDE (default), and enter 1, 0, and 0.6 into the fields c, a, and f respectively.
- Activate the Mesh Mode and initialize the triangular mesh.
- Now we can solve the PDE equation, for this select the Solve PDE line from the popup menu of the main menu Solve button. A 2D view of the solution appears with a colored bar.
- To format the graph according to the problem conditions, select the Plot button of the main menu and click there the "Parameters ..." option. In the "Plot Selection" panel that opens, check the Contour box (below the "Plot type" column name), type 5 in the "Contour plot levels" field, select "jet" in the popup list of the Colormap field, and click the "Plot" button. The resulting plot is presented in Figure 10.15.

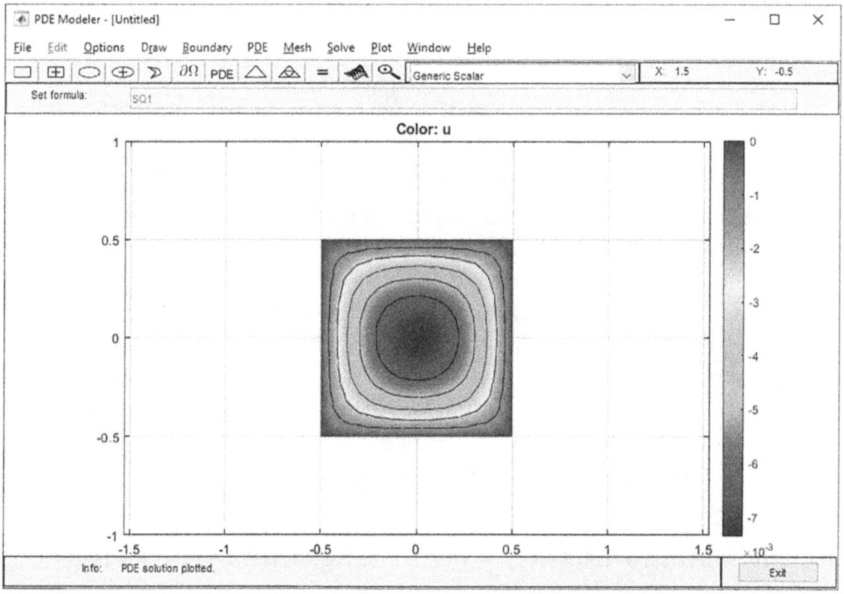

FIGURE 10.15 Membrane displacement in the 2D PDE Modeler representation.

To pass the solution and mesh parameters to the MATLAB® workspace, select the lines Export Mesh and Export Solution in the popup menus of the corresponding Solve and Mesh buttons of the main menu. Without changing the variable names, the solution is in the *u* matrix, and the mesh parameters are in the *p*, *e*, and *t* matrices. To obtain and display the *u*-values at the required points of the orthogonal grid, the following commands should be entered in the Command Window:

```
>> x=linspace(-0.5,0.5,7);y=x;
>> P=tri2grid(p,t,u,x,y)
P =
0        0        0        0        0        0        0
0  -0.0027  -0.0039  -0.0044  -0.0039  -0.0027        0
0  -0.0040  -0.0060  -0.0066  -0.0060  -0.0040        0
   -0.0000  -0.0043  -0.0066  -0.0073  -0.0066  -0.0044  0
0  -0.0039  -0.0060  -0.0066  -0.0060  -0.0039        0
0  -0.0027  -0.0039  -0.0044  -0.0040  -0.0027        0
0        0        0        0        0        0        0
```

To obtain the maximum membrane displacement value and closest to it the node coordinates, enter in the Command Window:

```
>> [M,i]=max(-u),p(:,i)
M =
0.0074
i =
41
ans =
      0.0262
      0.0041
```

The displacement values are negative, thus the "–u" values are used in the max command.

- The automatically generated program can be saved to a file. To perform this, select the "Save As ..." line in the popup menu of the File button of the main menu and enter the name ApExample_10_10 in the "File name" field of the panel that appears. To test the generated file, close the PDE Modeler window and then open the saved file in the MATLAB® Editor window and type ApExample_10_10 in the function definition line instead of the default name. Note that after running this file, the plot parameters should be adjusted as described above in order to generate the necessary contour lines and colormap.

10.4.3 Diffusion-Brusselator PDEs for Reagent Concentrations

The model of two unsteady dimensionless 2D PDEs describing concentration change of the two-reagent system in a reactor are as follows (based on Adomian 1995):

$$\frac{\partial u}{\partial t} = B + u^2 v - (A+1)u + \alpha\left(\frac{\partial^2 u}{\partial x^2} + \frac{\partial^2 u}{\partial y^2}\right)$$

$$\frac{\partial v}{\partial t} = Au - u^2 v + \alpha\left(\frac{\partial^2 u}{\partial x^2} + \frac{\partial^2 u}{\partial y^2}\right)$$

where u and v are the component concentrations of two individual components, t – time, x and y coordinates, α is a dispersion coefficient, A and B are positive real constants.

The system is solved for a rectangular tank and with the following ICs

$$u(0,x,y) = 2 + 0.25y$$

$$v(0,x,y) = 1 + 08x$$

and BCs

$$u(t,0,y) = 3.4$$

$$v(t,0,y) = 1$$

$$\frac{\partial u(t,W,y)}{\partial y} = \frac{\partial u(t,L,y)}{\partial x} = \frac{\partial u(t,x,0)}{\partial y} = 0$$

$$\frac{\partial v(t,W,y)}{\partial y} = \frac{\partial v(t,L,y)}{\partial x} = \frac{\partial v(t,x,0)}{\partial y} = 0$$

where L and W are the length and width of the tank, respectively.

Problem: Compose a live script named ApExample_10_11 that solves the above PDE system with given ICs and BCs. Consider A = 3.4, B = 1, α = 0.002, L = 20, W = 5, and 20t values in the range of 0 … 10. Present $v(x,y)$ and $v(x,y)$ at the final time on two 3D plots of one Figure window. Display two tables containing the final concentrations of each reagent at seven x and five y values, spaced at regular intervals.

The diffusion-Brusselator equations, rewritten to the standard form, are as follows:

$$\frac{\partial u}{\partial t} = \nabla\alpha\nabla u + B + u^2 v - (A+1)u$$

$$\frac{\partial v}{\partial t} = \nabla\alpha\nabla u + Au - u^2 v$$

Comparing these equations and the standard PDE (Section 6.3), we conclude that they are identical when the m, d, c, a, and f terms of PDEs are [0;0], [1;1], [α; α], [0; 0], and [$B + u^2 v - (A+1)u$; $Au - u^2 v$], respectively.

Steps to solve the above system:

8. Create a model object using commands

```
nEquations=2;
model = createpde(nEquations);
```

9. Build and plot the geometry of our object

```
L=20;W=5;
R1=[3,4,0,0,L,L,0,W,W,0]';                    % square
sf = 'R1';                               % formula set
ns = char('R1')';              %basic geometric form name
g=decsg(R1,sf,ns);                      % decomposition
geometryFromEdges(model,g);            % model geometry
pdegplot(model,'EdgeLabels','on',
'FaceLabels','on')% plotting
axis equal                         % for correct scaling
```

The resulting tank geometry is identical to that described in Section 10.4.1 and is shown in Figure 10.11.

10. BCs of both PDEs are of the Dirichlet type on the left E1 edge of the rectangle and of the Neumann type on all other edges. Considering this, the commands that specify the BCs in our model are

```
applyBoundaryCondition(model,'dirichlet','Edge',1, 'u',
[3.4;1]);
applyBoundaryCondition(model,'neumann','Edge',2:4, 'g',
[0;0], 'q', [0;0]);
```

Note that the coefficients are presented here as column vectors containing the boundary data for the first and second PDE.

The commands that specify the ICs are as follows:

```
u0=@(location,state) [2+0.25*location.y;1+0.8*location.x];
setInitialConditions(model,u0,'Face',1);
```

the argument u0 is presented here in the anonymous function as a column vector containing two equations for the first and second reactants.

11. The following PDE terms must be specified are $m = 0$, $d = 1$, $c = 2e{-}3$, $a = 0$, and $f = [B+u^2v-(A+1)u;\ Au-u^2v]$, so the commands that set these terms look like this:

```
f=@(location,state) [1+state.u(1,:).^2.*state.u(2,:)-(3.4+1)*
state.u(1,:);3.4*state.u(1,:)-state.u(1,:).^2.*state.u
(2,:)];
specifyCoefficients(model,'m',0,'d',1,'c',2e-3,'a',0,
'f',f);
```

Note that if the PDE term of each component is represented by the same number, it is not necessery to write both numbers in a column vector, only one number can be entered, and the computer uses that number for both components.

12. Generate a mesh with triangular elements using the command

```
generateMesh(model,'Hmax',0.25);
```

here the maximum value of the triangle edge length is 0.25.

13. After that, we can solve the PDE for a list of 20 times using the commands

```
tf=10;n_t=20;
tlist=linspace(0,tf,n_t);
res=solvepde(model,tlist);
u=res.NodalSolution;
```

the result is contained in the res.NodalSolution three-dimensional matrix and assigned to u. Each page of this matrix has two columns with the *u* and *x,y* node coordinates for each reagent and corresponds to a specific time from the time list.

14. The resulting graphs and the required table can be obtained using the following commands:

```
figure
subplot(2,1,1)
pdeplot(model,'XYData',u(:,1,n_t),'Zdata',u(:,1,n_t),
'Colormap','jet');
xlabel('x'),ylabel('y'),zlabel('u')
title(['Reagent 1, final t=',num2str(tlist(end))])
axis tight
subplot(2,1,2)
pdeplot(model,'XYData',u(:,2,n_t),'Zdata',u(:,2,n_t),
'Colormap','jet');
xlabel('x'),ylabel('y'),zlabel('v')
title(['Reagent 2, final t=',num2str(tlist(end))])
axis tight
xq=linspace(0,L,7);yq=linspace(0,W,5);
[x,y]=meshgrid(xq,yq);
uintrp=interpolateSolution(res,x,y,[1,2],length(tl
ist));
Reagent_1=reshape(uintrp(:,1),size(x)),Reagent_2=reshap
e(uintrp (:,2),size(x))
```

Here, the `interpolateSolution` command is used to retrieve the *u* and *v* values at seven *x* and five *y* grid points as required by the problem.

All of the above commands have been compiled into the following live script named ApExample_10_11.

```
%                              solves diffusion-Brusselator PDEs
clear, close all
nEquations=2;
model = createpde(nEquations);
L=20;W=5;
R1=[3,4,0,0,L,L,0,W,W,0]';                                % square
sf = 'R1';                                           % formula set
ns = char('R1')';                     % basic geometric form name
g=decsg(R1,sf,ns);                               % decomposition
geometryFromEdges(model,g);                     % model geometry
pdegplot(model,'EdgeLabels','on','FaceLabels','on') % plotting
axis equal                                for correct scaling
applyBoundaryCondition(model,'dirichlet','Edge',1,
'u',[3.4;1]);
applyBoundaryCondition(model,'neumann','Edge',2:4,
'g',[0;0],'q',[0;0]);
u0=@(location,state) [2+0.25*location.y;1+0.8*location.x];
setInitialConditions(model,u0,'Face',1);
f=@(location,state)
[1+state.u(1,:).^2.*state.u(2,:)-(3.4+1)*state.u(1,:);
3.4*state.u(1,:)-state.u(1,:).^2.*state.u(2,:)];
specifyCoefficients(model,'m',0,'d',1,'c',2e-3,'a',0,'f',f);
generateMesh(model,'Hmax',0.25);
tf=10;n_t=20;
tlist=linspace(0,tf,n_t);
res=solvepde(model,tlist);
u=res.NodalSolution;
figure
subplot(2,1,1)
pdeplot(model,'XYData',u(:,1,n_t),'Zdata',u(:,1,n_t),'Colorm
ap','jet');
xlabel('x'),ylabel('y'),zlabel('u')
title(['Reagent 1, final t=',num2str(tlist(end))])
axis tight
subplot(2,1,2)
pdeplot(model,'XYData',u(:,2,n_t),'Zdata',u(:,2,n_t),'Colorm
ap','jet');
xlabel('x'),ylabel('y'),zlabel('v')
title(['Reagent 2, final t=',num2str(tlist(end))])
axis tight
xq=linspace(0,L,7);yq=linspace(0,W,5);
[x,y]=meshgrid(xq,yq);
uintrp=interpolateSolution(res,x,y,[1,2],length(tlist));
Reagent_1 = reshape(uintrp(:,1),size(x)),Reagent_2 =
reshape(uintrp(:,2),size(x))
```

After running this script, the following graphs and tables are outputted "inline" (Figure 10.16).

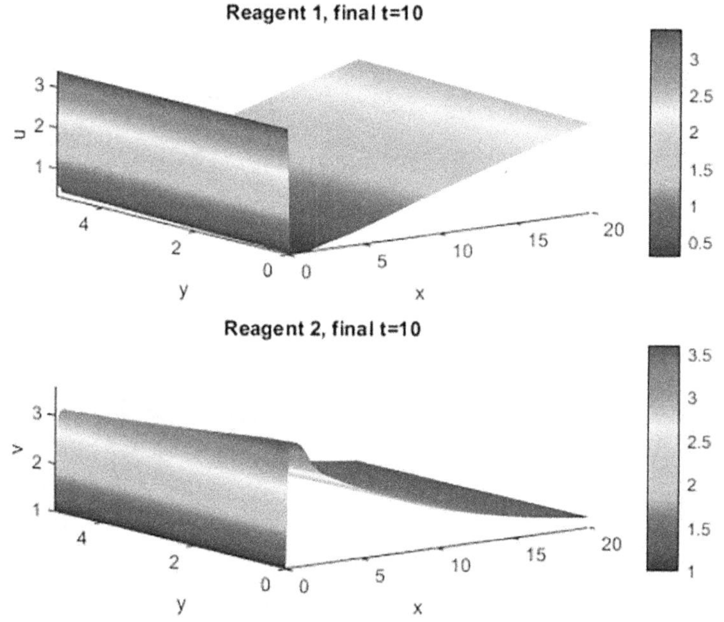

Reagent_1 = 5×7

3.4000	0.5497	0.9411	1.3638	1.7676	2.1495	2.4973
3.4000	0.5700	0.9716	1.3947	1.7974	2.1782	2.5250
3.4000	0.6006	1.0107	1.4333	1.8343	2.2134	2.5591
3.4000	0.6361	1.0526	1.4742	1.8730	2.2504	2.5948
3.4000	0.6699	1.0906	1.5107	1.9078	2.2835	2.6271

Reagent_2 = 5×7

1.0000	2.7298	2.2152	1.8360	1.5619	1.3591	1.2099
1.0000	2.6950	2.1836	1.8123	1.5443	1.3457	1.1992
1.0000	2.6452	2.1442	1.7834	1.5230	1.3294	1.1863
1.0000	2.5908	2.1033	1.7538	1.5011	1.3128	1.1731
1.0000	2.5418	2.0672	1.7276	1.4819	1.2982	1.1612

FIGURE 10.16 The live script solution of the diffusion-Brusselator equation is presented in two 3D plots and in two tables with final concentrations of the reagents.

Appendix A
Special Characters, Predefined Variables, Operators, and Commands Used in PDE ToolboxTM and MATLAB® Programming

The predefined variables, special characters, operators, and commands discussed, used in programs, or simply mentioned in this book are listed below in this appendix.

A.1 SPECIAL CHARACTERS, OPERATORS, ALTERNATIVE COMMANDS, AND PUNCTUATIONS

The symbols, special characters, and arithmetic operators with alternative commands (if exist) are shown in Tables A.1 and A.2.

TABLE A.1
Special Characters/Operators/Punctuations Used

No	Symbol	Description/Action	Symbol Name
1	@	designates an anonymous function or call to the user-defined function	At
2	=	assigns a value/set of values to a variable	Equal sign
3	%	introduces comments and specifies an output number format	Percent
4	()	is used for command input arguments, or elements in a vector/ matrix/array; sets the priority of the operation	Parentheses
5	[ ]	is used for input of the vector/matrix/array elements, or for output arguments of the user-defined functions	Brackets
6	{ }	delimits a block of comments that extend beyond one line; construct/access the contents in so-termed cell (the latter concept was not used explicitly in this book)	Curly brackets
7	(space)	separates elements of vectors/arrays/matrices; used also into output function specifications	Space

(*Continued*)

TABLE A.1 (*Continued*)
Special Characters/Operators/Punctuations Used

No	Symbol	Description/Action	Symbol Name
8	.	separates whole and decimal parts of a number; used in element-by-element array operations; refers to a structure field	Period or dot
9	:	creates vectors, used also for loops iterations and for selecting all array/matrix elements	Colon
10	,	separates commands on the same line and elements into array/matrix	Comma
11	;	prevents outputting when wrote after the command; separates matrix rows and commands on the same command line	Semicolon
12	...	denotes that a long statement to be continued on the next line; also the same as % when is written inside the command	Ellipsis
13	‘	transposes vector/array/matrix, and used for a text characters generation	Apostrophe
14	“	is used for text string generation	Double quotes
15	~	is used to ignore function argument or output; is used also as logical not	Tilde

TABLE A.2
Arithmetic Operators

No	Symbol	Description	Alternative Command Name
1	+	Addition	plus
2	-	Subtraction	minus
3	*	Scalar and matrix multiplication	mtimes
4	.*	Element-wise multiplication	times
5	/	Right matrix division	mrdivide
6	./	Element-wise right division	rdivide
7	\	Left matrix division	mldivide
8	.\	Element-wise left division	ldivide
9	^	Exponentiation or matrix power	mpower
10	.^	Element-wise power	power

A.2 PREDEFINED VARIABLES

Variables and values that are permanently stored in memory and are provided automatically without prior assignment are termed predefined. They are presented in Table A.3.

TABLE A.3
Predefined Variables

No	Variable	Value	Definition
1	ans	Last calculated value	Variable name, contains the most recent number obtained from the command line
2	eps	2.220446049250313·10–16	Smallest difference between two numbers
3	i	$\sqrt{-1}$	Imaginary unit
4	inf	∞	Infinity
5	j	$\sqrt{-1}$	the same as i
6	NaN	fff8000000000000 (IEEE code)	Not a number
7	pi	3.141592653589793	Number π

A.3 OPERATOR SYMBOLS USED IN RELATIONAL AND LOGICAL STATEMENTS

The operators used in the book in rational and logical commands are summarized in Table A.4 and accompanied by an alternative command that can be used in their place. When these commands are used for array or matrices, they are element-wise.

TABLE A.4
Relational and Logical Operators

No	Designation	Definition	Alternative Command Name
1	==	Equal (double)	eq
2	>	Greater than	gt
3	>=	Greater than or equal to	qe
4	<	Less than	lt
5	&	And (logical)	and
6	~	Not (logical)	not
7	|	Or (logical)	or
8	<=	Less than or equal to	le
9	~=	Not equal	ne

A.4 COMMANDS REPRESENTED IN THE BOOK

The commands implemented in this book are conditionally arranged below in three groups:

- commands used to manage the program and perform numerical calculations;
- graphic, drawing, and color commands;
- programming and interface commands of the PDE toolbox.

A.4.1 Commands for Program Management and Numerical Calculations

Commands for input/output, flow control, array/matrix processing, and mathematical calculations are called "non-graphical" here and are presented in Table A.5.

TABLE A.5
List of Non-graphical Commands

No	Command (Alphabetically)	Definition	Page
1	abs	Absolute value	14t
2	acos	Inverse cosine for angle in radians	14t
3	acosd	Inverse cosine for angle in degrees	14t
4	acot	Inverse cotangent for angle in radians	14t
5	acotd	Inverse cotangent for angle in degrees	14t
6	asec	Inverse second, result in radians	14t
7	asecd	Inverse second, result in degrees	14t
8	asin	Inverse sine for angle in radians	14t
9	asind	Inverse sine for angle in degrees	14t
10	atan	Inverse tangent for angle in radians	14t
11	atand	Inverse tangent for angle in degrees	14t
12	besselj	Bessel function of the first kind	15t
13	beta	Beta function	15t
14	bvp4c	Solves boundary value problem for ODEs	154
15	bvp5c	Solves boundary value problem for ODEs	157
16	ceil	Rounds toward infinity	15t
17	char	Creates an array of characters	42t
18	close	Closes the Figure/s window/s	109
19	clc	Clear the command window	13
20	clear	Remove variables from the Workspace	22
21	convforce	Converts force units	19
22	cos	Cosine for angle in radians	15t
23	cosd	Cosine for angle in degrees	15t
24	cosh	Hyperbolic cosine	15t
25	cot	Cotangent for angle in radians	15t
26	cotd	Cotangent for angle in degrees	15t
27	coth	Hyperbolic tangent	15t
28	cross	Calculates cross product of a 3D vector	42t
29	det	Calculates a determinant	42t
30	deval	Evaluates ODE solution values	155
31	diag	Creates a diagonal matrix from a vector	42t

(*Continued*)

TABLE A.5 *(Continued)*
List of Non-graphical Commands

No	Command (Alphabetically)	Definition	Page
32	diff	Calculates a difference, approximates a derivative	43t
33	disp	Display output	52
34	doc	Displays HTML documentation in the Help window	19
35	dot	Calculates scalar product of two vectors	43t
36	edit	Opens Editor window	73
37	else,elseif	Is used with if; conditionally executes if statement condition	63
38	end	Terminates scope of for, while, if statements, or serves as last index	63, 64
39	erf	Error function	15t
40	exp	Exponential	15t
41	eye	Creates a unit matrix	37
42	factorial	Factorial function	15t
43	find	Finds indices of certain elements of array	61
44	fix	Rounds toward zero	16t
45	floor	Rounds off toward minus infinity	16t
46	for	Repeats statements	64
47	format	Sets output number format	23
48	full	Converts a sparse matrix to a full matrix	43t
49	fprintf	Formates data and text	25,53
50	function	Defines an user-defined function	79
51	gamma	Gamma function	16t
52	global	Declares a global variable	81
53	help	Displays explanations in the Command Window	18
54	if	Executes conditionally	63
55	input	Prompts to user input	77
56	interp1	One-dimensional interpolation	44t
57	inv	Calculates the inverse matrix	37
58	length	Number of elements in vector	44t
59	linspace	Generates a linearly spaced vector	31
60	log	Natural logarithm	16t
61	log10	Decimal logarithm	16t
62	lookfor	Search for the word in all help entries	18
63	max	Returns maximal value	44t
64	mean	Calculates mean value	45t
65	median	Calculates median value	45t
66	min	Returns minimal value	45t
67	num2str	Converts numbers to a string.	46t
68	ode113	Solves nonstiff ODEs	152t

(Continued)

TABLE A.5 (*Continued*)
List of Non-graphical Commands

No	Command (Alphabetically)	Definition	Page
69	ode15s	Solves stiff ODEs	145
70	ode23	Solves nonstiff ODEs	152t
71	ode23s	Solves stiff ODEs	152t
72	ode23t	Solves stiff ODEs	152t
73	ode23tb	Solves stiff ODEs	153t
74	ode45	Solves nonstiff ODEs	145
75	odeset	Sets ODE options	150
76	ones	Creates an array with ones	40
77	pdepe	Solves 1D (spatially) PDE	240
78	polyfit	Fits the points with a polynomial	46t
79	polyval	Evaluates the polynomial value	46t
80	rand	Generates uniformly distributed random numbers	40
81	randi	Generates integer random numbers from uniform discrete distribution	46t
82	randn	Generates random numbers from normal distribution	46t
83	repmat	Duplicates a matrix	47t
84	reshape	Changes size of array/matrix	47t
85	rng	Controls random number generator	41
86	round	Rounds off toward nearest decimal or integer	16t
87	sign	Returns selectively –1 or 0 or 1	47t
88	sin	Sine	16t
89	sind	Sine for angle in degrees	16t
90	sinh	Hyperbolic sine	16t
91	size	Size of vector/array/matrix	48t
92	sort	Arranges elements in ascending or descending order	48t
93	sparce	Converts a full matrix to a sparce matrix	48t
94	sqrt	Square root	17t
95	std	Calculates standard deviation	49t
96	strvcat	Concatenates strings vertically	49t
97	sum	Calculates sum of elements together	50t
98	tan	Tangent for angle in radians	17t
99	tand	Tangent for angle in degrees	17t
100	tahh	Hyperbolic tangent	17t
101	trapz	Numerical integration with the trapezoidal rule	15t
102	ver	Displays versions of the MATLAB and toolboxes	21
103	while	Repeats execution of command/s	64
104	who	Displays stored variables	22
105	whos	Displays stored variables with additional information	22
106	zeros	Creates an array with zero	40

A.4.2 Commands for Plotting, Drawing, and Coloring

The studied commands of basic MATLAB® for plotting and formatting graphs are presented in Table A.6. This table does not include the PDE Toolbox™ plotting and object geometry commands, which are further included in the general list of the PDE tool command (see Section A.4.3).

TABLE A.6
Commands for Plotting and Formatting

No	Command (Alphabetically)	Definition	Page
1	axis	Controls axis scaling and appearance	110
2	bar	Draws a vertical bars on the plot	125
3	barh	Draws a horizontal bars on the plot	125
4	bar3	Generates 3D vertical bars on the plot	126t
5	clabel	Labels iso-level lines	126t
6	close	Closes one or more Figure Windows	109
7	colormap	Sets colors	121
8	contour	Creates a 2D-contour plot	127t
9	contour3	Creates a 3D-contour plot	127t
10	cylinder	Generates a 3D plot with cylinder	127t
11	errorbar	Creates a plot with error bounded points	128t
12	figure	Creates the Figure window	128t
13	fplot	Creates a 2D plot of a function	129t
14	grid on/off	Adds/removes grid lines	109
15	help graph2d	Displays list of 2D graph commands	124
16	help graph3d	Displays list of 3D graph commands	124
17	help specgraph	Displays list of specialized graph commands	124
18	legend	Adds a legend to the plot	110
19	loglog	Generates a 2D plot with log axes	129t
20	mesh	Creates a 3D plot with meshed surface	119
21	meshgrid	Creates X,Y matrices for further plotting	118
22	pie	Creates a 2D pie plot	129t
23	pie3	Creates a 3D pie plot	130t
24	plot	Creates a 2D line plot	97
25	plot3	Creates a 3D plot with points/lines	115
26	plottools	Opens the plot editing tools	111
27	polar	Generates a plot in polar coordinates	130t
28	semilogx	Creates a 2D plot with log-scaled x-axis	130t
29	semilogy	Creates a 2D plot with log-scaled y-axis	130t
30	sphere	Generates a sphere plot	131t
31	stairs	Creates star-like plot	131t
32	stem	Creates a 2D stem plot	132t

(Continued)

TABLE A.6 (*Continued*)
Commands for Plotting and Formatting

No	Command (Alphabetically)	Definition	Page
33	stem3	Creates a 3D stem plot	132t
34	subplot	Places multiple plots on the same page	113
35	surf	Creates a 3D surface plot	120
36	surfc	Generates surface and counter plots together	132t
37	text	Adds a text to the plot	110
38	title	Adds a caption to the plot	110
39	waterfall	Generates a mesh plot without column lines	133t
40	view	Specifies a viewpoint for 3D graph	122
41	xlabel	Adds a label to x-axis	110
42	ylabel	Adds a label to y-axis	110
43	yyaxis left/right	Activates left/right y-axes of the plot	133t
44	zlabel	Adds a label to z-axis	115

A.4.3 PDE Toolbox™ Commands for PDE Solution and Plotting

The commands for solving the PDEs and plotting the resulting graphs used in the book are presented in Table A.7.

TABLE A.7
PDE Toolbox Commands Used for PDE Solution and Plotting

No	Command (Alphabetically)	Definition	Page
1	applyBoundaryCondition	Adds boundary condition to PDE model	176
2	assempde	Solves elliptic PDE (not recommended)	290t
3	createpde	Creates model as PDE object	174
4	decsg	Decomposes 2D object geometry to minimum	175
5	extrude	Generates 3D geometry by extruding 2D geometry vertically	274
6	findBoundaryCondition	Finds the acting boundary conditins for a given region	289t
7	findCoefficients	Finds active PDE coefficients	291t
8	generateMesh	Generates triangular or tetrahedral mesh	178
9	geometryFromEdges	Creates 2D geometry matrix	175
10	geometryFromMesh	Generates 2D or 3D geometry from mesh	286
11	importGeometry	Imports 2D or 3D geometry from STL-file	285t
12	initmesh	Generates initial 2D mesh	288t
13	interpolateSolution	Interpolates solution to the desired point/s	179, 276

(Continued)

TABLE A.7 (*Continued*)
PDE Toolbox Commands Used for PDE Solution and Plotting

No	Command (Alphabetically)	Definition	Page
14	multicuboid	Generates cubic/multicubic geometry	270
15	multicylinder	Generates vertical cylinder/s geometry	272
16	multisphere	Generates sphere/s geometry	273
17	pdegplot	Plots body (object) geometry	175
18	pdeModeler	Opens graphical interface for solving 2D PDE	198
19	pderect	Draws a rectangle or square	211t
20	pdeellip	Draws an ellipse or circle	212t
21	pdecirc	Draws a circle	212t
22	pdepoly	Draws a polygon	212t
23	pdeplot	Plots 2D mesh or solution	178
24	plotpde3D	Plots 3D solution or mesh	276
25	pdesmesh	Calculates tensor function in Structural Mechanics application	229t
26	pdesurf	Plots 3D surface of PDE node data	292t
27	refinemesh	Refines triangular mesh	288t
28	rotate	Rotates object geometry	274
29	setInitialConditions	Sets initial conditions	176
30	solvepde	Solves PDE specified in a PDE model	178
31	solvepdeeig	Solves eigenvalue PDE specified in a PDE model	178
32	specifyCoefficients	Specifies PDE coefficients in a PDE model	176
33	squeeze	Removes the matrix siglrton dimensions	277
34	tri2grid	Converts solution from triangular to rectangular grid	210

Appendix B
List of Examples, Problems, and Applications Considered in the Book

The following is a nomenclature of examples, problems, and applications presented in the book. The list is divided by chapters and provided with a reference subsection of the book. Examples from summary tables containing additional commands and some small examples are not included in the list. The name of the problem, example, or application represents a brief definition of the problems considered in the book and may differ from the corresponding headings in the text.

B.1 CHAPTER 2

B.2 CHAPTER 3

B.3 CHAPTER 4

B.4 CHAPTER 5

B.5 CHAPTER 6

No	The Name of the Example, Problem, or Application	Section, Page
1	Elliptical PDE for electrostatic potential in a square with the rectangular slot	6.4.2, p. 179
2	Eigenvalue 2D PDE for square membrane clamped at its three edges	6.4.3, p. 184
3	Parabolic unsteady heating PDE for a small metallic plate	6.5.1, p. 187
4	Hyperbolic PDE for drumhead vibrations	6.5.2, p. 189
5	Eigenvalue 2D PDE for elliptic membrane	6.5.3, p. 192

B.6 CHAPTER 7

No	The Name of the Example, Problem, or Application	Section, Page
1	Elliptical PDE of the laminar flow for the axial velocity in the circular cross-section of a horizontal pipe	7.2.1, p. 198
2	Elliptical PDE for momentary pressure distribution in a lubricating film between two surfaces covered by hemispherical pores	7.4.1, p. 219
3	Parabolic thermal conduction PDE with temperature-dependent property of a material, square plate with rectangular hole in the center	7.4.2, p. 224
4	Structural Mechanics problem, plane stress in a rectangular plate with an elliptical hole	7.4.3, p. 227
5	Eigenvalue 2D PDE for the T-shaped membrane	7.4.4, p. 230

B.7 CHAPTER 8

No	The Name of the Example, Problem, or Application	Section, Page
1	A second order diffusion/adhesion 1D ODE; finite difference method illustration	8.2.4, p. 238
2	Transient 1D thermo-diffusion equation without a heat source for insulated wire or thin rod	8.3.1, p. 242
3	Transient 1D diffusion equation with Neumann boundaries and piecewise IC; substance concentration distribution	8.4.1, p. 247
4	Action potential model with diffusion term	8.4.2, p. 250
5	Velocity of fluid at startup flow in a pipe	8.4.3, p. 254
6	Fluid flow according to Bateman-Burgers PDE	8.4.4, p. 257

B.8 CHAPTER 9

No	The Name of the Example, Problem, or Application	Section, Page
1	Two elliptical PDEs for square plate under area load, PDE Modeler	9.2.1, p. 262
2	3D steady heat PDE for a hollow cuboid	9.3.5, p. 27
3	3D unsteady heat PDE for a disk	9.3.6, p. 278
4	Schnakenberg trimolecular reaction model; solving coupled 2D PDEs with the PDE Modeler	9.5.1, p. 284
5	3D wave PDE for vibration of slab with elliptical hole	9.5.2, p. 297
6	3D electric potential PDE for a plate with varying conductivity	9.5.3, p. 299

B.9 CHAPTER 10

No	The Name of the Example, Problem, or Application	Section, Page
1	First-order ODE rate equation for biomolecular reaction	10.2.1, p. 303
2	Second-order ODE for steady-state concentration distribution in a short tube, BVP problem	10.2.2, p. 305
3	A set of first-order ODEs for enzyme kinetics	10.2.3, p. 307
4	A set of first-order ODEs for two reactors in series	10.2.4, p. 311
5	Unsteady 1D PDE reactor model	10.3.1, p. 313
6	Coupled unsteady 1D PDEs for tumor growth model	10.3.2, p. 318
7	Coupled unsteady 1D PDEs for bacterial culture, diffusing, and reproducing	10.3.3, p. 321
8	Coupled unsteady 1D PDEs for a model of the reaction-diffusion system	10.3.4, p. 324
9	Steady-state 2D PDE for concentration distribution in a reactor	10.4.1, p. 328
10	Steady-state 2D PDE for homogenous membrane displacement	10.4.2, p. 332
11	Diffusion-Brusselator 2D PDEs for reagent concentrations	10.4.3, p. 334

References

Adomian, G., 1995. The diffusion – Brusselator equation. *Computers & Mathematics with Applications*. v. 29, issue 5, pp. 1–3.

Batchelor, G.K., 2012. *An Introduction to Fluid Dynamics*. Cambridge: Cambridge University Press.

Beentjes C. H. L., 2014. Pattern formation analysis in the Schnakenberg model. Mathematical Institute, University of Oxford. https://cbeentjes.github.io/files/Ramblings/PatternFormationSchnakenberg.pdf (accessed, May 24, 2021).

Beitz, W., and Kuttner, Eds., 1994. *Handbook of Mechanical Engineering*, NY: Springer-Verlag, 1994, p. C31.

Budinas, R. G., & Nisbett, G.K., 2011. *Shigley's Mechanical Engineering Design, 9th ed.*, NY: McGray-Hill, p.934.

Burr, A. H. & Cheatem, J. B., 1995. *Mechanical Analysis and Design, 2nd ed.*, Upper Saddle River, NJ: Prentice Hall, p.423.

Burstein, L., 2021a. *MATLAB® with Applications in Mechanics and Tribology*. Hershey, PA: IGI Global.

Burstein, L., 2021b. Load-carrying capacity of the lubricating film in the gap between surfaces textured by hemispherical pores. *International Journal of Surface Engineering and Interdisciplinary Material Science*, v. 9, issue 2, pp.1–19.

Burstein, L., 2020. *A MATLAB® Primer for Technical Programming in Material Science and Engineering*. Duxford, UK: Woodhead Publishing Series in Technology and Engineering, Elsevier Inc.

Chapra, S. C. and Canale, R. P., 2015. *Numerical Methods for Engineers. Seventh edition*. NY: McGraw-Hill Education.

Magrab, E. B., Azarm, S., Balachandral, B., Duncan, J. H., Herold, K. E., & Walsh, G. C., 2010. *An Engineer's Guide to MATLAB®. 3rd ed.* Boston, MA: Pearson.

Norton, R. L., 2010. *Machine Design. An Integral Approach. 4th ed.* Upper Saddle River, NJ: Prentice Hall, p.1003.

Orme, M. E. and Chaplain, M. A. J., 1996. A mathematical model of the first steps of tumor-related angiogenesis: capillary sprout formation and secondary branching. *IMA Journal of Mathematics Applied in Medicine & Biology*, v. 13, pp. 73–98.

Rooke, D. P., & Cartwright, D. J., 1976. *Compendium of stress intensity factors*. HMSO Ministry of Defense. Procurement Executive, https://en.wikipedia.org/wiki/Stress_intensity_factor#cite_note-rooke–4.

Sobhy, M. (n. d.) in *9 Piston Rings*. Retrieved from https://www.academia.edu/26077072/9_Piston_rings (accessed March 14, 2022).

Spotts, M. F., & Shoup, T. E., 1998. *Design of Machine Elements*, Upper Saddle River, NJ: Prentice Hall, p.613.

Index

Note: **Bold** page numbers refer to tables.

R

V

W

X

Y

Z

For Product Safety Concerns and Information please contact our EU
representative GPSR@taylorandfrancis.com
Taylor & Francis Verlag GmbH, Kaufingerstraße 24, 80331 München, Germany

www.ingramcontent.com/pod-product-compliance
Lightning Source LLC
Chambersburg PA
CBHW060757220326
41598CB00022B/2459
9781032060224